IntechOpen

Supporting open minds since 2005

Solvents, Ionic Liquids and Solvent Effects
http://dx.doi.org/10.5772/intechopen.82039
Edited by Daniel Glossman-Mitnik and Magdalena Maciejewska

Contributors
Ronewa Phadagi, Indra Bahadur, Angeles Dominguez, Begoña Gonzalez, Patricia F. Requejo, Sandra Corderí, Zhengping Liu, Dan He, Liyan Huang, Anna Tsegelskaya, Alexander A. Kuznetsov, Imed Gargouri, Fatma Omrane, Moncef Khadhraoui, Sanjoy Saha, Biswajit Sinha, Banothile Makhubela, Leah Matsinha, Novisi Oklu, Alwar Ramani, Suresh Iyer, Murugesan Muthu, Muhammad Saad Khan, Bhajan Lal, Raffaello Papadakis, Ioanna Deligkiozi, Paola Campodonico, Manuel Alberto Flores-Hidalgo, Diana Barraza-Jimenez, Azael Martínez De La Cruz, Sandra Iliana Torres-Herrera, María Leticia Saucedo-Mendiola, Adolfo Padilla Mendiola, Elva Marcela Coria-Quiñones, Raúl Armando Olvera-Corral, María Estela Frías-Zepeda

© The Editor(s) and the Author(s) 2020
The rights of the editor(s) and the author(s) have been asserted in accordance with the Copyright, Designs and Patents Act 1988. All rights to the book as a whole are reserved by INTECHOPEN LIMITED. The book as a whole (compilation) cannot be reproduced, distributed or used for commercial or non-commercial purposes without INTECHOPEN LIMITED's written permission. Enquiries concerning the use of the book should be directed to INTECHOPEN LIMITED rights and permissions department (permissions@intechopen.com).
Violations are liable to prosecution under the governing Copyright Law.

CC BY

Individual chapters of this publication are distributed under the terms of the Creative Commons Attribution 3.0 Unported License which permits commercial use, distribution and reproduction of the individual chapters, provided the original author(s) and source publication are appropriately acknowledged. If so indicated, certain images may not be included under the Creative Commons license. In such cases users will need to obtain permission from the license holder to reproduce the material. More details and guidelines concerning content reuse and adaptation can be found at http://www.intechopen.com/copyright-policy.html.

Notice
Statements and opinions expressed in the chapters are these of the individual contributors and not necessarily those of the editors or publisher. No responsibility is accepted for the accuracy of information contained in the published chapters. The publisher assumes no responsibility for any damage or injury to persons or property arising out of the use of any materials, instructions, methods or ideas contained in the book.

First published in London, United Kingdom, 2020 by IntechOpen
IntechOpen is the global imprint of INTECHOPEN LIMITED, registered in England and Wales, registration number: 11086078, 7th floor, 10 Lower Thames Street, London, EC3R 6AF, United Kingdom
Printed in Croatia

British Library Cataloguing-in-Publication Data
A catalogue record for this book is available from the British Library

Additional hard and PDF copies can be obtained from orders@intechopen.com

Solvents, Ionic Liquids and Solvent Effects
Edited by Daniel Glossman-Mitnik and Magdalena Maciejewska
p. cm.
Print ISBN 978-1-78985-281-3
Online ISBN 978-1-78985-282-0
eBook (PDF) ISBN 978-1-83968-530-9

For EU product safety concerns:
IN TECH d.o.o., Prolaz Marije Krucifikse Kozulić 3, 51000 Rijeka, Croatia, info@intechopen.com or visit our website at intechopen.com.

We are IntechOpen, the world's leading publisher of Open Access books

Built by scientists, for scientists

4,500+
Open access books available

119,000+
International authors and editors

135M+
Downloads

151
Countries delivered to

Our authors are among the
Top 1%
most cited scientists

12.2%
Contributors from top 500 universities

CLARIVATE ANALYTICS
BOOK CITATION INDEX
INDEXED

WEB OF SCIENCE™

Selection of our books indexed in the Book Citation Index
in Web of Science™ Core Collection (BKCI)

Interested in publishing with us?
Contact book.department@intechopen.com

Numbers displayed above are based on latest data collected.
For more information visit www.intechopen.com

Meet the editors

Dr. Daniel Glossman-Mitnik is a titular researcher at the Centro de Investigación en Materiales Avanzados (CIMAV), Chihuahua, Mexico as well as a level-III national researcher for the Consejo Nacional de Ciencia y Tecnología, Mexico. He also participates as a visiting professor at the Universitat de les Illes Balears, Palma de Mallorca, Spain. His research interest focuses on computational chemistry and molecular modeling of molecular systems useful for pharmacological, health, food and alternative energy applications by resorting to density functional theory (DFT) and conceptual DFT. He has authored and coauthored more that 230 peer-reviewed papers and 20 book chapters as well as having edited a book on DFT. He has delivered speeches at many international and domestic conferences, and he serves as a referee for more than 80 international journals, books and research proposals.

Dr. Magdalena Maciejewska is currently Assistant Professor at Lodz University of Technology, in the Institute of Polymer and Dye Technology, Lodz, Poland. She obtained a PhD in Chemical Technology and PhD DSc from Lodz University of Technology in 2008 and 2018, respectively. Her research work has been credited with 8 national patents, more than 50 research and review articles in peer-reviewed national and international journals, 14 book chapters and 33 proceedings. She has attended more than 30 international and domestic conferences. She has supervised 50 diploma theses and was an assistant supervisor for one of them. She is currently a principal supervisor of two doctoral dissertations. Her major areas of research are elastomer composites, ionic liquids in the elastomer technology, vulcanization of elastomers and thermal analysis of polymeric materials.

Contents

Preface

Solvents and ionic liquids are ubiquitous within our whole life since ancient times and their effects are actually being studied through basic sciences like Chemistry, Physics and Biology as well as being researched by a large number of scientific disciplines.

This book represents an attempt to present examples on the utility of old and new solvents and the effects they exercise on several fields of academic and industrial interest. The first section, Solvents, presents information on bio-solvents and their synthesis, industrial production and applications, about per and trichloroethylene air monitoring in dry cleaners in the city of Sfax (Tunsia) and on the synthesis of polyimides using molten benzoic acid as the solvent. The second section, Ionic Liquids, shows information about the synthesis, physicochemical characterization and exploration of antimicrobial activities of imidazolium ionic liquid-supported Schiff base and its transition metal complexes, the technology of heterogenization of transition metal catalysts towards the synthetic applications in an ionic liquid matrix, the progress in ionic liquids as reaction media, monomers, and additives in high-performance polymers, a pre-screening of ionic liquids as gas hydrate inhibitor via application of COSMO-RS for methane hydrates, the extraction of aromatic compounds from their mixtures with alkanes from ternary to quaternary (or higher) systems and a review on ionic liquids as environmental benign solvent for cellulose chemistry. The final section, Solvent Effects, displays interesting information on solvent effects on dye sensitizers derived from anthocyanidins for applications in photocatalysis, about the solvent effect on a model of SNAr reaction in conventional and non-conventional solvents, and on solvent effects in supramolecular systems.

The Editors would like to express their sincere gratitude to all authors who have contributed to this book: Novisi K. Oklu, Leah C. Matsinha, Banothile C. E. Makhubela, Fatma Omrane, Imed Gargouri, Moncef Khadhraoui, Alexander A. Kuznetsov, Anna Yu. Tsegelskaya, Diana Barraza-Jiménez, Azael Martínez-De la Cruz, Leticia Saucedo-Mendiola, Sandra Iliana Torres-Herrera, Adolfo Padilla Mendiola, Elva Marcela Coria Quiñones, Raúl Armando Olvera Corral, María Estela Frías-Zepeda, Manuel Alberto Flores-Hidalgo, Paola R. Campodónico, Raffaello Papadakis, Ioanna Deligkiozi, Dan He, Zhengping Liu, Liyan Huang, Alwar Ramani, Suresh Iyer, Murugesan Muthu, Biswajit Sinha, Sanjoy Saha, Muhammad Saad Khan, Bhajan Lal, Ángeles Domínguez, Begoña González, Patricia F. Requejo, Sandra Corderí, Ronewa Padagi and Indra Bahadur.

Daniel Glossman-Mitnik wants to express his warmest thanks to his beloved wife Carmen and his dear sister Perla for their support during the preparation of this book and to the memories of my late parents, Sofía and Miguel. He also thanks the financial support from the Consejo Nacional de Ciencia y Tecnología (CONACYT), Mexico and the Universitat de les Illes Balears (UIB), Palma de Mallorca, Spain, where part of this book has been edited.

Dr. Daniel Glossman-Mitnik
Laboratorio Virtual NANOCOSMOS,
Departamento de Medio Ambiente y Energía,
Centro de Investigación en Materiales Avanzados,
Chihuahua, Mexico

Magdalena Maciejewska
Lodz University of Technology,
Poland

Section 1

Solvents

Chapter 1

Bio-Solvents: Synthesis, Industrial Production and Applications

Novisi K. Oklu, Leah C. Matsinha
and Banothile C.E. Makhubela

Abstract

Solvents are at the heart of many research and industrial chemical processes and consumer product formulations, yet an overwhelming number are derived from fossils. This is despite societal and legislative push that more products be produced from carbon-neutral resources, so as to reduce our carbon footprint and environmental impact. Biomass is a promising renewable alternative resource for producing bio-solvents, and this review focuses on their extraction and synthesis on a laboratory and large scale. Starch, lignocellulose, plant oils, animal fats and proteins have been combined with creative synthetic pathways, novel technologies and processes to afford known or new bio-derived solvents including acids, alkanes, aromatics, ionic liquids (ILs), furans, esters, ethers, liquid polymers and deep eutectic solvents (DESs)—all with unique physiochemical properties that warrant their use as solvation agents in manufacturing, pharmaceutical, cosmetics, chemicals, energy, food and beverage industries, etc. Selected bio-solvents, conversion technologies and processes operating at commercial and demonstration scale including (1) Solvay's Augeo™ SL 191 renewable solvent, (2) Circa Group's Furacell™ technology and process for making levoglucosenone (LGO) to produce dihydrolevoglucosenone (marketed as Cyrene™), (3) Sappi's Xylex® technology and demonstration scale processes that aim to manufacture precursors for bio-solvents and (4) Anellotech's Bio-TCat™ technology and process for producing benzene, toluene and xylenes (BTX) are highlighted.

Keywords: bio-solvents, renewable resources, green chemistry, biorefinery, biomass

1. Introduction

Air quality deterioration, environmental, health and safety issues have raised serious concerns over continued processing of fossil-based feedstocks in producing chemical products such as fuels and solvents. As such, many efforts are being made to reduce the use of hazardous substances (particularly volatile organic solvents (VOCs)) and to eliminate or minimize waste generation in chemical processes. Switching from the currently widely used fossil-based solvents to greener ones derived from renewable resources constitutes a key strategy to drive sustainability as well as clean and safer chemical procedures in both industry and academia [1, 2].

Solvents are central to many chemical processes as they dissolve reagents and ensure sufficient interactions, at a molecular level, for chemical transformations to take place. For instance, the pharmaceutical industry is heavily reliant on solvents for

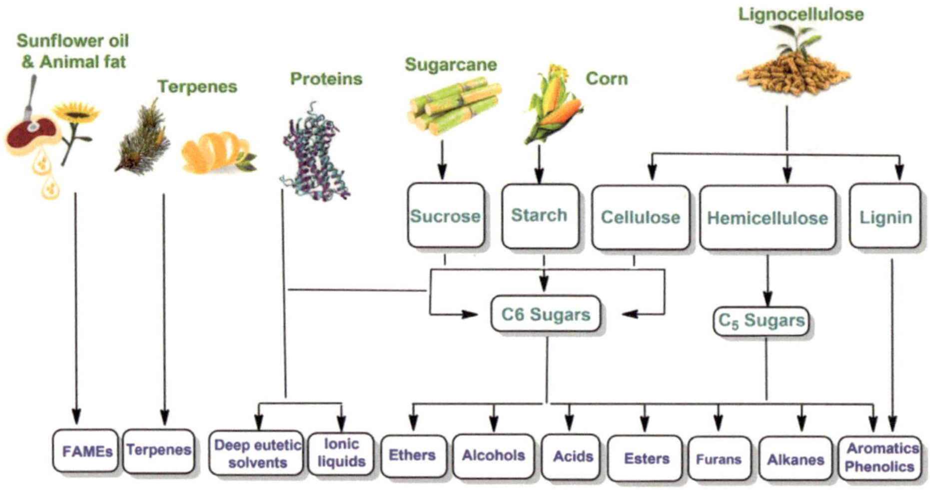

Figure 1.
The various sustainable solvents derived from biomass.

drug discovery, process development and drug manufacturing processes [3]. They are also used to extract and separate compounds from mixtures or natural products. Solvents are also important for performing compound purifications and form part of many manufacturing protocols as well as consumer products including fragrances, cleaning agents, cosmetics, paints, flavors, adhesives and inks, to name a few [1].

Many organic solvents are volatile, flammable and toxic; therefore, their use poses safety and health risks and impacts negatively on the environment. In order to address these issues, 'solvent-free' chemistry has been proposed, to mitigate against solvent exposure risks, and for chemical manipulations and formulations that absolutely require solvents, a system of ranking them by their environmental, safety and health (ESH) attributes has been introduced. This system aims to aid in selecting and using those solvents with minimum ESH risks and good green profiles [4–6].

Making use of renewable resources in producing solvents is a promising and important strategy to move towards sustainable chemical processing and to replace organic solvents derived from fossil raw materials. To this end, bio-based feedstocks such as carbohydrates, carbohydrate polymers, proteins, alkaloids, plant oils and animal fats have been used to produce bio-based solvents (**Figure 1**). This often requires prior processing of the raw materials (typical by thermochemical and biochemical conversion methods) to give familiar solvents or to provide completely new and innovative solvent entries [7, 8]. There are some, such as essential oils extracted from citrus peels (which are rich in terpenes), that are used directly.

The main processing methods include biochemical and thermochemical conversion [7]. Using one or a combination of these processing techniques, several classes of bio-based solvents (including alcohols [9–11], esters [12, 13], ethers [14], alkanes [15], aromatics [16] and neoterics [17, 18]) can be manufactured (**Figure 1**).

2. Alcohols

2.1 Methanol, ethanol, propanol and butanol

Bio-based ethanol is currently the most produced of all bio-solvents and is produced through biological transformation of sugars. These processes use either edible (sugarcane and corn) or nonedible (cellulose) feedstocks [19]. However, due to concerns over edible feedstock causing the rise of food prices, there has been a move towards optimizing processes that produce cellulosic ethanol [20, 21], and indeed the world's first cellulosic ethanol commercial-scale production plant was commissioned in 2013 by Beta Renewable. This plant is situated in Milan, Italy, and uses Proesa™ technology for the pretreatment of agricultural waste (such as rice straw, giant cane (*Arundo donax*) and wheat straw) for production of ethanol at 60,000 tons/year [9].

Ethanol's most common uses are as a biofuel and solvent in consumer products such as perfumes, food coloring and flavoring, alcoholic drinks and in certain mediation. The latter can be in both synthetic medicines and natural products. For example, highly efficient ethanol-assisted extraction of artemisinin, an active antimalarial drug, from *Artemisia annua*, has been demonstrated [22].

The presence of the hydroxyl group in ethanol makes it capable of hydrogen bonding, and it is therefore miscible with water and other solvents such as toluene, pentane and acetone.

Large-scale production of methanol is currently achieved using fossil-derived sources, by hydrogenation of carbon monoxide in the presence of a catalyst such as ZnO/Cr_2O_3 and $Cu/ZnO/Al_2O_3$ [23]. It can be obtained in small amounts in fermentation broths and from the gasification of biomass to produce bio-based syngas for subsequent conversion to methanol. The commercial viability of the latter is still under investigation [24].

Due to the structural similarity of methanol to ethanol, the former is often used in place of ethanol in synthetic procedures. However, its toxicity has limited its widespread application as a solvent in consumer products. Still, other useful applications where methanol features not only as a solvent but as a reagent and fuel do exist. For instance, methanol is used in methanol fuel cells and is a key reagent in the manufacture of fatty acid methyl esters (FAMEs) (biodiesel) through transesterification of triglycerides and in dimethyl ether (DME) (diesel substitute) production [24]. It is also used in acid-catalyzed formation of methyl levulinate from bio-derived furfural alcohol (FA) and 2,5-hydroxymethylfurfural (HMF) [25].

Thermochemical synthesis of alcohols such as methanol, ethanol and propanol (1-propanol and 2-propanol) from glycerol raw materials (**Figure 2**) is also known. This process is often promoted by a catalyst through a hydrogenolysis reaction [26–28].

n-Butanol production was first commercialized in the early 1900s, in the Weizmann process (ABE fermentation). Here, *n*-butanol, acetone and minute amounts of ethanol are produced from starch feedstock using *Clostridium acetobutylicum* [10, 29]. Later, fossil-based production of *n*-butanol became more cost-effective which led to abandonment of the Weizmann process. Today there are some ABE fermentation production units; however, most *n*-butanol is produced mainly from petroleum by either (1) the oxo synthesis which involves hydroformylation of propene and then hydrogenation of the afforded butraldehyde, in the presence of a rhodium or cobalt homogeneous catalysts, (2) the Reppe synthesis (which involves the treatment of propene with CO and H_2O) in the presence of a catalyst such as iron or (3) a multistep, catalytic hydrogen borrowing, cascade process involving self-aldol condensation of acetaldehyde, dehydration and then hydrogenation of the resultant croton aldehyde [30].

n-Butanol has a low order of toxicity and has an energy density of 29.2 MJ/L, which is comparable to that of gasoline (32.5 MJ/L). It is also miscible with gasoline, and so it is not surprising that this alcohol is sometimes blended with gasoline to improve its properties. *n*-Butanol is also used as a solvent for coatings (varnishes, resins and waxes), paints and cosmetics and is an intermediate for making solvents (e.g. butyl propanoate, dibutyl ether and butyl acetate), polymer monomers (e.g. butyl acrylate and butyl vinyl ether) and plasticizers (e.g. butyl phthalates) [31].

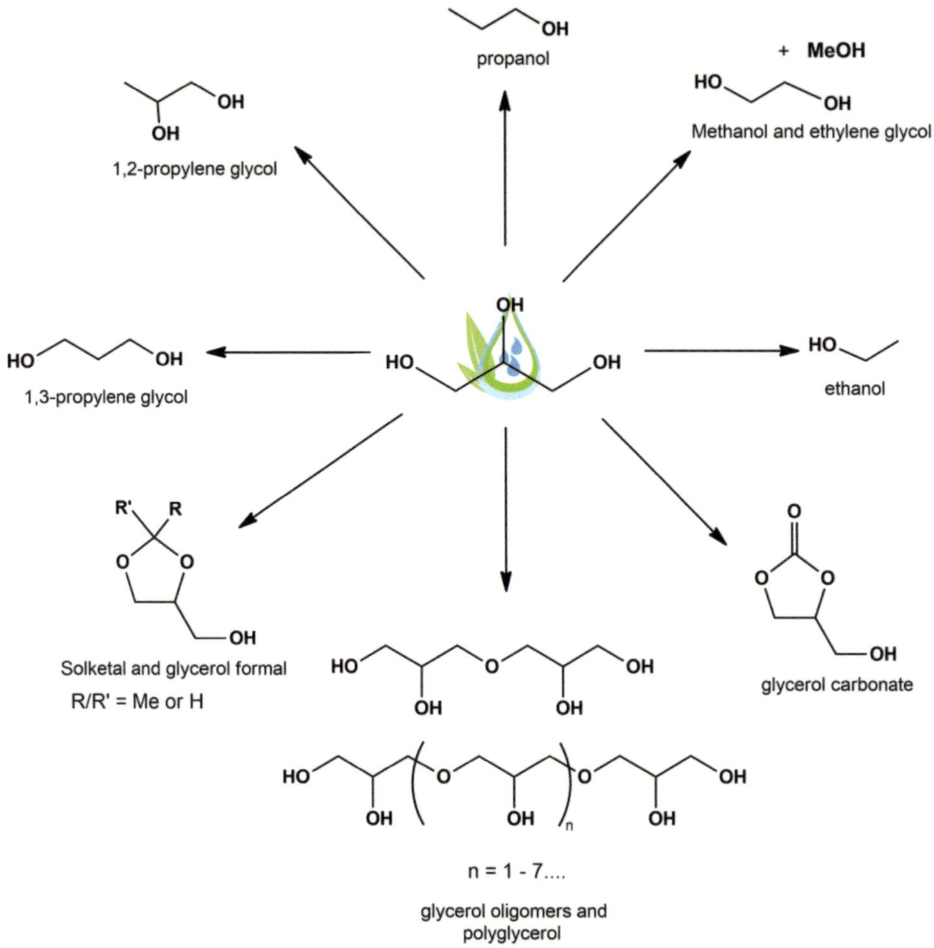

Figure 2.
Bio-based solvents derived from glycerol.

2.2 Furfuryl alcohol

The production of furfuryl alcohol involves two stages: (1) the pentosan is first hydrolysed to pentoses, e.g. xylose, and (2) cyclohydration of the pentoses into furfuryl alcohol. The reaction is catalysed by acids such as dilute sulfuric or phosphoric acid, and the furfuryl alcohol is recovered by steam distillation and fractionation. Yields are generally between 30 and 50% [32].

Furfuryl alcohol is a solvent used in the refining of lubricating oils [33]. During the production of butadiene, an extractive distillation step is required to remove impurities [34]. Furfuryl alcohol is one of the extractive solvents that have been used for this purpose on a commercial scale, for example, at TransFuran Chemicals, USA. Furfuryl alcohol is also used as a solvent during crystallization of anthracene oils and during the modification of high phenolic molding resins to improve the corrosion resistance of the cured resin [35].

2.3 Glycerol and its derivatives

Global biodiesel production has increased from 0.78 billion (in 2000) to 32.6 billion liters by 2016 [36]. This has made large quantities of glycerol widely available, since biodiesel production generates 10,000 liters of glycerol for every 100,000 liters of fuel produced. This oversupply of glycerol, and its low toxicity, has driven many efforts to increase the portfolio of glycerol applications both directly and indirectly (by using it as a raw material to access value-added products). Because of its sweet taste, it can be used directly as a sweetener in processed foods. It is also used as a thickener and stabilizer for foods containing water and oil. Glycerol functions well as a solvent for processing cosmetics (this is where most glycerol is used) and forms part of pharmaceutical formulations to name a few [37].

More recently, glycerol has been used as a sustainable replacement for fossil-derived monoethylene glycol in purification of bioethanol by extractive distillation. Using glycerol, up to 99% purity of bioethanol is recovered [38]. Although its high viscosity and boiling point complicate reaction workups, several examples of successful Aza-Michael addition and Suzuki-Miyaura and Mizoroki-Heck reactions in glycerol have been reported [38, 39].

The conversion of glycerol leads to a variety of other products with solvent properties such as solketal, glycerol formal, glycerol carbonate, glycerol oligomers and polymers, deep eutectic solvents (DESs) as well as diols (**Figure 1**) [37, 40].

Solketal and glycerol formal are the products of acid-catalyzed acetone and formaldehyde reactions with glycerol, respectively [41], and currently solketal is sold commercially under the name Augeo™ SL 191. It is used as a solvent in interior-scenting products and household cleaners [42]. Another route to solketal is by metal-catalyzed condensation of glycidol with acetone [43].

Solketal is polar and has a very high boiling point (188°C), which makes it useful as a coalescent solvent in paints and inks and in controlled release systems (where gradual release of species like pesticides and drugs is required). It is a good heat transfer fluid and fuel additive, especially in gasoline and biodiesel where it favorably reduces the latter's viscosity [41]. Its high sensitivity to acid has limited its use as a solvent for reactions.

2.3.1 Glycerol oligomers and polymers

The main route for preparing di-, tri-, tetra- and polyglycerol is by direct oligomerization or polymerization of glycerol using acid or base catalysts, such as potassium carbonate or sulfuric acid. The base catalyst is usually more efficient due to its better solubility in glycerol, albeit with poor selectivity [44]. Glycerol oligomers and polymers have hydrophilic heads which makes them useful for surfactant applications. By careful control of their length and branching during syntheses, desirable physiochemical properties can be built into them such that they can be used as lubricants, polymers and solvents. They have also been earmarked as replacement solvents for fossil-derived glycol ethers in paints, inks and cleaning agents [45].

2.3.2 Diols

In addition to methanol, ethanol and propanol [46], the hydrogenolysis of glycerol can lead to formation of ethylene glycol (EG) and propylene glycol (PG, 1,2-propanediol and 1,3-propanediol) (**Figure 1**) [47]. It is worth mentioning that diols can also be obtained from thermochemical conversion of fructose in the presence of homogeneous osmium and ruthenium catalysts [48] and from cellulose using a nickel-tungsten carbide catalyst [49]. Recently, Sappi Ltd. acquired Plaxica's Xylex® technology which they plan to utilize in valorizing the hemicellulose component of their pulp processing waste, in producing, furfural and xylitol [50]. Changchun Dacheng Industrial Group operates a commercial integrated biological and thermochemical process that manufactures EG, PG and 2,3-butanediol (2,3-BDO) from starch. Ten thousand tons of 2,3-BDO are produced annually [51].

Diols serve as dehumidifying and antifreeze agents. They have important applications as monomers (in polyester production) and work as solvents in the cosmetic and coating (varnishes and waxes) industries [46, 49].

3. Esters

3.1 Acyclic esters

Lactates are a class of non-toxic and green solvents obtained from treating lactic acid with various alcohols, such as ethanol, propanol and butanol. Lactic acid feedstock for making these solvents can be obtained *via* biochemical and thermochemical routes. The latter is economical but uses toxic hydrogen cyanide and gives a racemic mixture of lactic acid, while the biological process uses microbial fermentation technology and is more selective.

Ethyl lactate is the most common of the lactate esters, and Archer Daniels Midland Company, USA, operates a commercial production plant. Ethyl lactate has excellent physical properties, including a low vapor pressure and high boiling point (151–155°C) and solvent power (Kauri-butanol value > 1000). This makes it a good replacement for halogenated solvents, acetone and toluene and an excellent solvent for solubilizing resins and polymers. As such, it is used to dissolve plastics and to remove salts from circuit boards. Ethyl lactate also dissolves grease, inks and solder paste and strips paint [52].

Fatty acid methyl esters are produced by the transesterification of triglycerides, from vegetable oils or animal fats, with methanol for biodiesel applications. However, FAMEs can also be used as bio-based solvents and have been found to possess high solvent power. When mixed with ethyl lactate, evaporation is aided post utilization in dissolution or in cleaning industrial parts [53].

3.2 Glycerol carbonate and γ-valerolactone

Glycerol carbonate can be synthesized via direct or indirect routes. With direct routes, carbon monoxide and oxygen [54] or carbon dioxide [55] is treated with glycerol in the presence of metal catalysts such as Pd and Sn, respectively. The use of CO_2 is more desirable due to its abundance and lower toxicity in comparison to CO; however, this route is low yielding (7–35%). The indirect route involves carbonation of glycerol with an activated carbonation source such as urea, dimethyl carbonate or phosgene, and among these dimethyl carbonate is preferred from an industrial production perspective [56].

Due to its favorable properties which include high polarity, boiling point (110–115°C) and flash point (109°C) and low vapor pressure, glycerol carbonate has interesting applications as a polar protic solvent, electrolyte liquid carrier, detergent solvent, humectant and nail polish/gel stripper [57]. It has also been demonstrated to serve as a promising solvent in pretreatment of sugarcane bagasse [13].

γ-Valerolactone (GVL) can be produced from 5-hydroxymethylfurfural or furfural alcohol via their dehydration and hydrolysis, respectively [58]. The resultant levulinic acid can then be converted to GVL in the presence of hydrogen and a suitable catalyst [59, 60]. This therefore links the cost of GVL production to the production of hydrogen. This has resulted in the limited widespread production and use commercially. The increased number of hydrogen production plants by water electrolysis (or use of transfer hydrogenation techniques) should aid in improving the process and economics to favor GVL commercial-scale production in the near future.

The physical and chemical properties of GVL favor its application as a solvent and illuminating fuel. It has a low melting point, high boiling and flash point and low toxicity and has a characteristic herbal smell that can be used to identify leaks or spills [61]. GVL has been used as a solvent in many biomass conversion reactions. For example, it was used as a solvent in the pretreatment of lignocellulose, hydrolysis of lignocellulose and dehydration of carbohydrates to furans [61, 62]. It has also been used as solvent in cross-coupling reactions such as Sonogashira [63], Hiyama [64] and Mizoroki-Heck [65].

3.3 Cyrene™

Cyrene (dihydrolevoglucosenone or 6,8-dioxabicyclo[3.2.1]octanone) is a ketone functionality containing solvent that can be prepared in a two-step process from cellulose [66]. The common starting point in its synthesis is from levoglucosenone (LGO), which can be obtained from a variety of cellulosic starting feedstocks such as Bilberry presscake [67], corn cob [68], poplar wood [69] or bagasse [70]. LGO has been successfully obtained from cellulose [71–73], by the patented Furacell™ process discovered by Circa Group Ltd., Australia. The process gives 40% yield of LGO and has consequently been scaled up to a 50 ton/year production [74]. The next step after the production of LGO is its hydrogenation to afford Cyrene™ (**Figure 3**). The hydrogenation of levoglucosenone to Cyrene has been largely dominated by heterogeneous Pd catalysts [75, 76].

Concerns over the use of toxic solvents (such as dimethyl sulfoxide (DMSO), dimethylformamide (DMF) and *N*-methyl-2-pyrrolidone (NMP)) in the industry have led to the exploration of Cyrene™ as a potential replacement. These dipolar aprotic solvents have very similar properties to those of Cyrene™ [77]. As a result, more examples of its use as solvent in syntheses are becoming popular. For example, metal organic frameworks that were previously prepared in DMF solvent have been successfully synthesized in Cyrene™ [78]. Cyrene™ has also been used in amide bond formation reactions [79].

3.4 Alkanes and aromatics

3.4.1 Alkanes

Furfural and hydroxymethylfuran can react with acetone (derived from fermentative processes) resulting in aldol-condensed products with longer carbon chain lengths. Subsequent hydrodeoxygenation of the aldol condensation products using a supported Pd catalyst, H_2 (plus acetic and/or Lewis acid cocatalysts), produces alkanes as shown in **Figure 4** [80–83].

Alkanes are useful for reactions requiring nonpolar medium because they are generally unreactive. Consequently, they have been used as reaction medium during

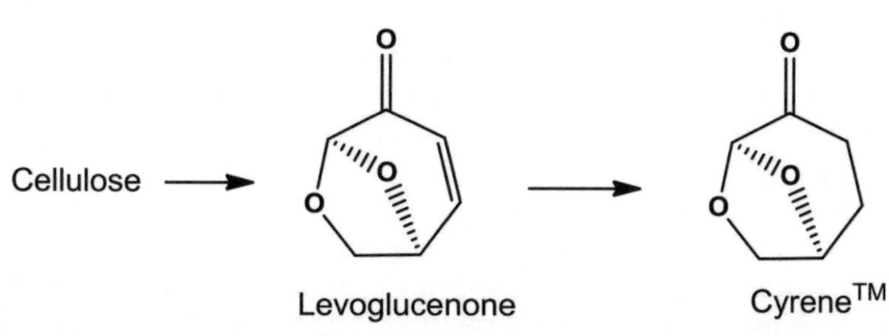

Figure 3.
Synthetic steps of Cyrene™ from cellulose.

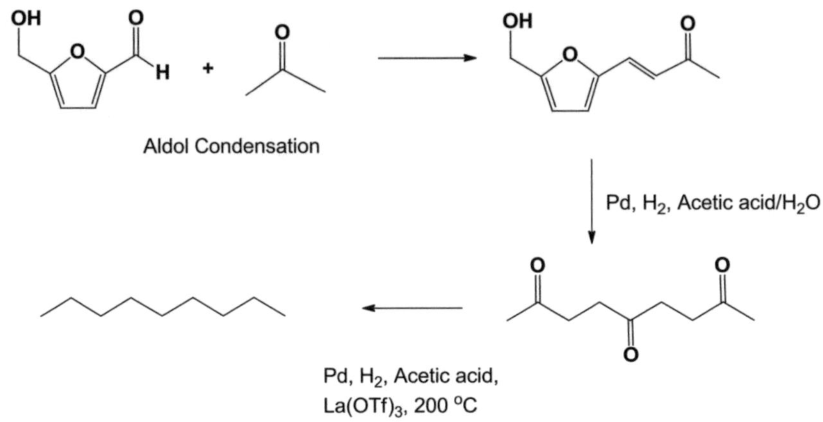

Figure 4.
Synthetic steps for straight-chain alkanes from bio-derived furan compounds.

the synthesis of drugs, pesticides and other chemicals. Their use as fuels varies depending on the length of the carbon chains. Generally C_3 and C_4 alkanes are used as liquefied petroleum gas for cooking and in cigarette lighters, respectively, while C_5–C_{18} alkanes are used in gasoline, diesel and aviation fuel.

3.4.2 Benzene, toluene and xylenes

Benzene, toluene and xylene (BTX) can be derived from wood and agricultural waste [84]. Anellotech, a US-based company, has scaled up a process that converts biomass, using their Bio-TCat™ technology, into BTX mixtures. Benzene, toluene and xylene have identical properties as their fossil-derived counterparts [85]. Benzene is used in the manufacture of resins, rubber lubricants, synthetic fibers, detergents, pesticides, drugs and plastics. Toluene is used in printing and leather tanning processes and as a solvent in paint, a nail polish remover, thinners and glues, while xylene finds use as a cleaning agent, ingredient in pesticide manufacturing, disinfectants, paints, paint thinners, polishes, waxes and adhesives.

An elegant multistep route to renewable *p*-xylene was proposed. It begins with (1) the conversion of glucose to fructose, (2) followed by dehydration of the fructose to HMF, (3) hydrodeoxygenation of HMF to 2,5-dimethylfuran (2,5-DMF) and (4) subjecting the resultant 2,5-DMF to Diels-Alder cycloaddition with ethylene to afford the product [86].

3.4.3 Terpenes

Terpenes are a class of natural solvents obtained from essential oils found in plants. They have C_5H_8 isoprene units and can be acyclic, bicyclic or monocyclic; therefore, they have varying physical and chemical properties. Extraction using water hetero-azeotropic distillation at low temperatures affords terpenes, and they in turn have been utilized as green alternative solvents for the extraction of oils from microalgae [87].

D-Limonene is extracted from citrus peels and pulp by steam distillation and alkali treatment [88]. Industrially, limonene is used as a solvent in place of various halogenated hydrocarbons. By 2023, the demand of D-limonene is expected to be 65 kilotons/year [88]. It is also used as solvent to degrease and grease wool and cotton wool and in the dissolution of cholesterol stone, where its performance is better than chloroform and diethyl ether (DEE) in the latter. The catalytic isomerization

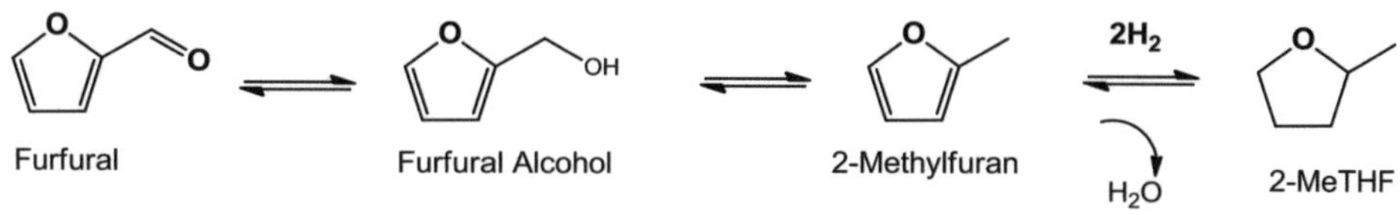

Figure 5.
The successive hydrogenation of furfural to 2-methyltetrahydrofuran.

and dehydrogenation of D-limonene give *p*-cymene, an aromatic hydrocarbon with equally good solvent properties [89].

α-Pinene is the most widely available terpenoid and is mostly found in essential oils of coniferous trees, or it can be recovered from paper pulping by-products, i.e. from crude sulphate turpentine [90]. Alpha-pinene is used as a household cleaning solvent and repellent for insects and in the production of perfumes. It also have medicinal properties as an anti-inflammatory and antibiotic agent.

3.5 Ethers

Bio-derived ethers are produced by the dehydration of bio-alcohols, mainly ethanol and methanol to form diethyl ether and dimethyl ether, respectively. These ethers are usually used as fuel additives to improve its octane rating and reduce emissions (NO_x and ozone) and engine wear.

2-Methyltetrahydrofuran (2-MeTHF) is a biodegradable, non-toxic, non-ozone-depleting, easy-to-recycle ether solvent with a good preliminary toxicology report [91]. It is produced from either furfural or levulinic acid *via* catalytic processes. For instance, the successive hydrogenation of furfural over Ni-Cu, Fe-Cu, Cu-Zn or Cu-Cr produces 2-MeTHF as illustrated in **Figure 5** [92, 93].

The first two hydrogenations have been reported to quantitatively convert furfural alcohol to 2-methylfuran using Ni-based catalysts followed by 2-methylfuran conversion over a Cu-Zn catalyst. 2-MeTHF can also be prepared from levulinic acid *via* consecutive hydrogenation and dehydrogenation steps. Here, Ru-based catalysts are the most active.

Pfizer, USA, has reported the application of 2-MeTHF as a solvent in two-phase reactions due to the poor solubility of water in 2-MeTHF [94]. This bio-derived solvent has also been used in place of dichloromethane (that has unfavourable environmental effects) because of its low boiling point, which makes it difficult to contain on a large scale. Using 2-MeTHF, high yields of product were afforded for amidations, alkylations and nucleophilic aromatic substitutions reactions.

Another pharmaceutical company, Actelion Pharmaceuticals Ltd. in Switzerland, uses 2-MeTHF as solvent in the synthesis of 5-phenylbicyclo-[2.2.2] oct-5-en-2-one [95], an intermediate during the preparation of an important pre-clinical candidate. Furthermore, the reaction was successfully upscaled to kilogram scale, and the reaction still gave excellent yields (98%) in 2-MeTHF.

4. Ionic liquids

Ionic liquids (ILs), defined here as materials that are made up of cations and anions (salts) which melt at or below 100°C, evolved from the nineteenth century. This field started with a report by Paul Walden, in which he studied the physical properties of ethylammonium nitrate ($[EtNH_3][NO_3]$), a salt which melted at around 14°C [96]. Ionic liquids are commonly formed through the combination of an organic cation (usually heterocyclic), such as dialkylimidazolium, and either an inorganic or organic anion (such as halides or methanesulfonate) [97]. The advantage with ionic liquids is their low vapor pressure, meaning the risk of atmospheric contamination and related health issues are reduced by using these solvents. It is for this particular reason that they are viewed as green solvents [98]; however, low volatility alone is not the only property that makes ionic liquids green. For instance, toxic and non-biodegradable ionic liquids will not be referred to as a green solvent even though they have negligible volatility [97]. In this regard, ionic liquids composed of bio-derived components have the added advantage and are outstanding candidates as sustainable and green solvents.

4.1 Sugar-based ionic liquids

Since the depolymerization of hard-to-dissolve carbohydrate polymers can be achieved using ILs, it would be beneficial to develop sugar-based ILs with the aim of employing these in a 'closed-loop' carbohydrate polymer depolymerization process [99]. Some sugar-based ionic liquids which have potential in this regard have already been synthesized, even though they have been used for various other processes. The glucose-linked 1,2,3-triazolium ionic liquids were synthesized by copper(I)-catalysed regioselective cycloaddition of a glucose azide to a glucose alkyne, followed by quaternization with methyl iodide (**Figure 6**) [100]. The ILs were used as reusable chiral solvents and ligands in copper(I)-catalysed amination of aryl halides with aqueous ammonia. The triazolium salt affords the compound in its liquid state at room temperature, and hence the compound can be used as a solvent, while the free hydroxyl groups of the glucose moiety aids in stabilizing the copper(I) species during the reaction.

Methyl-D-glucopyranoside has also been used to synthesize an ionic liquid with promising solvent potential [101]. The synthetic sequence involved uses thexyldimethylsilyl chloride (TDSCl) to protect the hydroxyl at the primary position. The other secondary hydroxyl groups were converted to methyl ethers followed by reduction of the anomeric position and further deprotection of the primary hydroxyl. After deprotection of the primary hydroxyl group, it was then converted to a triflate to form an intermediate. This triflate intermediate finally underwent a nucleophilic substitution reaction using diethyl sulphide to afford an ionic liquid, with a triflate anion and a sulfonium cation, which was a liquid at room temperature (**Figure 6**).

Handy et al. have used fructose to synthesize room temperature ionic liquids [102]. In their synthetic protocol, copper carbonate, ammonia and formaldehyde were used to ring-close fructose and form hydroxymethylene imidazole. This imidazole was then taken through a series of alkylation steps (with 1-bromobutane followed by iodomethane) to form an imidazolium cation. Anion metathesis was performed to give a series of room temperature ionic liquids (**Figure 7**), which were used as recyclable solvents for Mizoroki-Heck cross-coupling of aryl iodides with alkenes.

An arabinose-based imidazolium IL was synthesized in 2018 starting with 2,3,5-tri-*O*-benzyl-D-arabinofuranose [103]. The pentofuranoside starting material was prepared by benzylating all the hydroxyl groups of D-arabinose, except the hydroxyl group on the carbon adjacent the ring oxygen atom. The free hydroxyl group was then reacted with propane-1,3-diyldioxyphosphoryl chloride in the presence of 1-methylimidazole to form a mixture of anomeric phosphates. This was then reacted with 1-methylimidazolium chloride and trimethylsilyl triflate, in catalytic amounts, to give a pure anomeric ionic liquid with a chloride anion. Anion metathesis reactions gave two additional ionic liquids (**Figure 7**), which were used as co-solvents in the synthesis of alcohols from aromatic aldehydes.

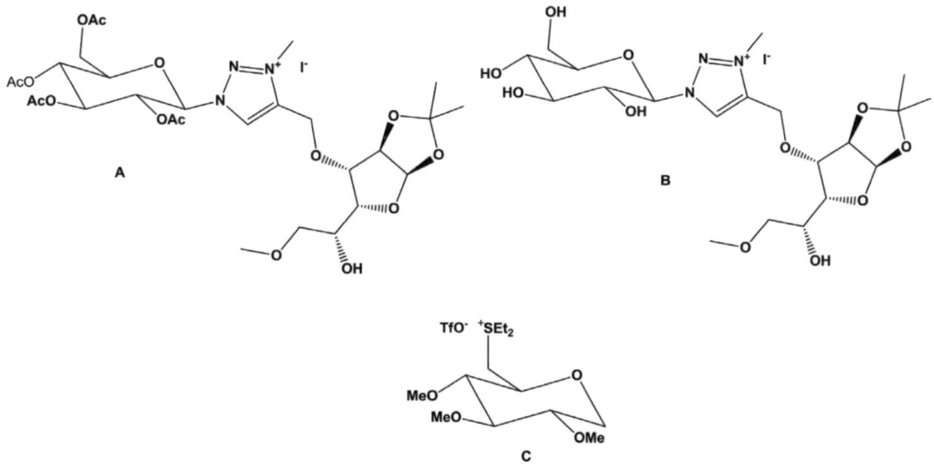

Figure 6.
Glucose-based ionic liquids [(A and B) glucose-tagged triazolium ILs; (C) glucopyranoside-based IL].

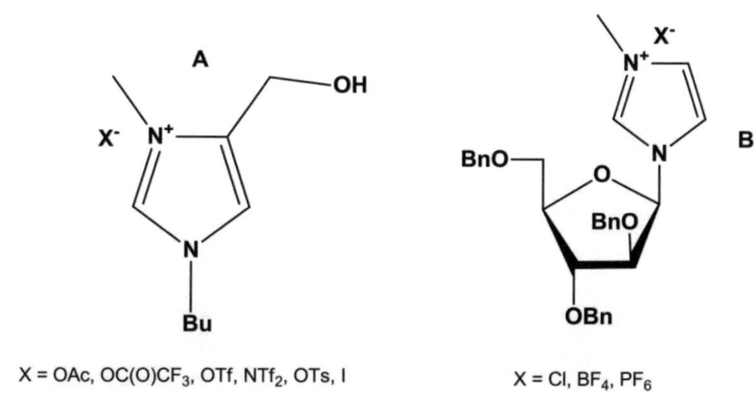

Figure 7.
ILs based on pentoses [(A) fructose-based ILs; (B) arabinose-based ILs].

4.2 Alkaloid-based ionic liquids

Ionic liquids which contain ampicillin as active pharmaceutical ingredients were developed by Ferraz et al., by neutralizing basic ammonia solutions of ampicillin with different organic cation hydroxides. These ampicillin-based ILs may be useful in the development of bioactive materials [104]. A chiral IL derived from ephedrine was synthesized by Wasserscheid et al. N-methylephedrine was synthesized from ephedrine and alkylated with dimethyl sulphate followed by ion exchange to form the IL, with a melting point of 54°C [105]. Heckel et al. reported the synthesis of ionic liquids derived from nicotine by the quaternization of the pyridine ring of nicotine with methyl iodide and ethyl bromide. The resulting chiral nicotine-based ILs were examined as chiral solvating agents (**Figure 8**) [106].

4.3 Lipid-based ionic liquids

Linoleate and oleate were used in the syntheses of four ILs with melting points below −21°C (**Figure 9**). Unsaturated fatty acids were initially taken through neutralization reactions with NaOH followed by ion exchange with either

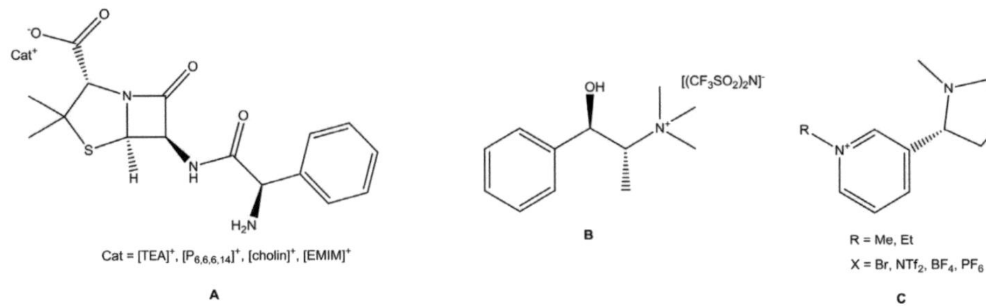

Figure 8.
Alkaloid-based ILs [(A) ampicillin-based ILs; (B) ephedrine-based IL; (C) nicotine-based ILs].

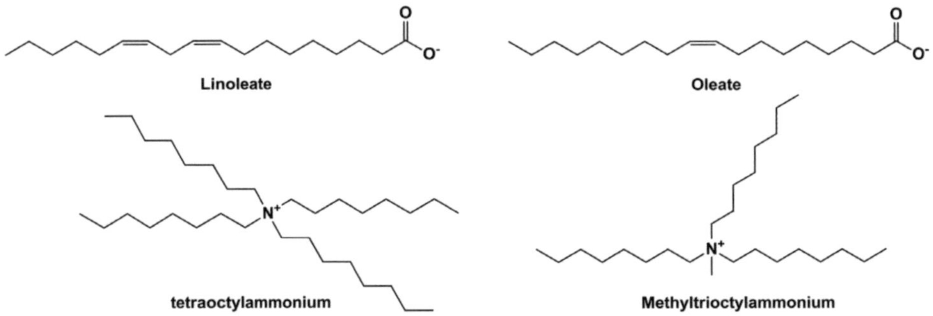

Figure 9.
Anions and cations used in the syntheses of lipid-based ILs.

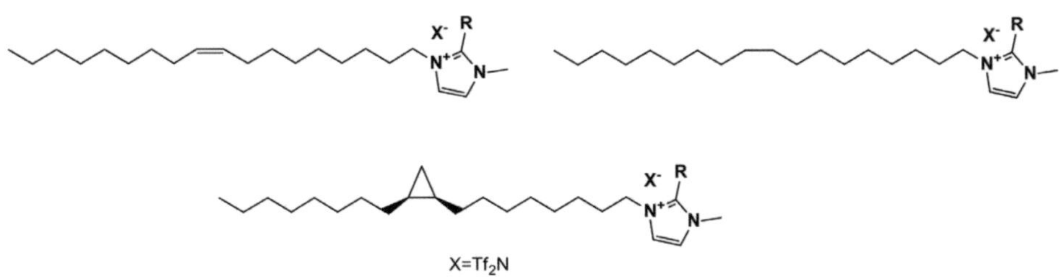

Figure 10.
Lipid-based imidazolyl ILs.

tetraoctylammonium or methyltrioctylammonium chloride. The ILs were then successfully used as solvents in the extraction of metal ions from aqueous solutions [107].

Kwan et al. also synthesized ILs using lipids through alkylating a tertiary amine (**Figure 10**). The lipids used in this instance were methyl oleate and methyl stearate. In a third instance, cyclopropanated oleic acid methyl ester synthesized the reaction of the double bond of the oleate with diiodomethane, and diethylzinc was used to alkylate the tertiary amine. After reacting the obtained alkyl iodide with the imidazoles, anion exchange of the iodide with bistriflimide gave imidazolium bistriflimide ILs [108].

5. Deep eutectic solvents

Deep eutectic solvents are analogous to ILs in the sense that they possess similar properties of low melting point and non-volatility, but they are, however, two different types of solvent systems. Unlike ILs, which are made up of one type of discrete cation and anion [17], DESs are formed from an eutectic mixture of a hydrogen bond acceptor (HBA), mostly quaternary ammonium salts, and a metal salt or hydrogen bond donor (HBD). The mixture forms a eutectic phase which has a lower melting point than the individual components [109]. There is a charge delocalization created through hydrogen bonding between, for example, a halide anion of the HBA and the HBD which causes the decrease in the melting point [18]. DESs are usually formed from relatively inexpensive bio-based components which makes them biodegradable and affordable [17]. Abbott et al. initially developed DESs based on choline chloride and zinc chloride as cheaper alternatives to imidazolium-based ionic liquids, which would make them readily accessible for bulk applications in syntheses. These new solvents were made by heating mixtures of the choline chloride and zinc chloride in molar ratios of 1:1, 1:2 and 1:3 until clear colorless liquids were obtained, with the 1:2 mixture giving the lowest freezing point of 25°C [110]. Choline is a provitamin which is derived from lecithin, found in plants and animal organs [111]. As such, choline-derived

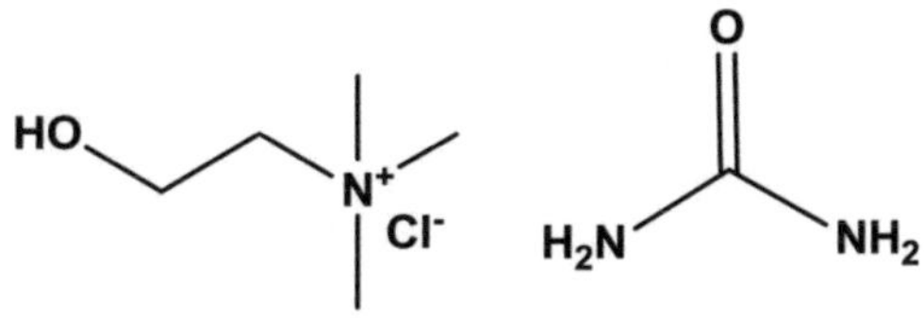

Figure 11.
DES based on choline chloride and urea.

Hydrogen bond donor (HBD)	ChCl/HBD ratio (mol/mol)	M.p. of HBD/°C
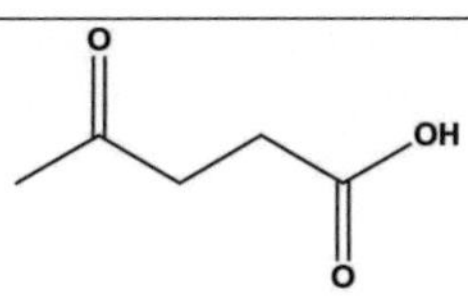 Levulinic acid	1:2	32
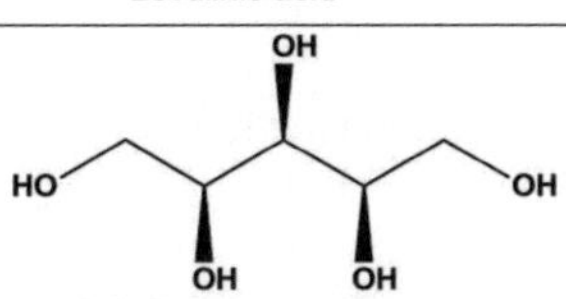 Xylitol	1:1	96
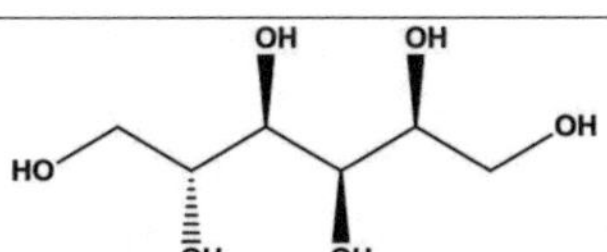 D-Sorbitol	1:1	99
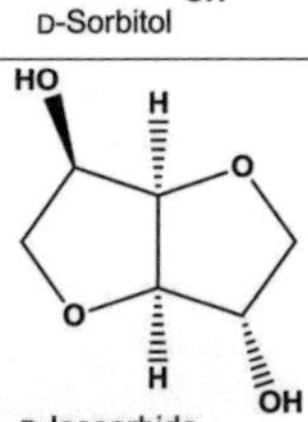 D-Isosorbide	1:2	62

Table 1.
Novel room temperature liquid DESs derived from renewable sources [119].

DESs are bio-based solvents, particularly where it is paired up with bio-sourced HBDs such as lactic acid, levulinic acid, glycerol and sugars. On a large scale, choline is produced by a single-step reaction between HCl, ethylene oxide and trimethylamine [17].

DESs based on urea (**Figure 11**) were prepared by mixing choline chloride (m.p. 302°C) and urea (m.p. 133°C) in different molar ratios. This showed that a eutectic phase occurs at a choline to urea ratio of 2. This DES has a very low freezing point of 12°C, which was significantly lower than its constituents, thus giving a room temperature solvent [112]. The choline chloride-urea DES has been used as both a catalyst and solvent for the selective *N*-alkylation of various aromatic primary amines. This method avoids the complexity of multiple alkylations, which is a problem encountered when polar volatile organic solvents are used. After the reaction, DES recycling by simple biphasic extraction with ethyl acetate was carried out, and the DES was reused at least five times (with just a slight loss in activity) [113]. This same group have used the choline chloride-urea DES as a solvent and catalyst for the bromination of 1-aminoanthra-9,10-quinone [114] and for the Perkin reaction [115].

DESs based on choline chloride and glycerol or ethylene glycol have also been used as extraction solvents to remove excess glycerol from biodiesel [116–118]. Also, new choline chloride-based DESs with levulinic acid and sugar-based polyols as renewable hydrogen bond donors were synthesized. The best ratios of choline chloride (ChCl) and HBD which gave liquid DESs at room temperature are shown in **Table 1** [119]. DESs based on different bio-based HBD (oxalic, lactic and malic acids) and HBA (choline chloride, betaine, alanine, glycine, histidine, proline and nicotinic acid) have also been prepared and tested as solvents for the dissolution of lignin, cellulose and starch. Majority of the resulting solvents exhibited high lignin solubility but poor cellulose solubility.

This was advantageous since DESs can be used in separating lignin from cellulose [120]. DESs can be viewed as a more environmentally friendly alternative to volatile organic solvents—they are affordable, easily prepared and scalable, inflammable and also biodegradable [121].

6. Conclusions

Rising concerns over depletion of fossil reserves and the drive towards sustainable and responsible consumption of resources have presented a challenge and opportunity for scientist and engineers to work together in developing methods, technologies and processes for producing of bio-based products. These include commodity chemicals, fuels and materials of which many of the chemicals have solvent properties. As such, they have been widely used in extractions and in solvating a range of physiochemical processes and numerous consumer goods (by either dissolving, stabilizing or modifying the latter's properties) across the manufacturing research and development landscapes. This review has contextualized and discussed some existing and emerging approaches in these areas.

Even though most bio-solvents are still volatile like their petroleum counterparts, the bio-solvents have the advantage of being biodegradable, derived from renewable resources, and their production often results in CO_2 emission savings. A novel array of ionic liquids and deep eutectic solvents based on inexpensive bio-derived components has been developed, and because these solvents have low volatility, it presents a solution to the volatile organic solvent concern. Additionally, they are usually recyclable after use, with the benefit of tunability to bring about smart solvents for tailored applications.

DOI: http://dx.doi.org/10.5772/intechopen.86502

Acknowledgements

This work was supported by the Royal Society and African Academy of Sciences, the Future Leaders—African Independent Researchers (FLAIR) Scheme (Fellowship ref.: 191779), the National Research Foundation (NRF) of South Africa (Grant numbers: 105559, 112809 and 117989) and the University of Johannesburg's Centre for Synthesis and Catalysis.

Conflict of interest

The authors declare no conflicts of interest.

Author details

Novisi K. Oklu, Leah C. Matsinha and Banothile C.E. Makhubela*
Department of Chemical Sciences, University of Johannesburg, Johannesburg, South Africa

*Address all correspondence to: bmakhubela@uj.ac.za

IntechOpen

© 2019 The Author(s). Licensee IntechOpen. This chapter is distributed under the terms of the Creative Commons Attribution License (http://creativecommons.org/licenses/by/3.0), which permits unrestricted use, distribution, and reproduction in any medium, provided the original work is properly cited.

References

[1] Kerton FM. Alternative Solvents for Green Chemistry. Cambridge: Royal Society of Chemistry; 2009. DOI: 10.1039/9781847559524

[2] Anastas PT, Levy IJ, Parent KE. Green Chemistry Education. Washington: American Chemical Society; 2009. DOI: 10.1021/bk-2009-1011.fw001

[3] Curzons AD, Constable DJC, Mortimer DN, Cunningham VL. So you think your process is green, how do you know?—Using principles of sustainability to determine what is green–a corporate perspective. Green Chemistry. 2001;**3**:1-6. DOI: 10.1039/b007871i

[4] Prat D, Wells A, Hayler J, Sneddon H, Mcelroy CR, Abou-shehada S, et al. CHEM21 selection guide of classical- and less classical-solvents. Green Chemistry. 2016;**18**:288-296. DOI: 10.1039/c5gc01008j

[5] Diorazio LJ, Hose DRJ, Adlington NK. Toward a more holistic framework for solvent selection. Organic Process Research and Development. 2016;**20**:760-773. DOI: 10.1021/acs.oprd.6b00015

[6] Capello C, Fischer U, Hungerbu K. What is a green solvent ? A comprehensive framework for the environmental assessment of solvents. Green Chemistry. 2007;**9**:927-934. DOI: 10.1039/b617536h

[7] Makhubela BCE, Darkwa J. The role of noble metal catalysts in conversion of biomass and bio-derived intermediates to fuels and chemicals. Johnson Matthey Technology Review. 2018;**62**(1):4-31. DOI: 10.1595/205651317X696261

[8] Clark JH, Hunt AJ, Topi C, Paggiola G, Sherwood J. Sustainable Solvents Perspectives from Research, Business and International Policy. Croydon: Royal Society of Chemistry; 2017. DOI: 10.1039/9781782624035

[9] Biochemtex, Crescentino Biorefinery. A New Era Begins: World's First Advanced Biofuels Facility. ETIP Bioenergy-SABS [Internet]. 2013. Available from: http://www.etipbioenergy.eu/images/crescentino-presentation.pdf [Accessed: March 25, 2019]

[10] Weizmann C. Production of Acetone and Alcohol by Bacteriological Processes 1315585; 1919

[11] Nguyen NPT, Raynaud C, Meynial-salles I, Soucaille P. Reviving the Weizmann process for commercial n-butanol production. Nature Communications. 2018;**9**:1-8. DOI: 10.1038/s41467-018-05661-z

[12] Opre JE, Bergemann EP, Henneberry M. Environmentally Friendly Solvent. 6191087 B1; 2001

[13] Zhang Z, Rackemann DW, Doherty WOS, Hara IMO. Glycerol carbonate as green solvent for pretreatment of sugarcane bagasse. Biotechnology for Biofuels. 2013;**6**(1):1. DOI: 10.1186/1754-6834-6-153

[14] Al-shaal MG, Dzierbinski A, Palkovits R. Solvent-free γ-valerolactone hydrogenation to 2-methyltetrahydrofuran catalyzed by Ru/C: A reaction network analysis. Green Chemistry. 2014;**16**:1358-1364. DOI: 10.1039/c3gc41803k

[15] Huber GW, Cortright RD, Dumesic JA. Renewable alkanes by aqueous-phase reforming of biomass-derived oxygenates. Angewandte Chemie International Edition. 2004;**43**:1549-1551. DOI: 10.1002/anie.200353050

[16] Anellotech Incorporated, New York, USA [Internet]. Available from: http://

anellotech.com/bio-tcat%E2%84%A2-renewable-chemicals-fuels

[17] Smith EL, Abbott AP, Ryder KS. Deep eutectic solvents (DESs) and their applications. Chemical Reviews. 2014;**114**:11060-11082. DOI: 10.1021/cr300162p

[18] Domínguez de María P. Recent trends in (ligno)cellulose dissolution using neoteric solvents: Switchable, distillable and bio-based ionic liquids. Journal of Chemical Technology and Biotechnology. 2014;**89**(1):11-18. DOI: 10.1002/jctb.4201

[19] Abengoa Bioenergy Corporation. Europe's Largest Bioethanol Producer [Internet]. Available from: http://www.abengoa.com/export/sites/abengoa_corp/resources/pdf/en/gobierno_corporativo/informes_anuales/2005/2005_Volume1_AnnualReport_Bioenergy.pdf

[20] Cellulosic Ethanol Technology. Ongoing Research and Novel Pathways', ETIP Bioenergy-SABS [Internet]. Available from: http://www.biofuelstp.eu/cellulosic-ethanol.html#ce6 [Accessed: March 30, 2019]

[21] Commercial Cellulosic Ethanol Plants in Brazil' and 'Cellulosic Ethanol in Canada', ETIP Bioenergy-SABS [Internet]. Available from: http://www.etipbioenergy.eu/value-chains/products-end-use/products/cellulosic-ethanol#brazil; http://www.etipbioener

[22] A Novel Commercial Method for Extracting Artemisinin: Extracting Artemisia Annua with Ethanol and Purifying Ethanolic Extracts. Scott Process Technology Ltd, [Internet]. Available from: https://www.mmv.org/sites/default/files/uploads/docs/artemisinin/06c_

[23] Hansen JB, Nielsen PEH. Methanol synthesis. In: Handbook of Heterogeneous Catalysis. 2nd ed. Vol. 1. Weinheim: Wiley-VCH Verlag GmbH & Co. KgaA; 2008

[24] Morone P, Cottoni L. Biofuels: Technology, economics, and policy issues. In: Handbook of Biofuels Production. 2nd ed. Duxford, UK: Woodhead Publishing; 2016

[25] Hu X, Westerhof RJM, Wu L, Dong D, Li CZ. Upgrading biomass-derived furans via acid-catalysis/hydrogenation: The remarkable difference between water and methanol as the solvent. Green Chemistry. 2014;**17**:219-224. DOI: 10.1039/c4gc01826e

[26] Bennekom JG, Venderbosch RH, Heeres HJ. Biomethanol from glycerol. In: Biodiesel-Feedstocks, Production and Applications. London, UK: IntechOpen; 2012. DOI: 10.5772/53691

[27] Speers AM, Young JM, Reguera G. Fermentation of glycerol into ethanol in a microbial electrolysis cell driven by a customized consortium. Environmental Science and Technology. 2014;**48**(11):6350-6358. DOI: 10.1021/es500690a

[28] Maglinao RL, He BB. Catalytic thermochemical conversion of glycerol to simple and polyhydric alcohols using Raney nickel catalyst. Industrial and Engineering Chemistry Research. 2011;**50**:6028-6033. DOI: 10.1021/ie102573m

[29] Ndaba B, Chiyanzu S, Marx I. n-Butanol derived from biochemical and chemical routes: A review. Biotechnology Reports. 2015;**8**:1-9. DOI: 10.1016/j.btre.2015.08.001

[30] Uyttebroek M, Van Hecke W, Vanbroekhoven K. Sustainability metrics of 1-butanol. Catalysis Today. 2015;**239**:7-10. DOI: 10.1016/j.cattod.2013.10.094

[31] Mascal M. Chemicals from biobutanol: Technologies and markets.

Biofuels, Bioproducts and Biorefining. 2012;**6**(4):483-493. DOI: 10.1002/bbb.1328

[32] Grand View Research, Marketing Research and Consulting, Global Furfural Market By Application (Furfuryl Alcohol, Solvents) Expected To Reach USD 1 200.9 Million By 2020. The Global Furfural Market By Application (Furfuryl Alcohol, Solvents) Expected To Reach USD 1,200.9 Million By 2020; 2015

[33] Available from: https://ihsmarkit.com/products/furfural-chemical-economics-handbook.html [Accessed: March 11, 2019–March 11, 2018], https://IhsmarkitCom/Products/Furfural-Chemical-Economics-HandbookHtml [Accessed: March 11, 2019]

[34] Available from: https://www.bechtel.com/services/oil-gas-chemicals/bhts/oil-processing/furfural-refining/ [Accessed: March 11, 2019], https://WwwBechtelCom/Services/Oil-Gas-Chemicals/Bhts/Oil-Processing/Furfural-Refining/ [Accessed: March 11, 2019]

[35] Available from: http://www.furan.com/furfural_applications_of_furfural.html [Accessed: March 11, 2018], http://WwwFuranCom/Furfural_applications_of_furfuralHtml [Accessed: March 11, 2018]

[36] World Bioenergy Association. Global Bioenergy Statistics. 2018. Available from: https://worldbioenergy.org/uploads/181203%20WBA%20GBS%20 2018_hq.pdf [Accessed: April 1, 2019]

[37] Pagliaro M, Rossi M. Glycerol: Properties and Production. The Future of Glycerol: New Uses of a Versatile Raw Material. Cambridge: Wiley-VCH; 2008. DOI: 10.1002/cssc.200800115

[38] Gu Y, Jerome F. Glycerol as a sustainable solvent for green chemistry. Green Chemistry. 2010;**12**:1127-1138. DOI: 10.1039/c001628d

[39] Wolfson A, Dlugy C, Shotland Y. Glycerol as a green solvent for high product yields and selectivities. Environmental Chemistry Letters. 2007;**5**:67-71. DOI: 10.1007/s10311-006-0080-z

[40] Pagliaro M, Rossi M. Aqueous Phase Reforming. The Future of Glycerol: New Uses of a Versatile Raw Material. Cambridge: Wiley-VCH; 2008. DOI: 10.1002/cssc.200800115

[41] Moity L, Benazzouz A, Molinier V, Nardello-Rataj V, Elmkaddem MK, De Caro P, et al. Glycerol acetals and ketals as bio-based solvents: Positioning in Hansen and COSMO-RS spaces, volatility and stability towards hydrolysis and autoxidation. Green Chemistry. 2015;**17**(3):1779-1792. DOI: 10.1039/c4gc02377c

[42] Solvay. Augeo™ SL 191 [Internet]. Available from https://www.solvay.us/en/markets-and-products/featured-products/augeo.html, http://www.youliao.com.cn/Uploads/Download/Document/tds/20398_tds.PDF [Accessed: April 1, 2019]

[43] Ellis JE, Lenger SR. A convenient synthesis of 3,4-dimethoxy-5-hydroxybenzaldehyde. Synthetic Communications. 1998;**28**(9):1517-1524. DOI: 10.1080/00397919808006854

[44] Salehpour S, Dube MA. Towards the sustainable production of higher-molecular-weight polyglycerol. Macromolecular Chemistry and Physics. 2011;**212**:1284-1293. DOI: 10.1002/macp.201100064

[45] Sutter M, Da Silva E, Duguet N, Raoul Y, Metay E, Lemaire M. Glycerol ether synthesis: A bench test for green chemistry concepts and technologies. Chemical Reviews. 2014;**115**(16):8609-8651. DOI: 10.1021/cr5004002

[46] Perosa A, Tundo P. Selective hydrogenolysis of glycerol with Raney

nickel. Industrial and Engineering Chemistry Research. 2005;**44**:8535-8537. DOI: 10.1021/ie0489251

[47] Bricker ML, Leonard LE, Kruse TM, Vassilakis JG, Bare SR. Methods for Converting Glycerol to Propanol. 8101807 B2; 2012

[48] Crabtree SP, Tyers DV. Hydrogenolysis of Sugar Feedstock; 2007

[49] Ji N, Zhang T, Zheng M, Wang A, Wang H, Wang X, et al. Direct catalytic conversion of cellulose into ethylene glycol using nickel-promoted tungsten carbide catalysts. Angewandte Chemie International Edition. 2008;**47**: 8510-8513. DOI: 10.1002/anie.200803233

[50] Sappi Limited. Sappi Invests in Sugar Separations and Clean-Up Technology to Strengthen its Renewable Bio-Chemicals Offering [Internet]. 2017. Available from: https://www.sappi.com/sappi-invests-sugar-separations-and-clean-technology-strengthe

[51] Ge L, Wu X, Chen J, Wu J. A new method for industrial production of 2,3-butanediol. Journal of Biomaterials and Nanobiotechnology. 2011;**2**:335-336. DOI: 10.1007/s002530000486

[52] Archer Daniels Midland Company. Fuels and Industrials Catalogue, ADM Solvents [Internet]. 2019. p. 16. Available from: https://assets.adm.com/Products-And-Services/Industrials/ADM-Fuels-and-Industrials-Catalog.pdf [Accessed: April 6, 2019]

[53] Gonnzalez YM, Thiebaud-roux YMG, De Caro P, Lacaze-Dufaure C. Fatty acid methyl esters as biosolvents of epoxy resins: A physicochemical study. Journal of Solution Chemistry. 2007;**36**:437-446. DOI: 10.1007/s10953-007-9126-5

[54] Hu J, Gu Y, Guan Z, Li J, Mo W, Li T, et al. An efficient palladium catalyst system for the oxidative carbonylation of glycerol to glycerol carbonate. ChemSusChem. 2011;**4**:1767-1772. DOI: 10.1002/cssc.201100337

[55] Aresta M, Dibenedetto A, Nocito F, Pastore C. A study on the carboxylation of glycerol to glycerol carbonate with carbon dioxide: The role of the catalyst, solvent and reaction conditions. Journal of Molecular Catalysis A: Chemical. 2006;**257**:149-153. DOI: 10.1016/j.molcata.2006.05.021

[56] Ochoa-Gomez RJ, Gomez-Jimenez-Abersturi O, Ramirez-Lopez C, Belsue M. A brief review on industrial alternatives for the manufacturing of glycerol carbonate, a green chemical. Organic Process Research and Development. 2012;**16**:389-399. DOI: 10.1021/op200369v

[57] Sonnati MO, Amigoni S, De Givenchy EPT, Darmanin T, Choulet O, Guittard F. Glycerol carbonate as a versatile building block for tomorrow: Synthesis, reactivity, properties and applications. Green Chemistry. 2013;**15**(2005):283-306. DOI: 10.1039/c2gc36525a

[58] Clarke CJ, Tu WC, Levers O, Bröhl A, Hallett JP. Green and sustainable solvents in chemical processes. Chemical Reviews. 2018;**118**(2):747-800. DOI: 10.1021/acs.chemrev.7b00571

[59] Amenuvor G, Makhubela BCE, Darkwa J. Efficient solvent-free hydrogenation of levulinic acid to γ-valerolactone by pyrazolylphosphite and pyrazolylphosphinite ruthenium(II) complexes. ACS Sustainable Chemistry & Engineering. 2016;**4**:6010-6018. DOI: 10.1021/acssuschemeng.6b01281

[60] Amenuvor G, Darkwa J, Makhubela BCE. Homogeneous polymetallic ruthenium(ii)^zinc(ii) complexes: Robust catalysts for the efficient hydrogenation of levulinic acid to

γ-valerolactone. Catalysis Science and Technology. 2018;**8**(9):2370-2380. DOI: 10.1039/c8cy00265g

[61] Zhang Z. Synthesis of γ-valerolactone from carbohydrates and its applications. ChemSusChem. 2016;**9**(2):156-171. DOI: 10.1002/cssc.201501089

[62] Zhang L, Yu H, Wang P, Li Y. Production of furfural from xylose, xylan and corncob in gamma-valerolactone using FeCl3·6H2O as catalyst. Bioresource Technology. 2014;**151**:355-360. DOI: 10.1016/j.biortech.2013.10.099

[63] Dougan L, Tych KM, Hughes ML. Article Type a Single Molecule Approach to Investigate the Role of Hydrogen Bond Strength on Protein Mechanical Compliance and Unfolding History2014. pp. 8-10. DOI: 10.1039/b000000x

[64] Ismalaj E, Strappaveccia G, Ballerini E, Elisei F, Piermatti O, Gelman D, et al. γ-Valerolactone as a Renewable Dipolar Aprotic Solvent Deriving from Biomass Degradation for the Hiyama Reaction2014. DOI: 10.1021/sc5004727

[65] Strappaveccia G, Ismalaj E, Petrucci C, Lanari D, Marrocchi A, Drees M, et al. A biomass-derived safe medium to replace toxic dipolar solvents and access cleaner heck coupling reactions. Green Chemistry. 2015;**17**(1):365-372. DOI: 10.1039/c4gc01677g

[66] Camp JE. Bio-available solvent cyrene: Synthesis, derivatization, and applications. ChemSusChem. 2018;**11**(18):3048-3055. DOI: 10.1002/cssc.201801420

[67] Zhou L, Lie Y, Briers H, Fan J, Remón J, Nyström J, et al. Natural product recovery from bilberry (*Vaccinium myrtillus* L.) presscake via microwave hydrolysis. ACS Sustainable Chemistry & Engineering. 2018;**6**(3):3676-3685. DOI: 10.1021/acssuschemeng.7b03999

[68] Zhuang G, Bai J, Wang X, Leng S, Zhong X, Wang J, et al. Role of pretreatment with acid and base on the distribution of the products obtained via lignocellulosic biomass pyrolysis. RSC Advances. 2015;**5**(32):24984-24989. DOI: 10.1039/c4ra15426f

[69] Lu Q, Wang TP, Wang XH, Dong CQ, Zhang ZB, Ye XN. Selective production of levoglucosenone from catalytic fast pyrolysis of biomass mechanically mixed with solid phosphoric acid catalysts. BioEnergy Research. 2015;**8**(3):1263-1274. DOI: 10.1007/s12155-015-9581-6

[70] Wang Z, Zhang Y, Liao B, Guo QX, Sui XW. Preparation of levoglucosenone through sulfuric acid promoted pyrolysis of bagasse at low temperature. Bioresource Technology. 2011;**103**(1):466-469. DOI: 10.1016/j.biortech.2011.10.010

[71] Halpern Y, Riffer R, Briodo A. Levoglucosenone. Journal of Organic Chemistry. 1973;**38**(2):204-209

[72] Sarotti AM, Spanevello RA, Suárez AG. An efficient microwave-assisted green transformation of cellulose into levoglucosenone. Advantages of the use of an experimental design approach. Green Chemistry. 2007;**9**(10):1137-1140. DOI: 10.1039/b703690f

[73] Zhang H, Meng X, Liu C, Wang Y, Xiao R. Selective low-temperature pyrolysis of microcrystalline cellulose to produce levoglucosan and levoglucosenone in a fixed bed reactor. Fuel Processing Technology. 2017;**167**(August):484-490. DOI: 10.1016/j.fuproc.2017.08.007

[74] Available from: https://ilbioeconomista.com/2018/02/19/an-interview-with-tony-duncan-ceo-

circa-group-the-most-innovative-bioeconomy-ceo-2017/; https://IlbioeconomistaCom/2018/02/19/an-Interview-with-Tony-Duncan-Ceo-circa-Group-the-Most-Innovative-Bioeconomy-Ceo-2017/ [Accessed: 2018]

[75] Krishna SH, McClelland DJ, Rashke QA, Dumesic JA, Huber GW. Hydrogenation of levoglucosenone to renewable chemicals. Green Chemistry. 2017;**19**(5):1278-1285. DOI: 10.1039/c6gc03028a

[76] Kudo S, Goto N, Sperry J, Norinaga K, Hayashi JI. Production of levoglucosenone and dihydrolevoglucosenone by catalytic reforming of volatiles from cellulose pyrolysis using supported ionic liquid phase. ACS Sustainable Chemistry & Engineering. 2017;**5**(1):1132-1140. DOI: 10.1021/acssuschemeng.6b02463

[77] Sherwood J, De Bruyn M, Constantinou A, Moity L, McElroy CR, Farmer TJ, et al. Dihydrolevoglucosenone (Cyrene) as a bio-based alternative for dipolar aprotic solvents. Chemical Communications. 2014;**50**(68):9650-9652. DOI: 10.1039/c4cc04133j

[78] Zhang J, White GB, Ryan MD, Hunt AJ, Katz MJ. Dihydrolevoglucosenone (Cyrene) as a green alternative to N,N-dimethylformamide (DMF) in MOF synthesis. ACS Sustainable Chemistry & Engineering. 2016;**4**(12):7186-7192. DOI: 10.1021/acssuschemeng.6b02115

[79] Wilson KL, Murray J, Jamieson C, Watson AJB. Cyrene as a bio-based solvent for HATU mediated amide coupling. Organic and Biomolecular Chemistry. 2018;**16**(16):2851-2854. DOI: 10.1039/c8ob00653a

[80] Sutton AD, Waldie FD, Wu R, Schlaf M, Pete'Silks LA, Gordon JC. The hydrodeoxygenation of bioderived furans into alkanes. Nature Chemistry. 2013;**5**(5):428-432. DOI: 10.1038/nchem.1609

[81] Simakova IL, Murzin DY. Transformation of bio-derived acids into fuel-like alkanes via ketonic decarboxylation and hydrodeoxygenation: Design of multifunctional catalyst, kinetic and mechanistic aspects. Journal of Energy Chemistry. 2016;**25**(2):208-224. DOI: 10.1016/j.jechem.2016.01.004

[82] Xia QN, Cuan Q, Liu XH, Gong XQ, Lu GZ, Wang YQ. Pd/NbOPO4 multifunctional catalyst for the direct production of liquid alkanes from aldol adducts of furans. Angewandte Chemie International Edition. 2014;**53**(37):9755-9760. DOI: 10.1002/anie.201403440

[83] Song HJ, Deng J, Cui MS, Li XL, Liu XX, Zhu R, et al. Alkanes from bioderived furans by using metal triflates and palladium-catalyzed hydrodeoxygenation of cyclic ethers. ChemSusChem. 2015;**8**(24):4250-4255. DOI: 10.1002/cssc.201500907

[84] Li Z, Lepore AW, Salazar MF, Foo GS, Davison BH, Wu Z, et al. Selective conversion of bio-derived ethanol to renewable BTX over Ga-ZSM-5. Green Chemistry. 2017;**19**(18):4344-4352. DOI: 10.1039/c7gc01188a

[85] Available from: http://www.anellotech.com/technology.BTX.pdf [Accessed: March 26, 2019]

[86] Williams CL, Chang CC, Do P, Nikbin N, Caratzoulas S, Vlachos DG, et al. Cycloaddition of biomass-derived furans for catalytic production of renewable p-xylene. ACS Catalysis. 2012;**2**(6):935-939. DOI: 10.1021/cs300011a

[87] Tanzi CD, Vian MA, Ginies C, Elmaataoui M, Chemat F. Terpenes as green solvents for extraction of oil from microalgae. Molecules.

2012;**17**(7):8196-8205. DOI: 10.3390/molecules17078196

[88] John I, Muthukumar K, Arunagiri A. A review on the potential of citrus waste for D-limonene, pectin, and bioethanol production. International Journal of Green Energy. 2017;**14**(7):599-612. DOI: 10.1080/15435075.2017.1307753

[89] Martin-Luengo MA, Yates M, Rojo ES, Huerta Arribas D, Aguilar D, Ruiz Hitzky E. Sustainable p-cymene and hydrogen from limonene. Applied Catalysis A: General. 2010;**387**(1-2):141-146. DOI: 10.1016/j.apcata.2010.08.016

[90] Deepthi Priya K, Petkar M, Chowdary GV. Bio-production of aroma compounds from alpha pinene by novel strains. International Journal of Biological Sciences and Applications. 2015;**2**(2):15-19

[91] Antonucci V, Coleman J, Ferry JB, Johnson N, Mathe M, Scott JP, et al. Toxicological assessment of 2-methyltetrahydrofuran and cyclopentyl methyl ether in support of their use in pharmaceutical chemical process development. Organic Process Research and Development. 2011;**15**(4):939-941. DOI: 10.1021/op100303c

[92] Zhu YL, Xiang HW, Li YW, Jiao H, Wu GS, Zhong B, et al. A new strategy for the efficient synthesis of 2-methylfuran and γ-butyrolactone. New Journal of Chemistry. 2003;**27**(2):208-210. DOI: 10.1039/b208849p

[93] Teng BT, Zhu YL, Li Y, Xiang HW, Zheng HY, Zhao GW, et al. Effects of calcination temperature on performance of Cu–Zn–Al catalyst for synthesizing γ-butyrolactone and 2-methylfuran through the coupling of dehydrogenation and hydrogenation. Catalysis Communications. 2004;**5**(9):505-510. DOI: 10.1016/j.catcom.2004.06.005

[94] Brown Ripin DH, Vetelino M. 2-Methyltetrahydrofuran as an alternative to dichloromethane in 2-phase reactions. Synlett. 2003;**34**(15):2353. DOI: 10.1055/s-2003-42091

[95] Funel JA, Schmidt G, Abele S. Design and Scale-Up of Diels À Alder Reactions for the Practical Synthesis of 5-Phenylbicyclo [2.2.2]oct-5-en-2-one2011. pp. 1420-1427

[96] Wilkes JS. A short history of ionic liquids—From molten salts to neoteric solvents. Green Chemistry. 2002;**4**: 73-80. DOI: 10.1039/b110838g

[97] Plechkova NV, Seddon KR. Applications of ionic liquids in the chemical industry. Chemical Society Reviews. 2008;**37**:123-150. DOI: 10.1039/b006677j

[98] Sheldon RA. Green solvents for sustainable organic synthesis: State of the art. Green Chemistry. 2005;**7**:267-278. DOI: 10.1039/b418069k

[99] Hulsbosch J, De Vos DE, Binnemans K, Ameloot R. Biobased ionic liquids: Solvents for a green processing industry? ACS Sustainable Chemistry & Engineering. 2016;**4**:2917-2931. DOI: 10.1021/acssuschemeng.6b00553

[100] Jha AK, Jain N. Synthesis of glucose-tagged triazolium ionic liquids and their application as solvent and ligand for copper (I) catalyzed amination. Tetrahedron Letters. 2013;**54**(35):4738-4741. DOI: 10.1016/j.tetlet.2013.06.114

[101] Poletti L, Chiappe C, Lay L, Pieraccini D, Russo G. Glucose-derived ionic liquids: Exploring low-cost sources for novel chiral solvents. Green Chemistry. 2007;**9**:337-341. DOI: 10.1039/b615650a

[102] Handy ST, Okello M, Dickenson G. Solvents from biorenewable sources: Ionic liquids based on fructose. Organic Letters. 2003;**5**(14):2513-2515. DOI: 10.1021/ol034778b

[103] Chiappe C, Marra A, Mele A. Synthesis and applications of ionic liquids derived from natural sugars. In: Rauter AP, Vogel P, Queneau Y, editors. Carbohydrates and Sustainable Development II—A Mine for Functional Molecules and Materials. Berlin: Springer-Verlag Berlin Heidelberg; 2010. DOI: 10.1007/128_2010_47

[104] Ferraz R, Branco C, Marrucho IM, Rebelo PN, Nunes M, Prud C. Development of novel ionic liquids based on ampicillin. Medicinal Chemistry Communications. 2012;**3**:494-497. DOI: 10.1039/c2md00269h

[105] Wasserscheid P, Bolm C. Synthesis and properties of ionic liquids derived from the 'chiral pool'. Chemical Communications. 2002;**1**(3):200-201. DOI: 10.1039/B109493A

[106] Heckel T, Winkel A, Wilhelm R. Chiral ionic liquids based on nicotine for the chiral recognition of carboxylic acids. Tetrahedron: Asymmetry. 2013;**24**(18):1127-1133. DOI: 10.1016/j.tetasy.2013.07.021

[107] Parmentier D, Metz SJ, Kroon MC. Tetraalkylammonium oleate and linoleate based ionic liquids: Promising extractants for metal salts. Green Chemistry. 2013;**15**:205-209. DOI: 10.1039/c2gc36458a

[108] Kwan ML, Mirjafari A, Mccabe JR, Brien RAO, Essi DF, Baum L, et al. Synthesis and thermophysical properties of ionic liquids: Cyclopropyl moieties versus olefins as Tm-reducing elements in lipid-inspired ionic liquids. Tetrahedron Letters. 2013;**54**(1):12-14. DOI: 10.1016/j.tetlet.2012.09.101

[109] Calvo FG, María F, Monteagudo J. Green and bio-based solvents. Topics in Current Chemistry. 2018;**376**(18):1-40. DOI: 10.1007/s41061-018-0191-6

[110] Abbott AP, Capper G, Davies DL, Munro HL, Rasheed RK. Preparation of novel, moisture-stable, Lewis-acidic ionic liquids containing quaternary ammonium salts with functional side chains. Chemical Communications. 2001;**1**(19):2010-2011. DOI: 10.1039/b106357j

[111] Zeisel SH. A brief history of choline. Annals of Nutrition and Metabolism. 2012;**61**:254-258. DOI: 10.1159/000343120

[112] Abbott AP, Capper G, Davies DL, Rasheed RK, Tambyrajah V. Novel solvent properties of choline chloride/urea mixtures. Chemical Communications. 2003;**1**(1):70-71

[113] Singh B, Lobo H, Shankarling G. Selective N-alkylation of aromatic primary amines catalyzed by bio-catalyst or deep eutectic solvent. Catalysis Letters. 2011;**141**:178-182. DOI: 10.1007/s10562-010-0479-9

[114] Phadtare SB, Shankarling GS. Halogenation reactions in biodegradable solvent: Efficient bromination of substituted 1-aminoanthra-9,10-quinone in deep eutectic solvent (choline chloride: Urea). Green Chemistry. 2010;**12**:458-462. DOI: 10.1039/b923589b

[115] Pawar PM, Jarag KJ, Shankarling GS. Environmentally benign and energy efficient methodology for condensation: An interesting facet to the classical Perkin reaction. Green Chemistry. 2011;**13**:2130-2134. DOI: 10.1039/c0gc00712a

[116] Shahbaz K, Mjalli FS, Hashim MA, ALNashef IM. Using deep eutectic solvents for the removal of glycerol from palm oil-based biodiesel. Journal of

Applied Sciences. 2010;**10**(24):3349-3354. DOI: 10.3923/jas.2010.3349.3354

[117] Shahbaz K, Mjalli FS, Hashim MA, Alnashef IM. Eutectic solvents for the removal of residual palm oil-based biodiesel catalyst. Separation and Purification Technology. 2011;**81**(2):216-222. DOI: 10.1016/j.seppur.2011.07.032

[118] Ho KC, Shahbaz K, Mjalli FS. Removal of glycerol from palm oil-based biodiesel using new ionic liquids analogues. Journal of Engineering Science and Technology. 2015;**10**(1):98-111

[119] Maugeri Z, Domı P. Novel choline-chloride-based deep-eutectic-solvents with renewable hydrogen bond donors: Levulinic acid and sugar-based polyols. RSC Advances. 2012;**2**:421-425. DOI: 10.1039/c1ra00630d

[120] Francisco M, Van Den Bruinhorst A, Kroon MC. New natural and renewable low transition temperature mixtures (LTTMs): Screening as solvents for lignocellulosic biomass processing. Green Chemistry. 2012;**14**:2153-2157. DOI: 10.1039/c2gc35660k

[121] Ge X, Gu C, Wang X, Jiangping T. Deep eutectic solvents (DESs)-derived advanced functional materials for energy and environmental applications: Challenges, opportunities, and future vision. Journal of Materials Chemistry A: Materials for Energy and Sustainability. 2017;**5**:8209-8229. DOI: 10.1039/C7TA01659J

Chapter 2

Per- and Trichloroethylene Air Monitoring in Dry Cleaners in the City of Sfax (Tunisia)

Fatma Omrane, Imed Gargouri and Moncef Khadhraoui

Abstract

The use of chlorinated solvents in dry cleaning poses risks to human health. The occupational health exposure assessment to these volatile chemicals is conducted through quantification of airborne concentrations inside the facilities. Indeed, the lack of such measurements in Tunisia pushed us to conduct the study. After identifying dry cleaners in Sfax city, we conducted door-to-door canvassing in 47 facilities. Then, the levels of perchloroethylene (PCE) and trichloroethylene (TCE) in the indoor air are measured in two sampling positions: fixed and individual. The pollutants are adsorbed with charcoal sorbent tubes where their amounts correspond to given air volumes that are suctioned through the pump. It is later used to calculate their mean concentrations. These solvents are desorbed using carbon disulfide and analyzed by gas chromatography—flame ionization detection. After the analytical validation of the protocol, 19 air samples were quantified. The measured concentrations of TCE are close to the occupational exposure limit value in almost all facilities, whereas the PCE concentrations are about half of the OELV. The overall results showed that the working environment in dry cleaning in Sfax city are concerning and can lead to many adverse health effects up to several types of cancers.

Keywords: air monitoring, trichloroethylene, perchloroethylene, exposure assessment, occupational health

1. Introduction

In the modern world, synthetic chemicals are a big part of the human life. Among these products, the organic solvents represent a group of diverse chemical substances with a generally high volatility and solubilization ability that allow their use in broad range of applications. In this respect, there are about a thousand different solvents involving a hundred common uses, especially in industrial sectors.

Depending on their properties, these organic solvents can be used as degreasers (cleaning textiles and metals), additives and thinners (paints, varnishes, inks, glues, and pesticides), strippers (removal of organic products), and even purifiers (perfumes) [1].

Nevertheless, all these solvents have negative impacts on human health and the environment in case of noncaution. The health effects are variable depending on the solvent and the exposure duration and intensity [1]. Due to their volatility, humans can be typically exposed by the three routes: (i) inhalation; (ii) dermal contact, whatever the state of the skin; and (iii) ingestion through accidental absorption or contaminated water or food.

In occupational settings where organic solvents are used or processed, prevention measures have to be established and followed. Their establishment should obey the general principles of prevention and also be based on chemical risk prevention methods [1].

Among the main chemical substance families of organic solvents, we can distinguish the family of the halogenated hydrocarbon. The latter includes a subgroup named the chlorinated hydrocarbons. Since the 1920s, chlorinated hydrocarbons have been widely used for their stripping property. In fact, because they are nonflammable compounds, chlorinated solvents were used in degreasing, notably in the cleaning of clothes. However, since the 1970s, their use has declined steadily because of the growing awareness of their harmful effect on humans and the environment [2].

Once in the human body, their effects are multiple. Some effects are common to all halogenated solvents, and others are distinct and specific to the solvent depending on toxicological proprieties [2]. The common effects include irritation mainly of the skin and mucous membranes (ocular and respiratory) and neurological disorders especially neurobehavioral difficulties. Many chlorinated solvents cause liver or kidney damages [2]. Ultimately, some of them can even induce cancers.

In addition, the recent literature reports with certainty that several volatile organic solvents including the chlorinated ones are deemed to be harmful to the environment by contributing to the production of tropospheric ozone via photochemical reactions. This is consequently causing respiratory disorders for suffering from asthma or respiratory failure. Besides, due to their low solubility in water and their limited biodegradability, these solvents can also lead to soil pollution and in some cases can lead to groundwater contamination [2].

Due to their potential hazardous properties, in recent years, the use of organic solvents in industry and laboratories has progressed considerably. The precaution instructions, which fall under what is called "green chemistry," led to the suspension or a significant limitation of some solvents (chlorinated solvents, glycol ethers, aromatic solvents, etc.) [3].

The perchloroethylene (PCE) and trichloroethylene (TCE) are considered as the second and the third most used chlorinated solvents, respectively. One of the industrial sectors that are largely using these two solvents is the dry cleaning industry.

Actually, PCE and TCE are currently well-known to induce many adverse health effects [2, 4–7]. However, in Tunisia, dry cleaners generally use them by dint of their important cleaning properties. In case of noncompliance with the standardized prevention and industrial hygiene measures, their consumption may lead to the contamination of the workplaces' atmospheres, especially nearby solvent handlers, which can lead to health quality degradation in case of long-term exposure.

In this context, for the purpose of protection of solvent handlers' health, the control of the occupational exposure is obligatory in Tunisia according to Law 94-28 of 21 February 1994 [8] and Law No. 95-56 of 28 June 1995 [9]. However, the actual indoor air monitoring is almost absent; besides, there is a shortage of specialized laboratories in the field of air and biomonitoring.

In contrast, for instance, in the United States and Japan, PCE is automatically removed if frequent and intense exposure is confirmed in the workplaces. In France, a national project of gradual cessation of its use in dry cleaning industry is ongoing. In 2013, the installation of new PCE dry cleaning machines is prohibited in buildings contiguous to dwellings [5]. In January 2022, the substitution of all PCE machines is going to take place [7].

Actually, the occupational exposure monitoring of chemical pollutants has an important place in today's strategies for chemical risk prevention [10]. Moreover, the technical advances in the analytical chemistry field have allowed the development of sufficiently sensitive techniques in order to detect pollutants in low concentrations.

Therefore, given the absence of occupational air monitoring standards and the lack of industrial hygiene control in our country, we are especially interested in the current study in occupational exposure assessment by means of the indoor air monitoring of PCE and TCE in dry cleaning facilities in Tunisia (case of Sfax City). This study aims also to assess the human health risks associated to the long-term exposure to these solvents.

2. Material and methods

2.1 Selection of the organic solvents

Dry cleaning is clothes and textiles cleaning process that uses a solvent other than water. Most of the time, the conventional and common technique consumes chlorinated hydrocarbons, primarily PCE [7]. In the current study, we decided to also include TCE because it was traditionally used in dry cleaning. Even though TCE is no longer used in France since the 1960s [7], we suspect its presence in dry cleaners as it is not explicitly banned by law in Tunisia. Thus, considering its toxicity and proved carcinogenicity, we decided to monitor it as well.

These solvents incur serious adverse effects threatening the health of workers following chronic exposure [11, 12]. They are also proved to be carcinogenic and probably carcinogenic to humans, for TCE and PCE, respectively [13].

Table shows the main exposure routes of these two solvents and their health effects and carcinogenicity.

2.2 Identification and selection of dry cleaning facilities

In order to explore the exhaustive list of dry cleaning facilities in the city of Sfax, we contacted the Sfax Chamber of Commerce and Industry and the Regional Union of Industry, Commerce and Handicrafts in Sfax. However, they neither had sufficient information about this industrial sector nor an updated list.

Therefore, we decided to carry out a door-to-door canvassing in the study area. We only included facilities on a radius of 5 km from the city center. We excluded facilities using solvents other than TCE and PCE. Besides, we concomitantly requested their acceptance to participate in our study. After that, we randomly selected half number of the facilities who freely and voluntarily accepted to collaborate.

Afterward, information was collected using a questionnaire with the aim of:

i. identifying all the handled substances in the facilities in order to get complete qualitative and quantitative inventories and

ii. describing the working environments and the ventilation quality in each facility.

2.3 Indoor air measurements

There are two sampling methods of gas and vapors: passive and active air monitoring of TCE and PCE [14–18]. In this study, we adopted the active sampling of vapors by pumping onto an activated charcoal tube [19, 20]. This method is suitable for the quantification of high exposure levels of air pollutants and of many chlorinated compounds simultaneously. It has also the advantage of allowing the control of sampling conditions (air flow rate and duration).

Solvents	Trichloroethylene (TCE)	Perchloroethylene (PCE)
Main exposure route [11, 12]	General population: inhalation, ingestion, and dermal contact: depending on the exposure media Occupational exposure: inhalation and/or dermal contact depending on the job process	General population: inhalation and ingestion: depending on the exposure media Occupational exposure: inhalation
Reported effects: chronic exposure via inhalation [11, 12, 25, 26]	• Neurological effects: o Neuropsychic disorders o Neuromotor function alteration o Cranial nerve damage o Risk of Parkinson's disease • Developmental effects: • Risk of congenital heart defects • Hepatic effects: o Liver damage and diseases • Immunological and lymphoreticular effects: o TCE hypersensitivity syndrome o Immune system alterations o Autoimmune disease: Scleroderma • Gastrointestinal effects: o Anorexia, nausea, and vomiting	• Neurological effects: o Neurobehavioral difficulties o Visual alterations • Ocular effects o Irritation • Respiratory effects o Respiratory hypersensitivity and pulmonary edema o Respiratory irritation • Hepatic effects o Subclinical liver effects • Renal effects o Risk of hypertensive end-stage renal disease o Increase in kidney markers
IARC classification: cancer site [13, 38]	• Group 1: o Carcinogenic to humans (kidney) • Limited evidence in humans: leukemia and/or lymphoma and liver	• Group 2A: o Probably carcinogenic to humans: urinary bladder

IARC: International agency for research on cancer.

Table 1.
The main exposure routes of trichloroethylene and perchloroethylene, and their adverse health effects due to chronic inhalation and carcinogenicity.

In order to assess the exposure level of the workers, we chose the sampling mode according to the workers' mobility inside the workplace. A fixed monitoring station characterizing the workplace atmosphere was carried out when employees are working in a steady position. The sampling device is fixed near employees and at the level of their airways in order to measure the exposure level (**Figures 1** and **2**) [14, 21]. Elsewhere, a personal air sampling was carried out in the worker's breathing zone in order to quantify the individual exposure when the job task requires the worker's mobility (**Figure 3**) [14, 21]. Samples were taken in the middle of the week for each facility (from Wednesday to Friday).

The sampling device consists of an activated charcoal sorbent tube with two sorbent sections (SKC® tube 226-16) [22]. The tube is connected by a flexible hose to a sampling pump: pocket pump (SKC® 210-1002 TX). The sampling takes place at a regular rate of 100 (±5%) cm^3/min for a 4-h shift (**Figures 1** and **2**). So, the indoor air is pumped out through two sorbent layers that are separated with foam and glass wool (800 and 200 mg). The first one is the sampling section and the second is a backup section that detects sample breakthrough [21].

After sampling, the sorbent tubes were stored in unpolluted areas at 4°C. Even though it is possible to store them at ambient temperature [20, 21] (limited to 8 days for PCE [19]), we preferred to avoid any minimal loss [17].

Figure 1.
Active air monitoring using fixed sampling station.

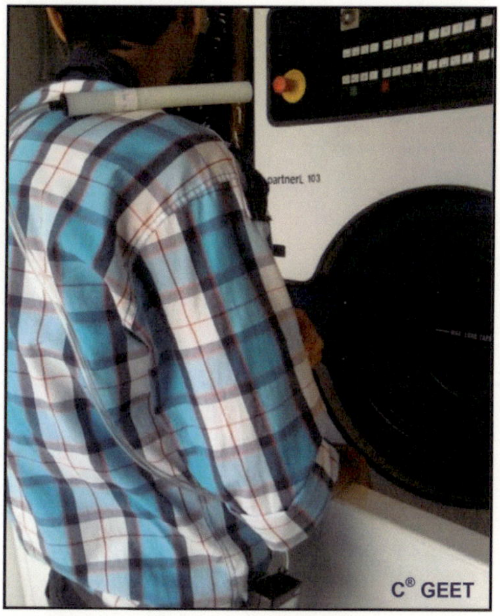

Figure 2.
Active air monitoring using personal sampling.

2.4 Sample analysis

Desorption of both solvents from activated charcoal was achieved using carbon disulfide. In fact, after breaking the tube glass tips, the content was put in a glass flask already placed in an ice bath to avoid any vaporization. Then, the content was stirred for 30 min to promote desorption [19, 20].

After the separation of the organic phase using filtration, it was directly injected (1 μL) and analyzed by gas chromatography with flame ionization detector (GC-FID). The used instrument is a SHIMADZU® chromatograph GC-2014. The capillary

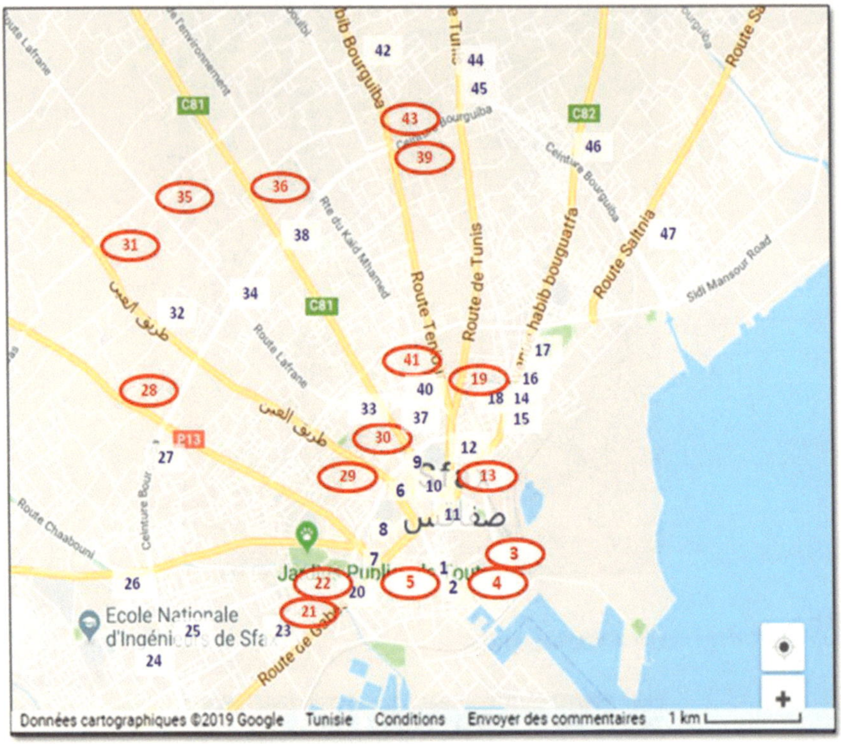

Figure 3.
Location of dry cleaning facilities in a radius of 5 km from the city center of Sfax. The facilities that profited from indoor air monitoring are encircled in red.

OELV	Solvents			
	TCE		PCE	
	(mg/m³)	(ppm)	(mg/m³)	(ppm)
France (INRS, ANSES) [17, 18, 25–27]	405	75	138	20
Europe (SCOEL) [17, 18, 27, 28]	55	10	138	20
USA (ACGIH) [25, 26]	55	10	170	25
Selected for this study	55	10	138	20

Table 2.
Occupational exposure limit values developed from different countries for trichloroethylene and perchloroethylene.

column (DB-23) features are length = 60 m, inner diameter = 0.32 mm, and film thickness = 0.25 μm. The temperature of injection is 200°C and the oven program is the following: 35°C (3 min), 7.5°C/min, 115°C (0 min), 15°C/min, 200°C (2 min). The vector gas is nitrogen with a flow rate of 2 ml/min. We adopted external calibration using the same solvent as for the samples [19, 20, 23]. Then, the air concentrations of both solvents were calculated using the airflow rates and sampling durations [24].

2.5 Occupational exposure limit values

The fixed amount on a sorbent tube over a work shift corresponds to average concentrations of solvents and subsequently will be compared with occupational exposure limit values OELVs that are time-weighted averages on the basis of 8 h a day.

In the absence of Tunisian guide values [8, 9], several OELVs are available internationally, for instance in France [17, 18, 25–27], Europe [27, 28], and the United States of America [25, 26]. The values are presented in **Table 2** OELVs are fixed for the purpose of helping employers to protect workers from occupational diseases and adverse health effects due to the exposure to hazardous chemicals within their job tasks.

In this study, we chose to select the most protective values among the explored ones.

3. Results

Following the door-to-door canvassing, 47 dry cleaning facilities were located in the city of Sfax within a radius of 5 km. **Figure 3** shows the locations of facilities in each major road using the map of Sfax city from "Google Maps." The facilities' locations on the map show their congestion in the city center with 19 dry cleaners which represents more than 40%. No facility has been identified in the medina quarter or the Sidi Mansour road.

After the meetings with the facilities' managers, only 33 dry cleaning companies accepted to participate to our study. Thus, we had an agreement with them about the confidentiality of their names, contacts and any personal specific data. The identification questionnaire was carried out for all of them. Then, we randomly selected 16 facilities. The selected ones are highlighted on the map in **Figure 3**.

It was noticed that the majority of dry cleaning establishments are located next to habitation. Moreover, they are exclusively naturally ventilated through open doors/windows. The 33 explored facilities have small surfaces with a mean value of 31.3 m^2 (95% CI [25.6-37]) where there are one or two machines. Despite all these exposure circumstances, we highlighted the almost total lack of personal protective equipment (PPE). In fact, employees never use masks even when they are directly manipulating the solvents, only one facility provides gloves, and three facilities use protective clothing.

The work activity is semiindustrial and the dry cleaners reported that they are solely using PCE with a mean quantity of 177.6 L/year (95% CI [159.9-195.3]). The number of workers in the facilities varied from 1 to 6, yet about 70% of them were women.

The chromatograms of our analysis show sharp, narrow, and well-separated peaks of both solvents. Interestingly, TCE is detected in all samples (fixed and personal), whereas the PCE is detected only in 6 workplaces among the 16 (**Table 3**).

TCE concentrations in the workplaces' atmospheres are almost equal in all facilities except for F3 (coefficient of variation (CV) is equal to 12.5%). In fact, their mean value is nearly the same as their median: ~43.2 mg/m^3 (95% CI [40.4-46]). Only in F3, the TCE concentration in the workplace's atmosphere is 40% lower than the mean concentration in all facilities. This could be due to the large surface of F3 compared to the other workplaces, which is 43% higher than the mean surface value. Moreover, we noticed that dry cleaners in F3 are using the lowest quantity of solvents, which is 60% lower than the mean quantity used in all facilities.

As for PCE concentrations in the workplaces' atmospheres, they are remarkably variable with a CV higher than 100%, which is significantly higher than the CV of TCE concentrations in the workplaces' atmospheres. PCE concentrations in the workplaces' atmospheres have no relationship with the used quantity of solvents. This may be explained by the differences in the working behaviors between the facilities, since all of them are not following the same standard prevention measures.

As for the personal sampling, TCE concentrations are almost the same as the workplace atmosphere levels, except for F3 where it increased by about 25%. However,

	Fixed monitoring				Personal sampling			
	TCE		PCE		TCE		PCE	
F	mg/m^3	ppm	mg/m^3	ppm	mg/m^3	ppm	mg/m^3	ppm
F3	26.35	4.98	12.60	1.89	35.38	6.69	25.08	3.76
F4	40.58	7.67	7.86	1.17	—	—	—	—
F5	44.67	8.45	ND	ND	—	—	—	—
F13	42.06	7.95	ND	ND	41.60	7.87	—	—
F19	49.61	9.38	ND	ND	—	—	—	—
F21	43.57	8.24	ND	ND	—	—	—	—
F22	49.72	9.40	ND	ND	—	—	—	—
F28	42.22	7.98	ND	ND	—	—	—	—
F29	41.10	7.77	54.34	8.14	44.06	8.33	65.40	9.80
F30	41.45	7.84	ND	ND	—	—	—	—
F31	48.89	9.25	7.98	1.19	—	—	—	—
F35	43.21	8.17	8.47	1.27	—	—	—	—
F36	45.86	8.67	ND	ND	—	—	—	—
F39	46.21	8.74	ND	ND	—	—	—	—
F41	43.74	8.27	12.27	1.84	—	—	—	—
F43	42.27	7.99	ND	ND	—	—	—	—

ND, not detected; F, dry cleaning facility.

Table 3.
Indoor air measured concentrations of TCE and PCE in dry cleaning facilities.

it still is 18% lower than the mean value of TCE concentrations in the workplaces' atmospheres, which confirms its connection with the lowest consumed quantity.

Regarding PCE, the personal sampling concentrations are higher than workplaces' atmospheres, particularly in P3 where the PCE level has doubled. Thus, the employees who are carrying personal sampling devices (particularly in F3 and F29) are exposed to greater levels of solvents. This is obviously due to their specific movement patterns where they could be moving closer to the emission sources. This consequently justifies our judgment to use the personal sampling in these job tasks.

As we can notice in **Table 3**, almost all TCE concentrations are close to the selected OELV (**Table 2**). These results are quite surprising because all facilities have declared to only use PCE. In fact, we suspected to have found TCE in trace amounts as impurity content. But, these high concentrations are an alarming finding. Thus, even though all PCE concentrations are lower than one-tenth of the OELV (except for both measurements in F29 and for F3 personal sampling level), we believe that urgent corrective actions should be carried out in all facilities.

4. Discussion

4.1 Location of dry cleaning facilities in Sfax and residential exposure

First of all, since there was no list of dry cleaning facilities in Sfax city, we encountered some difficulties during the identification and location steps. That is why, it was decided to conduct a door-to-door canvassing in the study area.

The lack of dry cleaners in the medina quarter may be explained by its transformation into a commercial zone with plenty of stores. However, this is not the case in the Sidi Mansour road, which is predominantly an industrial zone. It is a less populated area and seems like its inhabitants did not see the need for dry cleaning, which could be explained by their relatively low standard of living. Finally, we noticed that the facilities are generally located in densely populated areas of Sfax city. Indeed, the downtown area that includes 40% of the total number has a high demographic density. Actually, this could represent a major issue for the health of people living close to the dry cleaning facilities. In fact, they could be exposed to higher concentrations of solvents in comparison with the rest of the general populations [12].

Several studies have explored many adverse health effects related to the exposure of people residing near dry cleaning facilities and TCE- or PCE-emitting sites. A study showed that the maternal age increased the risk of fetal heart defects when mothers were living around a trichloroethylene-emitting site [11, 18]. Many epidemiological studies have reported that visual functions, such as contrast sensitivity, were reduced when residents were exposed to low perchloroethylene levels from dry cleaning facilities in their neighborhood [12, 29, 30]. This association was also confirmed in children in further studies but with even lower levels [12]. These Visual alterations are rather considered as neurological effects [12, 29]. Actually, other studies showed the increase of neurobehavioral difficulties, such as memory, cognitive, and attention impairments, in people living in residential buildings that include a dry cleaning facility [12, 29]. These neurological effects were also noticed in occupationally exposed populations, but according to a meta-analysis study, they were more significant in the general population [12]. This could be explained by the fact that the general populations include all ages and genders, which increases the worry about them.

On the other hand, a study conducted in North Carolina demonstrated that the dry cleaning facility was the main source of drinking water contamination with trichloroethylene [11]. Since many Sfaxians are using rainwater harvesting, this could induce a great concern about possible supplementary adverse health effects related to other routes of exposure through water.

4.2 Occupational exposure limit values and threshold effects

OELVs are established to prevent occupational illnesses in general but they are not specific to a health effect or target organ. Thus, even though the TCE and PCE concentrations are below the OELVs, some serious adverse health effects may occur [31]. Furthermore, since the general population is also involved, people, including the occupationally exposed ones, are more effectively protected if the air concentrations are below the toxicological reference values (TRV). These TRVs are exclusively established based on scientific considerations [31]. In this section, we are only discussing TRVs that are protecting populations from threshold effects.

Since there are no Tunisian values, we are checking TRVs that are conceived from French and American agencies. The French values are constructed by the French Agency for Food, Environmental and Occupational Health & Safety (ANSES) [32]. The TCE chronic TRV via inhalation is constructed based on noncarcinogenic renal effect because of a well-established nephrotoxicity mechanism [33]. The TRV is fixed on 3.2 mg/m^3. The mean value of all TCE concentrations is 13.4 times the French TRV, which means that there is a great risk of nephrotoxicity for workers. Among the American TRVs, we can mention, for instance, the ones that are established by the Environmental Protection Agency (US EPA) [34]. The TCE chronic Reference concentration (RfC) is fixed on 2 10^{-3} mg/m^3 [35] based on both developmental and immune effects. The mean value of all TCE concentrations is 21,385 times the RfC, which means that there is also an enormous risk related to the increase of congenital cardiac malformations and a decrease of thymus weight [36]. This developmental effect could be also alarming because 70% of the workers are women.

Regarding PCE, the chronic French TRV via inhalation is developed based on neurological effects. Among them, the ANSES distinguished the visual alterations as the most sensitive effect toward the lowest exposure levels [30]. The TRV is fixed on 0.4 mg/m^3 [30, 32], which is over 60 times the mean value of all PCE concentrations. Since the PCE concentration distribution is highly dispersed, the ratio, and consequently the risk, is even higher (108 times) if we use the upper bound of the 95% confidence interval for mean. As for the RfC, it is set at the tenth of the French TRV (0.04 mg/m^3 [37]) based also on neurological effects, notably neurobehavioral effects and color vision impairment induced by neurotoxicity [29]. Thus, the risk of neurotoxicity could be even 10 times higher for workers. It is worth to note that these risks maybe also extrapolated to the general population with lower intensity, but with great significance and likely adverse consequences.

4.3 Carcinogenic effects

Chronic exposure can also induce carcinogenic effects. Thus, there are non-threshold TRVs as well. Each value is established for a specific tumor site or sites based on a lifetime exposure.

As shown in **Table 1**, according to World Health Organization (WHO) [13, 38], TCE was proved to cause kidney cancer to humans. Indeed, the French carcinogen TRV via inhalation, established by ANSES, is the ERU "excès de risque unitaire," and is equal to 10^{-6} $(\mu g.m^{-3})^{-1}$ [32]. It was constructed based on kidney cancer, specifically the renal cell carcinoma [33].

Yet, other solid scientific evidences indicate other potential carcinogenic effects of TCE. In fact, the US EPA established the Inhalation Unit Risk for three different cancer sites: the hematologic, liver, and kidney cancers. Its value is more protective and equal to 4.1 10^{-6} $(\mu g/m^3)^{-1}$ [35, 36]. According to the latter value, the lifetime cancer risk, which is defined as the product of multiplying the carcinogen TRV by the exposure concentration, is equal to 17.5 10^{-2}, when compared with the mean value of all TCE concentrations. This means that, more than 17 additional cases of hematologic, hepatic, and renal cancers are estimated to occur during a lifetime exposure to TCE in a population of 100 people. This value is considerably high even for an occupationally exposed population.

Furthermore, other international agencies have also demonstrated the carcinogenic effect of TCE for other target organs, such as in testes or in lungs [33].

As for PCE, according to WHO [38], it is probably inducing urinary bladder cancer to humans.

Elsewhere, ERU established by ANSES is based on liver cancer (hepatocellular adenoma and carcinoma) [30] and is equal to 2.6×10^{-7} $(\mu g/m^3)^{-1}$ [32]. Additionally, the Inhalation Unit Risk by US EPA has the same value, but only for the hepatic cancer. According to this carcinogen TRV, the lifetime cancer risk is 6.3×10^{-3}. It means that six additional cases of liver cancer are expected to take place during a lifetime exposure in a population of 1000 people. Due to the high variability of PCE concentrations, this value could be even higher (more than 11 cases if the upper bound of the 95% confidence interval for mean is used) in some cases. Although these findings are slightly better than those for the TCE, they are still high and troubling. In fact, the inhalation cancer risk could be worse if we consider a total cancer risk due to the cumulative exposure to both solvents, especially when the target organ is the same.

4.4 Health risk assessment and chemical risk prevention

In Tunisia, chemical risk assessment studies have started in many industrial sectors that are handling solvents, such as adhesive [39] and shoe manufacturing [40] industries. Yet, the occupational exposure in dry cleaning industry is not explored. The exposure assessment was achieved using questionnaires and indoor air measurements. In fact, air samplings were carried out for the first time in the region of Sfax. We consider that this study will help for better understanding the dry cleaning industrial sector in Tunisia, which will lead to further improvement in health risk assessment studies in this sector.

All the 33 dry cleaning facilities in Sfax have announced to use PCE as a dry cleaning product; however, due to the high TCE concentrations, we assume the following:

- The dry cleaning products are contaminated by TCE but in significant amounts, so either the product manufacturers did not inform their customers about TCE or solvent handlers are mixing PCE with TCE to reduce cost since it is cheaper.

- TCE could be coming from a different emission source, as it could be used as a degreaser in maintenance operations.

- Both managers and workers are not aware of the chemical risks and the adverse health effects of both solvents, especially TCE.

It is worth to note that in our study, the air monitoring was conducted during the warm season (in June and the temperature was ~30°C) when the workplaces were relatively highly ventilated; doors and windows were open. However, in winter season, they are rather closed. So, we suppose that the exposure levels will be even higher in poor ventilation conditions.

The suggested corrective measures could start with the substitution. It consists of eliminating the use of the hazardous products by replacing them with less dangerous ones, especially because they are suspected to contain a high amount of TCE or by switching to a different process. Among the substitution processes, there are the (i) wet cleaning that consists of using a mixture of water, detergents, and surfactants whose risks are little or currently unknown, (ii) hydrocarbon dry cleaning that involves solvents that are less volatile than PCE, and (iii) siloxane D5 dry cleaning that are using the latter product as a liquid solvent that is barely volatile. These alternative machines may be quite costly; however, other easier corrective actions could be carried out.

For the second level of the corrective actions, we highly recommend the implementation of collective protection measures (CPM). In every facility, general mechanical ventilation ought to be installed immediately, with fume extraction systems positioned toward every job task. It should be noted that the polluted air needs to be rejected after purification [7]. In addition, the mechanical ventilation has to be supplied by outdoor fresh air and the reuse of the same air after gas scrubbing is interdicted [7].

The third level of the corrective actions involves the mandatory use of PPE in some job tasks. The wearing of protective clothing is advisable for every worker. Elsewhere, we noticed that the loading of solvents in the cleaning machines is generally introduced by manual pouring on the back or through the porthole, which increases the risk of exposure by direct contact with solvents. Thus, in such job tasks where workers are directly handling solvents for short duration, the use of gloves, safety glasses, and appropriate respiratory protective device is compulsory. The used gloves have to be made of polyvinyl alcohol or Viton® or Téflon® [25, 26]. Latex, butyl rubber, and polyethylene gloves are not suitable to be used for PCE [26]. The respiratory equipment should be equipped with gas filter type A [25, 26].

5. Conclusion

The current study is the first one in Tunisia and Sfax city that aimed to assess the occupational exposure and health risks of PCE and TCE in the dry cleaning industry.

The exposure assessment was achieved by means of the quantification of airborne concentrations of the chlorinated solvents via active air monitoring and their chemical analyses.

Our results revealed and responded to many interrogations and suspicions about the qualitative and quantitative exposure conditions and consequently the health status of dry cleaning workers in Sfax. In fact, all facilities are not following the standardized prevention and industrial hygiene measures. Moreover, they have declared to exclusively work with PCE products; however, TCE was detected in all facilities, and its concentrations were high and concerning.

Due to the inhalation exposure levels, many adverse effects are probably threatening the occupationally exposed population and even the general one, because of the location of all facilities in residential settings.

Among the threshold effects, we can distinguish high risks of neurological, nephrotoxic, developmental, and immune effects.

As for the carcinogenic effects, considerably high cancer risks were noticed if the lifetime exposure to these solvents would have the current average levels. Actually, 17 additional cases of hematologic, liver, and kidney cancers are expected to take place in population of 100 people.

Taking this into account, PCE and TCE air concentrations have to be reduced. Thus, some corrective measures were suggested in order to improve the working conditions.

We believe that the implementation of this study is very significant, at the Tunisian level, for better understanding of the dry cleaning industrial sector and for the improvement of future risk assessment studies in this field. Indeed, this pilot study provides the first occupational exposure data to TCE and PCE emissions from randomly selected dry cleaning facilities in Sfax city.

Acknowledgements

We acknowledge the cooperation and efforts of all the facility managers and workers.

Conflict of interest

The authors declare that there is no conflict of interest.

Author details

Fatma Omrane, Imed Gargouri* and Moncef Khadhraoui
Laboratory of Environmental Engineering and EcoTechnology, LR16ES19,
National School of Engineering, University of Sfax, Tunisia

*Address all correspondence to: imed.gargouri@fmsf.rnu.tn

IntechOpen

© 2020 The Author(s). Licensee IntechOpen. This chapter is distributed under the terms of the Creative Commons Attribution License (http://creativecommons.org/licenses/by/3.0), which permits unrestricted use, distribution, and reproduction in any medium, provided the original work is properly cited.

References

[1] INRS. Dossier Solvants. Prévenir les risques liés aux solvants [Internet]. 2017. Available from: http://www.inrs.fr/risques/solvants/ce-qu-il-faut-retenir.html

[2] INRS. Fiche solvants: Les hydrocarbures halogénés (chlorés, fluorés, bromés)—ED 4223 [Internet]. 2011. Available from: http://www.inrs.fr/media.html?refINRS=ED%204223

[3] Nun P, Colacino E, Martinez J, Lamaty F. Chimie sans solvant [Internet]. 2008. Available from: https://www.techniques-ingenieur.fr/base-documentaire/biomedical-pharma-th15/production-des-medicaments-procedes-chimiques-et-biotechnologiques-42610210/chimie-sans-solvant-k1220/synthese-sans-solvant-k1220niv10001.html

[4] INRS. Nettoyage à sec: Fiche d'aide au repérage de produit cancérogène [Internet]. 2013. Available from: http://www.inrs.fr/media.html?refINRS=FAR%2028

[5] INRS. Brochure: Perchloroéthylène. Nettoyage à sec [Internet]. 2013. Available from: http://www.inrs.fr/media.html?refINRS=FAS%202

[6] INRS. Brochure: Trichloroéthylène. Nettoyage, dégraissage [Internet]. 2015. Available from: http://www.inrs.fr/media.html?refINRS=FAS%201

[7] INRS. Le pressing: Aide-mémoire technique—ED 6308 [Internet]. 2018. Available from: http://www.inrs.fr/media.html?refINRS=ED%206308

[8] Official Journal of the Republic of Tunisia. Law n° 94-28 "portant régime de réparation des préjudices résultant des accidents du travail et des maladies professionnelles". 1994. Available from: http://www.ilo.org/dyn/natlex/natlex4.detail?p_lang=en&p_isn=38574

[9] Official Journal of the Republic of Tunisia. Law no 95-56 "portant régime particulier de réparation des préjudices résultant des accidents du travail et des maladies professionnelles dans le secteur public". 1995. Available from: http://ilo.org/dyn/natlex/natlex4.detail?p_lang=fr&p_isn=41295

[10] Officiel Prevention. Dossiers CHSCT: La métrologie des expositions professionnelles [Internet]. 2011. Available from: http://www.officiel-prevention.com/sante-hygiene-medecine-du-travail-sst/appareils-de-mesure/detail_dossier_CHSCT.php?rub=37&ssrub=152&dossid=285

[11] ATSDR. Toxicological Profile: Trichloroethylene (TCE) [Internet]. 2014. Available from: https://www.atsdr.cdc.gov/toxprofiles/TP.asp?id=173&tid=30

[12] ATSDR. Toxicological Profile: Tetrachloroethylene (PERC) [Internet]. 2014. Available from: https://www.atsdr.cdc.gov/toxprofiles/TP.asp?id=265&tid=48

[13] WHO (World Health Organization). IARC Monographs-Classifications [Internet]. 2018. Available from: http://monographs.iarc.fr/ENG/Classification/latest_classif.php

[14] INRS. Principe général et mise en œuvre pratique du prélèvement. (In French). 2018

[15] INRS. Préparation des dispositifs de prélèvement en vue d'une intervention en entreprise. (In French). 2018

[16] INRS. Les dispositifs de prélèvement actif pour le prélèvement de gaz ou vapeurs. (In French). 2015

[17] ANSES. Valeurs limites d'exposition en milieu professionnel: Le perchloroéthylène. (In French). 2010

[18] ANSES. Valeurs limites d'exposition en milieu professionnel Le trichloroéthylène [Internet]. 2017. Available from: https://www.anses.fr/fr/system/files/VLEP2007SA0432Ra.pdf

[19] INRS. Fiche Metropol: Tétrachloroéthylène M-405. 2019

[20] INRS. Fiche Metropol: Trichloroéthylène M-410. 2018

[21] INERIS. Trichloréthylène, tétrachloréthylène et chlorure de vinyle dans l'air Sources, mesures et concentrations [Internet]. 2005. Available from: https://www.ineris.fr/sites/ineris.fr/files/contribution/Documents/AIRE_05_0094.pdf

[22] Inc S. Sorbent Tubes Anasorb CSC Coconut Charcoal 10 × 110-mm size 2 Sections. SKC, Inc. [Internet]. Available from: http://www.skcinc.com/catalog/product_info.php?products_id=610

[23] INRS. Méthodes d'étalonnage pour la quantification des polluants. 2015

[24] INRS. Calcul de la concentration en polluants—Métropol. 2015

[25] INRS. Fiche Toxicologique: Trichloroéthylène (FT 22) [Internet]. 2011. Available from: http://www.inrs.fr/publications/bdd/fichetox/fiche.html?refINRS=FICHETOX_22

[26] INRS. Fiche Toxicologique: Tétrachloroéthylène (FT 29) [Internet]. 2012. Available from: http://www.inrs.fr/publications/bdd/fichetox/fiche.html?refINRS=FICHETOX_29

[27] HCSP. Commission specialisee risques lies a l'environnement. Valeurs repères d'aide à la gestion dans l'air des espaces clos: le trichloroéthylène. 2012

[28] Latvian Environment, Geology and Meteorology Centre. Substance Evaluation Conclusion Document as required by REACH Article 48 for Tetrachloroethylene [Internet]. 2014. Available from: https://echa.europa.eu/documents/10162/4b963c47-906a-4456-a8ea-8d9c352abe26

[29] US EPA. Integrated Risk Information System (IRIS), Chemical Assessment Summary: Tetrachloroethylene (Perchloroethylene) [Internet]. 2012. Available from: https://cfpub.epa.gov/ncea/iris/iris_documents/documents/subst/0106_summary.pdf#nameddest=rfc

[30] ANSES. Élaboration de VTR par voie respiratoire pour le perchloroéthylène [Internet]. 2018. Available from: https://www.anses.fr/fr/system/files/VSR2016SA0116Ra.pdf

[31] Péry A, Bonvallot N, Yamani ME, Boulanger G, Karg F, Mosqueron L, et al. Valeurs limites d'exposition professionnelles (VLEP), valeurs toxicologiques de référence (VTR): objectifs et méthodes. Environnement, Risques & SantÃ©. 2013;**12**:442-449

[32] ANSES—Agence nationale de sécurité sanitaire de l'alimentation, de l'environnement et du travail. List of Toxicity Reference Values (TRVs) Established by ANSES [Internet]. Available from: https://www.anses.fr/en/content/list-toxicity-reference-values-trvs-established-anses

[33] ANSES. Proposition de VTR par voie respiratoire pour le trichloroéthylène [Internet]. 2018. Available from: https://www.anses.fr/fr/system/files/VSR2016SA0117RA.pdf

[34] US EPA. IRIS Assessments; Integrated Risk Information System [Internet]. 2019. Available from: https://cfpub.epa.gov/ncea/iris2/atoz.cfm

[35] US EPA O. Trichloroethylene CASRN 79-01-6 | IRIS | US EPA, ORD [Internet]. 2011. Available

from: https://cfpub.epa.gov/ncea/iris2/chemicalLanding.cfm?&substance_nmbr=199

[36] US EPA. Integrated Risk Information System (IRIS), Chemical Assessment Summary: Trichloroethylene [Internet]. 2011. Available from: https://cfpub.epa.gov/ncea/iris/iris_documents/documents/subst/0199_summary.pdf#nameddest=rfc

[37] US EPA O. Tetrachloroethylene CASRN 127-18-4 | IRIS | US EPA, ORD [Internet]. Available from: https://cfpub.epa.gov/ncea/iris2/chemicalLanding.cfm?&substance_nmbr=106

[38] WHO (World Health Organization). IARC List of classifications by cancer site. 2018

[39] Gargouri I, Khadhraoui M, Nisse C, Leroyer A, Larbi Masmoudi M, Elleuch B, et al. Case study: Occupational assessment of exposure to organic solvents in an adhesive producing company in Sfax, Tunisia. Journal of Occupational and Environmental Hygiene. 2012;**9**:D71-D76

[40] Gargouri I, Khadhraoui M, Elleuch B. What are the health risks of occupational exposure to adhesive in the shoe industry? In: Rudawska A, editor. Adhesives—Applications and Properties. London: InTech; 2016 Available from: http://www.intechopen.com/books/adhesives-applications-and-properties/what-are-the-health-risks-of-occupational-exposure-to-adhesive-in-the-shoe-industry-

Chapter 3

Synthesis of Polyimides in the Melt of Benzoic Acid

Kuznetsov Alexander Alexeevich
and Tsegelskaya Anna Yurievna

Abstract

Review of the authors' works on the synthesis of polyimides (PIs) by the method of one-stage high-temperature polycondensation in an unusual solvent—molten benzoic acid (BA). Compared with a known synthesis in inert high-boiling solvents, synthesis in BA takes place under mild conditions (140°C, 1–2 hour) to give completely imidized PIs. The approach has a number of advantages. Due to catalysis of the first reversible stage of amic acid (AA) formation and low equilibrium constant (K = 10–20 l/mol), the first stage disappears kinetically, and the imidization reaction becomes limiting. The process becomes less sensitive to the basicity of diamines; therefore, low reactivity diamines can be involved. Water is easily removed from the melt by evaporation, which makes the whole process irreversible. Specific features of the method are successfully used to control the microstructure of the chain copolyimides (statistical to multiblock) and to synthesize hyperbranched PIs and star-shaped PIs with narrow molecular weight distribution.

Keywords: Polycondensation, catalysis, benzoic acid, polyimides, block copolymers, hyperbranched polymers, star-shaped polymers

1. Introduction

Aromatic polyimides (PIs) are a class of polymers with a unique combination of properties. Their main advantages are high physical and mechanical characteristics in a wide range from cryogenic temperatures up to 250–300°C and heat resistance up to 400–450°C, etc. Factors that provide such a set of properties of PIs—the presence of conjugated heteroaromatic fragments, strong intermolecular interaction, and chain stiffness—at the same time create difficulties in their processing. Several approaches to the synthesis of PIs are known. The most widespread are the two-stage and one-stage high-temperature methods for the synthesis of PIs from diamines and aromatic tetracarboxylic acids dianhydrides. The general reaction scheme leading to polyimide formation from diamines and dianhydrides consists in two discrete steps: polyacylation to give polyamic acid and imidization (cyclodehydration). The fundamental aspects of the two-stage process of PI synthesis have been the subject of detailed study of many researchers [1–3]. One of the fundamentally important results obtained was the conclusion that the formation of polyamic acids (PAA) is a reversible reaction [4, 5]; the equilibrium constant and degree of polymerization are controlled by the solvent basicity and temperature. This means that to obtain a PAA with a high degree of polymerization, the process is

IntechOpen

to be carried out in highly basic (amic) solvents at 20–40°C. In the case of one-stage process in high-boiling solvents, both reactions proceed simultaneously. In the process of one-stage high-temperature polycondensation (HTPC) of diamines and tetracarboxylic acids dianhydrides, the reactions of acylation and cyclization occur simultaneously in a high-boiling solvent at 180–210°C. The synthesis of PIs according to this scheme was first reported by Korshak, Vinogradova, and Vygodskii in 1967 [6–8]: the synthesis of so-called cardo-PIs. This approach has found wide application.

PI synthesis in molten benzoic acid (BA) described in this review should be considered as a variant of the said PI synthesis by HTPC approach. The advantages of using molten BA compared to other high-boiling solvents used in this process (nitrobenzene, m-cresol, o-dichlorobenzene, etc.) are a strong catalytic effect, the lack of solvent toxicity, and easy isolation of polymer due to crystallization of solvent on cooling. In contrast to other novel ecologically improved ("green") solvents for polyimide synthesis such as ionic liquids [9] and supercritical CO_2 [10], this approach does not require any special chemicals and equipment.

2. General features of the process

2.1 Rate of the process and molecular weight

The process of synthesis of PIs in molten BA can be carried out under relatively mild conditions (140°C, 1–2 hour) at slow inert gas flow [11–13]. Within 1 hour after the start of synthesis, fully cyclized polyetherimides (PEIs) and PIs with logarithmic viscosity values η_{log} = 0.4–1.2 dL*g^{-1} (N-methyl-2-pyrrolidone (N-MP), 25°C), depending on the structure of the monomers, were isolated. In the IR spectrum of the products, there are typical absorption bands of the imide cycle in the regions of 720, 1370, 1720, and 1780 cm^{-1}; no peaks of amic acid fragments (1660 cm^{-1}, 3600 cm^{-1}) were observed. Imide peaks in IR spectrum did not change after additional heat treatment of polymers at 300°C for 0.5 hour; this allows us to conclude that during the synthesis, almost 100% conversion of imidization has been achieved. Molecular weight Mw = 5–15•10^{4} (GPC) is sufficient for the subsequent processing of PIs into films, semifinished products, and bulk products by extrusion and injection molding or for use as binders for composite materials. The rate of PEI molecular weight increase, when synthesized in the BA melt at 140°C, is significantly higher than that in m-cresol at the same temperature.

The ability of carboxylic acids as additives to accelerate the process of one-pot synthesis of PIs was first demonstrated for synthesis of PIs with cardo fragments in nitrobenzene [14] at 160–210°C. However, it was noted that with an increase in the concentration of BA in a system above 2.5 mol BA per 1 mole of repeating unit (about 10%-weight solution in reaction mixture), the total rate of the process decreases—up to complete inhibition—probably due to the fact that the excess BA deactivates the amino group by the mechanism of acid–base interaction. In our works, we used concentration up to 95% in reaction mixture BA as a solvent. Under these conditions, the (BA/repeating unit) mole ratio is about 30–35. The absence of inhibition by excess BA can be explained by the fact that the melt of BA is a nonpolar liquid—by analogy with 100% acetic acid (AcOH) which has dielectric constant $k \sim 6$; therefore, the dissociation constant BA in its own melt can be significantly lower than that in nitrobenzene.

The general scheme of the process of obtaining PIs by high-temperature polycondensation of diamines and dianhydrides can be represented as follows (**Figure 1**). **Figure 1** includes four conjugated reactions, including two main ones, acylation and

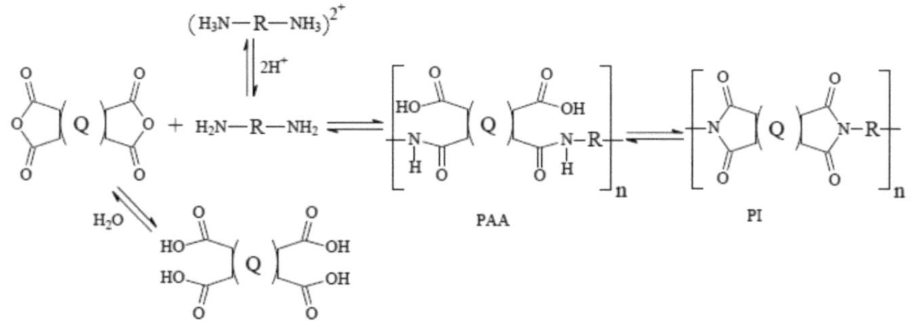

Figure 1.
Scheme of PIs obtaining by polycondensation of diamines and dianhydrides.

imidization, and two side reactions, binding of amino groups by acidic medium and hydrolysis of anhydride groups by water released during imidization. The degree of reversibility of each reaction depends on the temperature, acidity of solvent, and rate of water vapor removal. The replacement of dianhydride for corresponding tetracarboxylic acid does not influence on the overall rate and final molecular weight. This allows suggestion that dehyderatation of diphthalic acids is reversible and fast.

2.2 Leveling of monomer reactivity

In the course of experiments with different monomer pairs (**Table 1**), it was found that the chain growth rate in molten BA showed a rather weak dependence on the basicity of the diamines used [13, 15, 16]. This observation is in stark contrast to regularities of the low-temperature polycondensation in amic solvents in which the basicity of the diamines has a strong influence on the rate of polycondensation [1–3]. Such phenomenon of partial "reactivity leveling" in molten BA is of interest as a method of obtaining new high molecular weight copolyimides from low reactive diamines. In **Table 2**, the logarithmic viscosity (η_{log}) values are given of polyimides synthesized in molten BA from dianhydrides and bridged aromatic diamines having different basicities expressed as pK_b values.

It was observed that the change in basicity index of aromatic diamines in a range from $pK_b = 5.5$ (ODA) to $pK_b = 2.5$ (SDA) did not result in a significant difference in η_{log} of final PIs (**Table 2**). It allows a conclusion that conversion of amino groups reached for 1.5 h was at least 90–95% in both cases. In other words, the difference in apparent reactivity of diamines of both the low and the high basicities is not so large. Such a behavior is quite different from the results reported for low-temperature polycondensation of diamines and dianhydrides in amide solvents. For the latter reaction, changing the type of bridge substituent in diamines in a row —O—; —CH_2—; —SO_2— results in about three orders of magnitude decrease in the rate constant of polycondensation with pyromellitic dianhydride [1–3].

2.3 Mechanism and kinetics

Initially, we suggested that low sensitivity of reaction rate to basicity of diamines is caused by interaction with acid medium, i.e., the higher the basicity of amino group, the higher its deactivation by acid medium. To check this supposition, we studied a character of interaction of different diamines with BA by the method of the phase diagrams.

In **Figure 2a, b**, the phase diagrams are shown of binary system BA-diamine constructed on the basis of DSC thermograms for BA-diamine mixtures of different compositions [15]. It is seen that highly basic 1,12-dodecamethylene diamine with

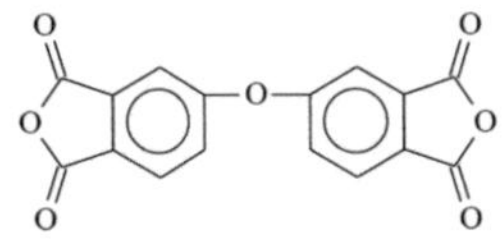	ODPA	Oxydiphthalic dianhydride
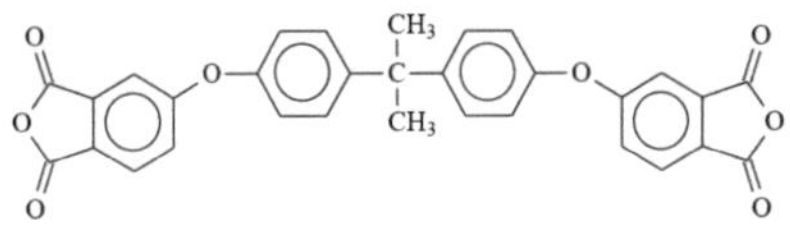	BPADA	Dianhydride of 2,2-bis-[(3,4-dicarboxyphenoxy)phenyl]propane
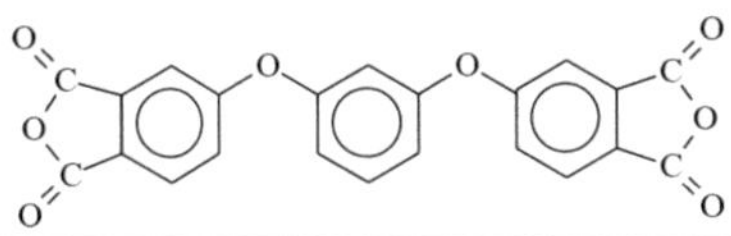	RDA	1,3-Phenylenedioxy-bis-4-phthalic anhydride
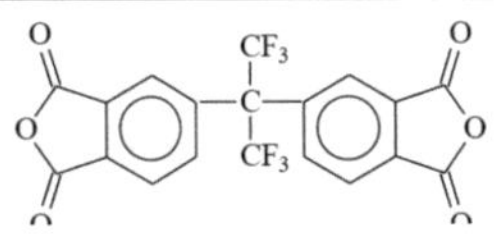	6F	2,2-Hexafluorpropylidene-diphthalic anhydride
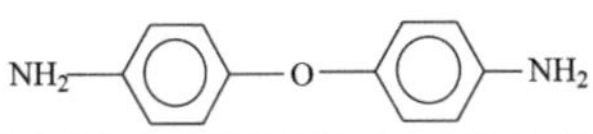	ODA	4,4'-Oxydianiline
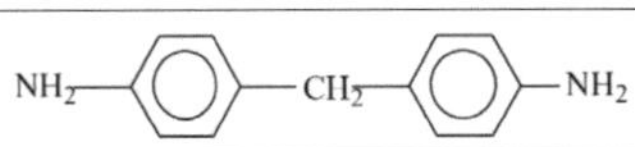	MDA	4,4′-Methylenedianiline
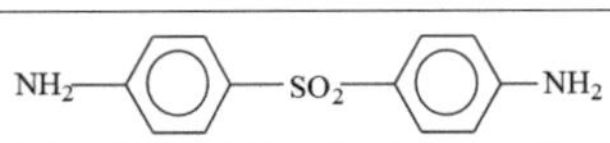	SDA	4,4'-Sulfonyldianiline
$H_2N–(CH_2)_6–NH_2$	HMDA	1,6-Hexamethylenediamine
$H_2N–(CH_2)_{12}–NH_2$	DDA	1,12-Dodecametylene diamine
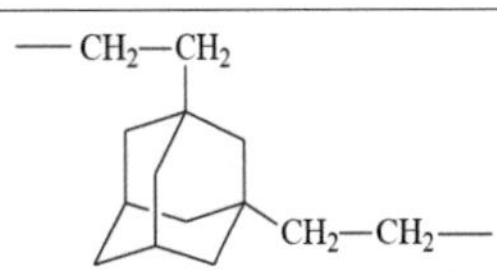	ADA	1,3-Bis-(2-aminoethyl)-adamantane
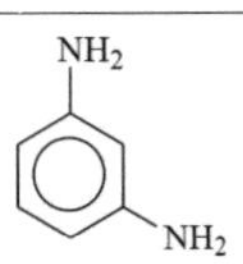	m-PDA	m-Phenylene diamine
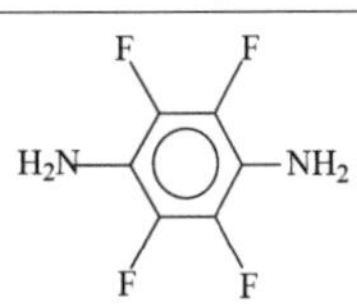	TFPDA	Tetrafluoro-p-phenylene diamine

Table 1.
Monomers used for polycondensation.

Diamine	Dianhydride	$\eta_{log}(dL\ g^{-1})^a$
ODA (pK_b = 5.5)	ODPA	0.57 (H_2SO_4)
SDA (pK_b = 2.5)	ODPA	0.44 (N-MP)
ODA	BPADA	1.0 (N-MP)
SDA	BPADA	0.45 (N-MP)

[a]N-MP, 0.5 g/dL, 25° C

Table 2.
Characterization of polyimides obtained in molten BA, at 140° C (1.5 hour) [12].

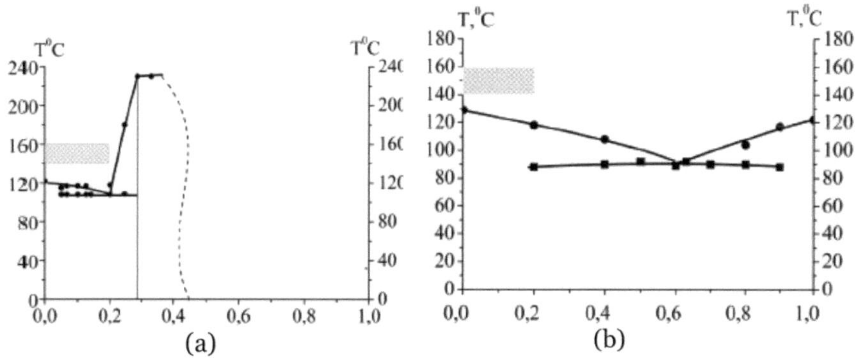

Figure 2.
Phase diagrams of binary system BA-diamines: DDA (a) and ODA (b); point (0.0) corresponds to 100%-Mol BA and point (1.0) to 100%-Mol diamine [15].

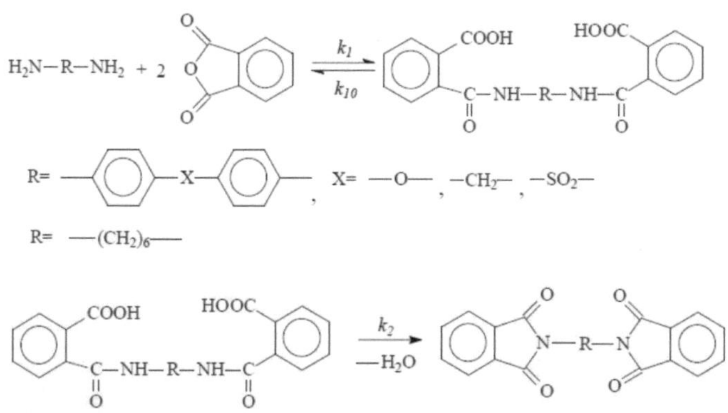

Figure 3.
Model reactions of amic acid formation and imidization.

low melting point (Tm = 67°C) interacts with BA to form alkylene-bis-ammonium benzoate (**Figure 2a**) with Tm = 236°C, whereas less basic ODA does not form such a salt; the phase diagram of BA-ODA system has an appearance of ordinary physical mixtures of two substances with limited solubility in each other (**Figure 2b**).

We also investigated the kinetics of model reaction—acylation of diamines ODA, MDA, SDA, and HMDA by phthalic anhydride (PhA) in glacial AcOH (**Figure 3**).

Kinetic data were obtained by potentiometric titration of amino groups with solution of perchloric acid in AcOH after reaching an equilibrium state [16, 17]. Results are presented in **Table 3** [16, 17]. It is seen that at the initial period of acylation, kinetics obeys the equation of second-order reaction (**Figure 4**); considerable difference is observed in the rate constants of bridged

Diamine	k_1 (mole * min)$^{-1}$ (22°C)	K_p (mole)$^{-1}$ (22°C)	E_a (kJ*mole^{-1})	-ΔH (kJ* mole^{-1})
ODA	490	8670	34.4	44.0
MDA	175	5800	33.1	41.5
SDA	6.9	1130	28.1	40.6
HMDA	0.054[a]	95.6[a]	77.1	13.0

[a]*Extrapolated value.*

Table 3.
Parameters of the acylation reaction of different diamines in glacial acetic acid.

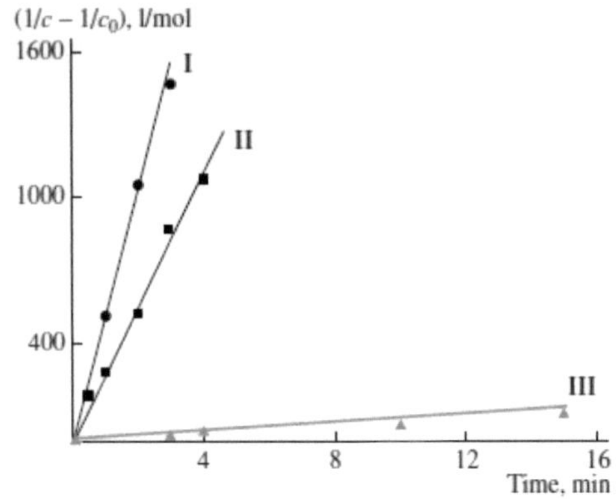

Figure 4.
Acylation kinetics of diamines ODA (I), MDA (II), and SDA (III) with phthalic anhydride in AcOH at 22°C (second-order reaction plot). Starting concentration of amino groups: ODA, 0.005; MDA, 0.01; and SDA, 0.03 Mol L^{-1} *[16].*

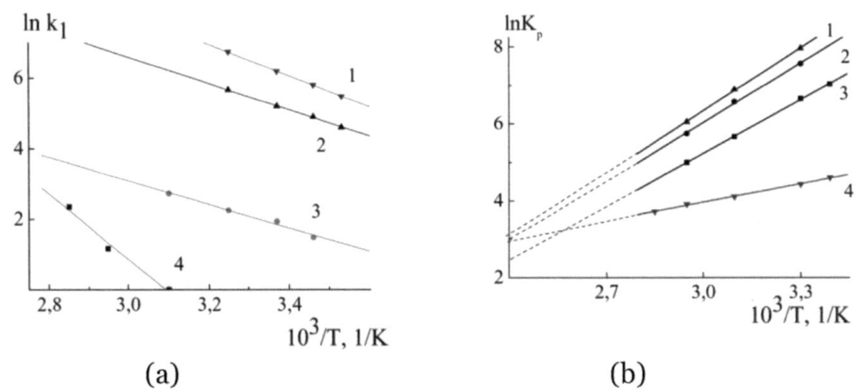

Figure 5.
Arrhenius plots (a) and logarithm of the equilibrium constant vs. the reciprocal temperature (b) for the acylation of diamines ODA (1), MDA (2), SDA (3), and HMDA (4) with phthalic anhydride [16, 17].

—O—; **—CH$_2$—**; **—SO$_2$—** diamines. So, partial effective reactivity leveling in a row of aromatic diamines (not able to form salt) observed experimentally in one-pot PI synthesis in molten BA at 140°C hardly can be explained by difference in acid–base-type interaction of amino groups with BA.

Alternative explanation of this phenomenon is the change of limiting stage from amic acid moiety formation for amic acid moiety imidization; the latter is much less sensitive to chemical structure of starting reagent.

From Arrhenius plots (**Figure 5a**), the values of activation energy (E_a) of acylation in glacial acetic acid (AcOH) were determined (**Table 3**). At elevated temperatures (50°C and higher), the reversible character of the acylation was established. On the basis of experimentally measured equilibrium amino group concentration, the equilibrium constants (K_p) were determined at different temperatures. In **Figure 5b**, temperature dependences of equilibrium constants for acylation reaction different diamines with PhA are shown.

From these data, using the vant' Hoff equation, the enthalpy change (ΔH) values for the acylation reaction of diamines in AcOH were determined (**Table 2**). On the basis of the values of the equilibrium constants and the rate constants for direct reaction (k_1), the first-order rate constants for the back-reaction (k_{-1}) were calculated for each temperature. In **Table 4**, the values of rate constants at 140°C obtained by extrapolation are given.

Diamine	k_1 (l*mole^{-1} min^{-1})^a	k_{-1} (min^{-1})^a	K_p (l*mole^{-1})^a	k_2 (min^{-1})^b
ODA	24,200	960	25.2	0.8
MDA	8100	370	22.0	—
SDA	150	15.6	9.6	0.4
HMDA	430	20	21.4	0.6

[a]*Extrapolated.*
[b]*Experiment in closed system.*

Table 4.
Parameters of acylation and imidization in carboxylic acid media (140°C).

It should be noted that a very low acylation rate was observed in the case of aliphatic diamines at 22°C (**Table 3**). The reason is occurrence of the concurrent reaction of salt formation with AcOH. This conclusion is confirmed by the appearance of adsorption peaks of benzoate anion and alkylene-bis-ammonium cation in IR spectrum.

Formation results in increase in effective activation energy of acylation. Due to this, the acylation rate increases sharply with increasing temperature (**Figure 5a**, line 4, and **Tables 3** and **4**).

We also estimated the value of effective rate constants for the imidization step [16, 17]. Low molecular weight model amic acids were synthesized from ODA, SDA, and HMDA and PhA. Kinetics of their imidization in the melt of BA at 140°C was followed by FTIR. First-order reaction rate constants were determined and corrected taking into account the conjugated reactions of decay and resynthesis of amic acid. Corresponding set of kinetic equations was written and solved numerically to give best fitting with experimental data on kinetics of imide cycle accumulation. From the analysis of the kinetic data, the following conclusions are apparent:

1. Imidization of amic acid moieties at 140°C in molten BA acid medium is a first-order reaction with a very fast pre-equilibrium stage (**Figure 3**).

2. Due to catalysis of the acylation step in molten BA in combination with low equilibrium constant, this stage becomes kinetically insignificant, and imidization becomes the rate-determining step of PI formation. In comparison with acylation reaction, imidization is less sensitive for chemical structure of reagents, so in one-pot PI synthesis, partial effective reactivity leveling of the low and high reactive diamines is observed.

3. Synthesis of random and multiblock copolymers

Multiblock (MB) copolymers attract big attention due to their ability to self-organize and form continuous two-phase morphology with controlled characteristic size of the phase particle in the micro- or nanoscale range. MB copolyimides (MB CPIs) are also known as promising materials for design of gas separation and proton-conductive membranes for fuel cells.

Conventionally, polycondensation-type MB copolymers are synthesized by polycondensation of two preliminary synthesized oligomers containing terminal reactive groups. MB CPIs can also be obtained by transimidization reaction of oligoimides with pyrimidine end groups. The limitation of this approach is the fact that only few oligoimides are soluble in organic solvents.

Vasnev and Kuchanov [18] investigated theoretically the regularities of copolymer chain microstructure formation in the course of copolycondensation in a system (A_2 + B_2 + C_2). Here A_2 and B_2 are the same type bifunctional monomers *of* different reactivities; C_2 is bifunctional "intermonomer" which reacts with A_2 and B_2; monomer A_2 does not react to B_2. According to this theory, MB copolymer can be formed from the system (A_2 + B_2 + C_2) in a regime of slow loading of intermonomer C_2 to the mixture of comonomers A_2 + B_2, but only in the case if polycondensation process meets the following requirements (so-called "ideal" interbipolycondensation): (1) reaction is irreversible; (2) any difference in reactivity of A_2 and B_2 occurs; any by-reactions are absent; and (3) reaction system is homogenous.

We synthesized a series of CPI samples from BPADA (intermonomer) and different pairs of diamines (CPI-1 series, **Figure 6**) in the melt of BA at 140°C with variable order of intermonomer loading [19]. In [20], two more CPI series (CPI-2 and CPI-3) were synthesized using dianhydrides ODPA and RDA as intermonomer and AFL and DDA as comonomers (**Figure 7**). Starting comonomers in all cases were chosen taking into account the solubility of copolymers in $CDCl_3$.

Chain microstructure of CPIs was studied by high-resolution NMR ^{13}C. In **Figure 8**, selected "sensitive" regions of NMR ^{13}C spectra (134–136 and 161–162 ppm) are shown for samples prepared of CPI-2 series. Designation *aa*, *bb*, and *ab* renders to monomer moiety consequences in chain: ACA, BCB, and ACB, correspondingly.

In **Figure 8**, curves 3–5 the inner two signals in 134–135 and 160–161 ppm regions correspond to *aa* and *bb* triads, and the outer two corresponds to *ab* triads. In the 113–114-ppm region, only one *ab* signal is observed. Analogous signal attribution was executed for every other CPI samples. Distribution of comonomer moieties in copolymer chain was characterized quantitatively by the coefficient of chain microheterogeneity (K_m; Eq. (1)) introduced by Yamadera and Murano [21]:

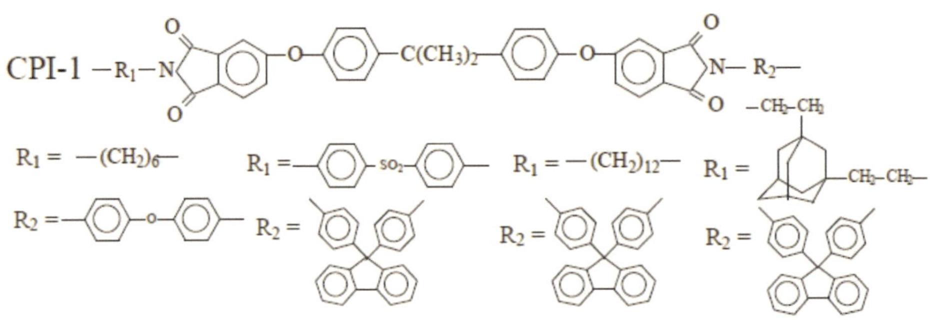

Figure 6.
Chain structure of copolyimides of CPI-1 series.

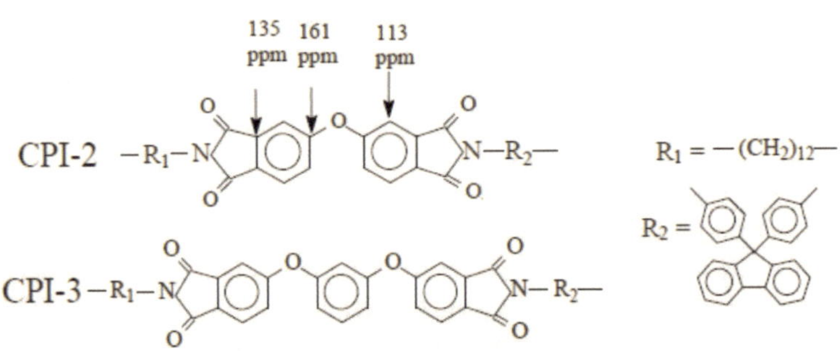

Figure 7.
Chain structure of copolyimides of CPI-2 and CPI-3 seria.

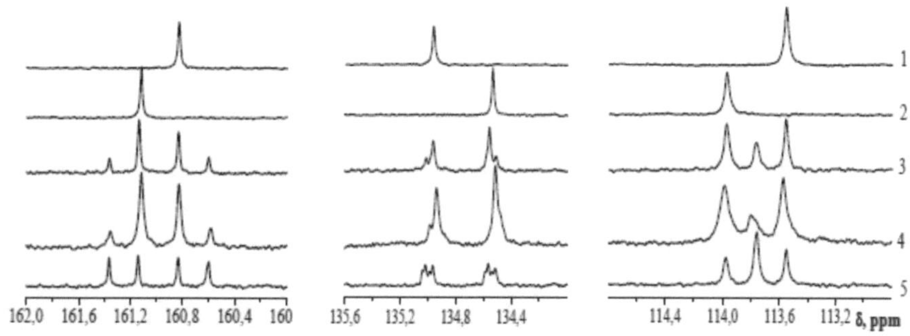

Figure 8.
*NMR ^{13}C spectra of CPI-2 series and corresponding homopolyimides in the structure-sensitive regions. Numbers to the right of curves correspond to the experiment number in **Table** 5. Signal attribution to the aa, bb, and ab triads was executed by comparison with signals of homopolyimides (**Table 6**).*

Experiment	Sample (order of loading, mole parts)	Triad ratio *aa/ab/bb*	K_m
1	Homopolyimide DDA-ODPA	1/0/0/	—
2	Homopolyimide AFL-ODPA	0/0/1	—
3	Slow (30 min) addition of ODPA to the mixture DDA/AFL (0.5/0.5)	1.7/1/1.6	0.48
4	Slow (30 min) addition of the mixture ODPA/DDA (1.0/0.5) to AFL (0.5)	2/1/1/1.9	0.39
5	One-shot loading of DDA/AFL/ODPA (0.5/0.5/1.0)	0.55/1/0.62	0.92

Table 5.
Loading condition and characteristics of CPI-2.

$$K_m = \frac{P_{ab}}{P_{ab} + 2P_{aa}} + \frac{P_{ab}}{P_{ab} + 2P_{bb}} \tag{1}$$

where P_{aa}, P_{bb}, and P_{ab} are the fractions of corresponding *aa*, *bb*, and *ab* triads.

The average block length l_A, l_B can be calculated as a unit divided by the first and second term in Eq. (1), correspondingly. In such a description, values $K_m = 0$; l_A, $l_B = \infty$ correspond to the mixture of two homopolymers; $K_m = 1$; l_A, $l_B = 2$, for random copolymer, and $K_m = 2$; l_A, $l_B = 1$, for strict alternation of moieties in copolymer. K_m values calculated from experimental NMR ^{13}C data for the samples of CPI-1–3 are shown in **Table 5**.

As it is seen from **Table 5**, at slow addition of intermonomer, in all experiments, we have obtained block CPIs with five-membered imide cycles. In the case of simultaneous loading, only random CPI was obtained. These results differ from data obtained in a work [22], in which only random CPI with five-membered imide cycles was obtained when conventional high-boiling solvent was used at any character of intermonomer addition to the mixture of diamines.

In **Figure 9**, NMR ^{13}C spectra are given for sample CPI-4 series, in which 1,3-bis (2-aminoethyl)adamantane (ADA) was used as intermonomer, and two anhydrides of different reactivities—BPADA and 6F (**Table 1**)—as comonomers (**Figure 10**) [23]. Curves 1 and 2 refer to homopolymers; curves 3 and 4 to the CPI-4 samples, obtained with simultaneously monomers loading and slow intermonomer loading, correspondingly. Attribution to triads is the following (ppm): 46.51 (*aa*), 46.63 (ab), and 46.71 (*bb*). The values of K_m and average block length calculated from the NMR ^{13}C spectra are $K_m = 0.91$ (l_A, $l_B = 2.2$) for simultaneous comonomer loading and $K_m = 0.76$ (l_A, $l_B = 2.63$) for slow loading of intermonomer. This result is

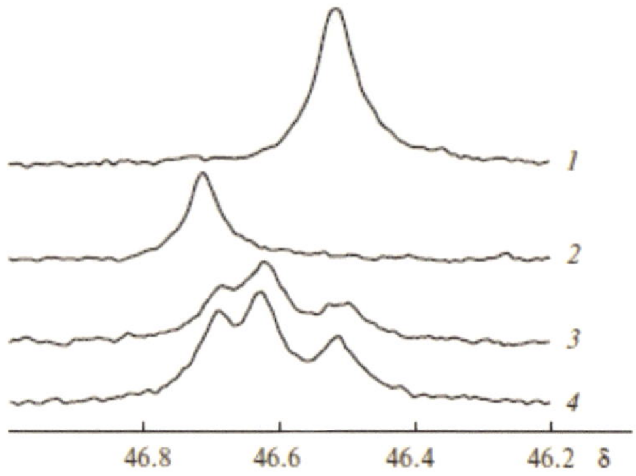

Figure 9.
A structure-sensitive region of the NMR^{13}C spectra of CPI-4 series.

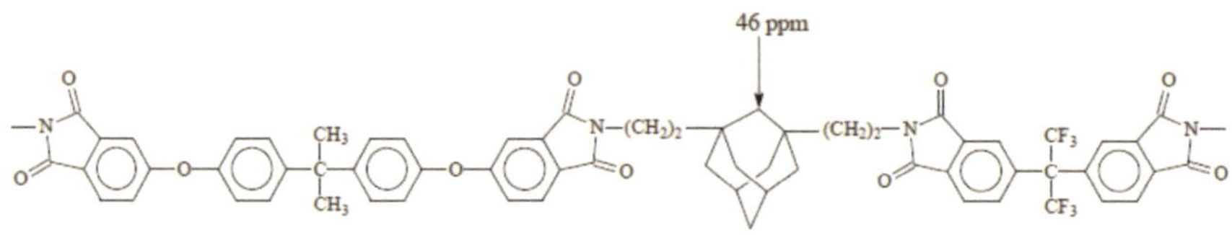

Figure 10.
Chain structure of copolyimides of CPI-4.

indicative of trend of formation of a multiblock chain microstructure at slow intermonomer loading.

This trend is not very pronounced in the comonomers chosen and shows rather weak influence of chemical structure of anhydride component on reactivity in molten BA.

In [24, 25], mathematical model has been developed by us for chain microstructure formation in copolyimide (CPI) synthesis (CPI) from two diamines A and B (comonomers) and one dianhydride C (intermonomer) in molten BA different regimes of intermonomer loading. The kinetic scheme was analyzed involving acylation of both diamines with anhydride fragment, decomposition, and imidization of two intermediate amic acid fragments.

Kinetic constants of acylation and imidization stages necessary for calculations were taken from our earlier experiments with model reactions described in Part 2.3 of this review. By numerical solution of the system of kinetic equations for different regimes of intermonomer loading, we calculated dependences of the change in time of the average block length l_A, l_B, the current concentrations of amino and anhydride groups, amic acid fragments, imide cycles, and triads *aa*, *bb*, and *ab* for CPI-1–3. The calculated values of the average block length and the chain microheterogeneity parameter (K_m = 0.5–0.6) for several comonomer pairs at slow intermonomer loading are in good agreement with the experimental values obtained from NMR ^{13}C data.

The kinetics of the block length (l_A and l_B) growth for regime of the slow intermonomer loading (for 30 min) is given in **Figure 11**. The length of block (l_A) containing the moieties of more active comonomer reaches its final value l_A = 4 already to the end of intermonomer loading, whereas the block length l_B goes on to increase. In the end, block copolymer forms. The difference in times of block formation is the sequence of difference in comonomer reactivity.

Typical consumption kinetics of amino and anhydride groups is presented in **Figure 12**. Consumption rates of amino groups belonging to the first and the second comonomers differ considerably. Concentration of transient amic acid fragments is

aa	ab	bb
160.82	160.60; 161.37	161.12
134.96	135.01; 134.51	134.53
113.56	113.77	113.97

Table 6.
Attribution of signals in NMR^{13}C spectra of CPI-1 series to triads.

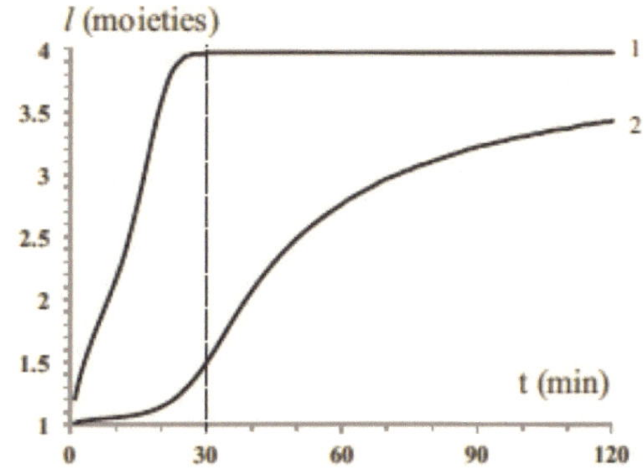

Figure 11.
Typical curve of change in time of the average length of blocks l_A (1) and l_B (2) in experiment with slow loading of the intermonomer (30 min).

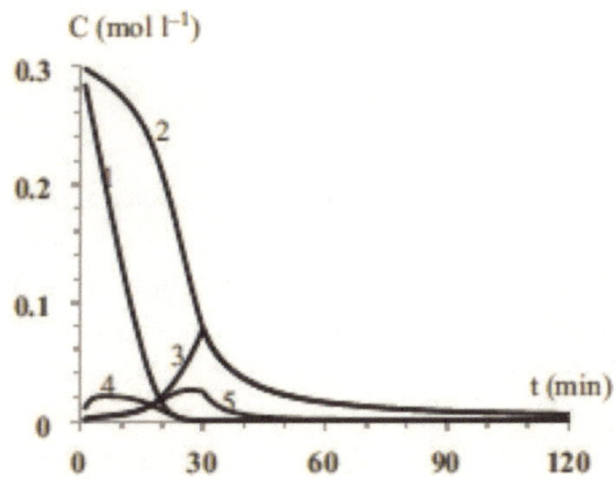

Figure 12.
Typical curves of change in time of the concentration of amino groups of A(1) and B (2), anhydride groups C (3), and amic acid fragments (4, 5).

low, less than 10% of starting amino groups. These fragments react rapidly to give imide cycles.

Kinetics of accumulation of imide cycles and different types of triads (*aa*, *bb*, and *ab*) are presented in **Figure 13**.

Accumulation of imide cycles and *aa* triads from the more reactive comonomer occurs faster than that for the less reactive monomer and forms a block consisting of several triads *aa*. Concentration of imide cycles and triads *bb* from less reactive comonomer increases with conversion more slowly. So, the model developed by us can be used to predict microstructure of the CPI chains at any conversion and at any loading order of intermonomer and comonomers. Dependence of parameter K_m for the final CPI on the duration of intermonomer BPADA loading for system AFL-DDA is presented in **Figure 14**. It should be noted that the fact of influence of intermonomer loading order on chain microstructure is very important for

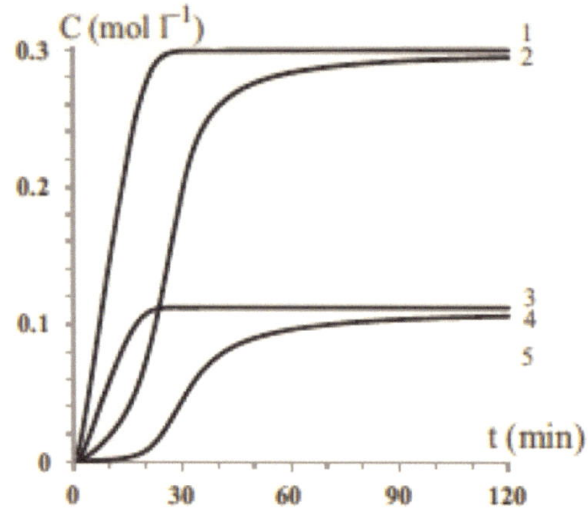

Figure 13.
Typical curves of change in time of the concentration of imide cycles (1,2) and triads: aa (3), bb (4), ab (5).

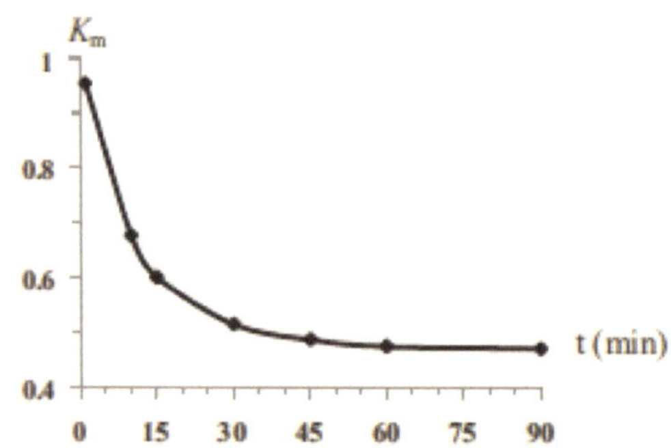

Figure 14.
Dependence of parameter K_m for CPI (120 min) on duration of intermonomer loading.

understanding the mechanism of PI synthesis in molten BA. The process shows symptoms typical for ideal interbipolycondensation. This means that the basic reaction—imidization—is practically irreversible at these conditions, i.e., the rate of evacuation with inert gas flow of the vapor of water released in the course of imidization is high. Otherwise, formation of long blocks would be impossible.

4. Synthesis of (hyper)branched (HB) polyimides

Investigation of hyperbranched (HB) polymers is a new rapidly developing field of polymer chemistry. HB polyimides (HB PIs) are of special interest for development of new functional materials as they can combine unique characteristic properties of polyimides (thermal and chemical stability, photostability barrier properties, etc.) with some common characteristic properties of HB polymers (solubility, possibility of placing many functional groups in one macromolecule, etc.). Examples are described in applying HB PI as proton-conductive or gas separation membranes, photosensible materials, etc. The study of these objects is largely constrained by multistage synthesis and a number of difficulties encountered in the synthesis process. Therefore, the creation of a simple convenient synthesis methodology is an important task.

In our work [26], we used the method of one-pot high-temperature polycondensation in molten BA at 140°C for synthesis of HB PIs with reactive anhydride groups via scheme $A_3 + B_2$, the monomer B_2 being AFL, and monomer A_3—trianhydride of hexacarboxylic acid prepared by reaction of 1,3,5-triaminotoluene sulfate with excess of BPADA. The A_3/B_2 ratio was chosen 1:1-mol. HB PI prepared was found to have wide MWD, and the glass temperature was $T_g = 235°C$, which is 55°C less than that for corresponding linear PI AFL-BPADA. In a frame of $A_2 + B_3$ approach, HB PI with terminal amino groups was prepared from 2,4,6-tris-(4-aminophenoxy)toluene and BPADA [27]. New HB polyimide with terminal amino groups was prepared also in molten BA via $A_2 + B_4$ scheme by polycondensation of 1,4-bis(3,5-diaminophenoxy) benzene (BDAPB, B_4) dialkyl semi-ester derivative of BPADA (precursor of A_2) obtained by treatment of BPADA with boiling ethyl alcohol (**Figure 15**) [28]. Structure of the products was confirmed by IR and NMR spectroscopy.

IR spectrum of synthesized HB PI sample (**Figure 16a**, curve 2) contains characteristic absorption bands at 1720 and 1780 cm^{-1} (antisymmetric and symmetric stretch C=O vibrations of imide cycle). The absorption band retains at 3300–3050 cm^{-1} which is observed also in the spectrum of starting B4 (stretch N-H vibrations of amino groups) (**Figure 16a**, spectrum 1).

The final polymer product was obtained as beige color powder soluble in THF, DMSO, and amic solvents. Mw of HB PI obtained is of about $2*10^4$. In NMR ^1H

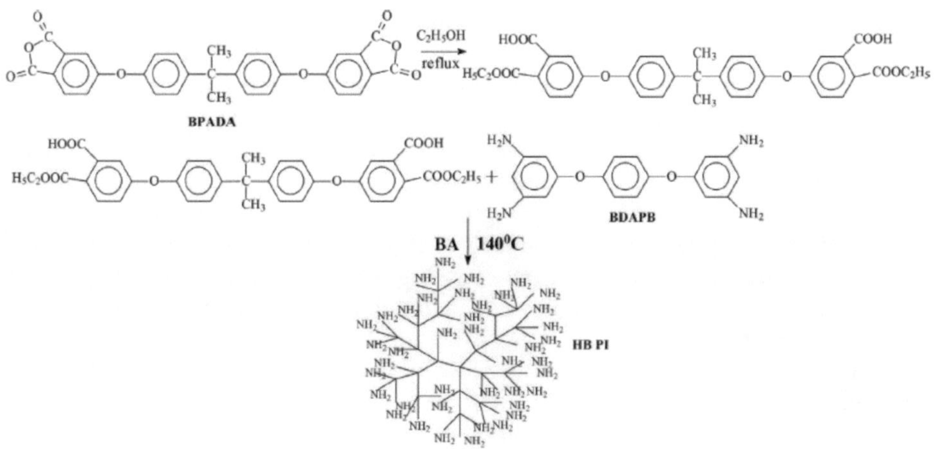

Figure 15.
Synthesis of hyperbranched polyimide.

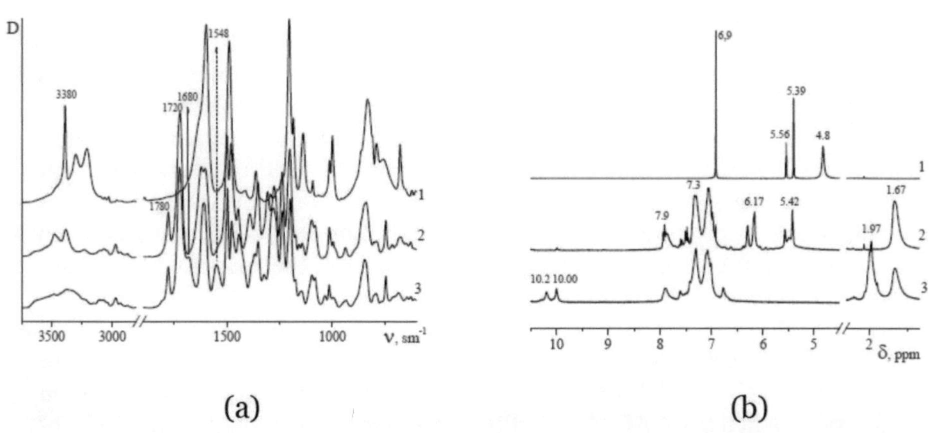

Figure 16.
Review IR spectra (a) and NMR ^1H spectra (b) of monomer B4 (1), HB PI (2), and HB PIac (3).

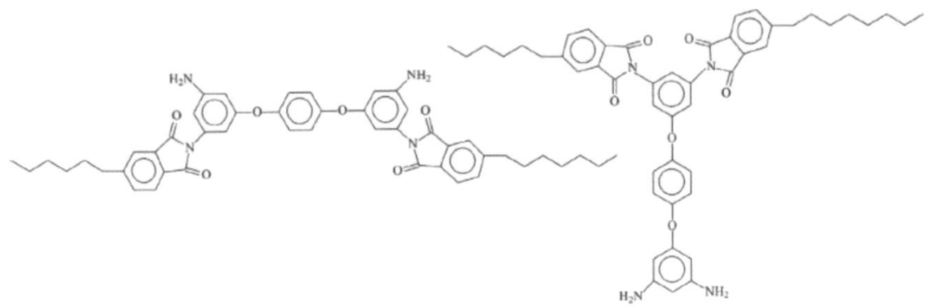

Figure 17.
Isomeric "linear" fragments in hyperbranched polyimide.

spectra of the starting B4 monomer (**Figure 16a**, spectrum 1), there are signals of protons of amino groups at 4.8 ppm, two signals of protons of aromatic fragments nearest to amino group at 5.58 and 5.42, and one signal of protons of central aromatic cycle at 6.9 ppm. In a spectrum of HB PI (**Figure 16b**, spectra 2), the signals of amino groups are shifted to 5.4 and 5.8 ppm due to appearance of imide fragments. Also, new signals of aromatic protons appear at 7–8 ppm belonging to protons of aromatic nuclei of dianhydride fragment as well as signals of isopropylidene fragment at 1.67 ppm.

With the presence of reactive terminal amino groups in HB PI, we carried out its reaction with acetic or phthalic anhydrides (at 25°C in dimethylacetamide and in molten BA, respectively). In IR spectra of the product isolated after treatment with acetic anhydride (HB PIac) (**Figure 16a**, curve 3), an absorption band at 3300–3500 cm^{-1} (NH_2 group) disappears, and new bands at 1680 and 1550 cm^{-1} (deformation N-C vibrations, "amide 1" and "amide 2") appear.

In NMR ^1H spectrum of HB PIac (**Figure 16b**, spectra 3), the signals at 5.4 and 5.8 ppm (NH_2) disappear, and new signals at 10.0 and 10.2 ppm appear related to two isomeric "linear" fragments of HB PIac (**Figure 17**). From the fact of equal intensity of these signals, it should be concluded that reactivity of all amino groups is about the same.

This observation is of importance because on its basis it can be concluded that all four amino groups of B4 monomer participate in chain growth with nearly the same probability, i.e., the scheme of reaction indeed is A2 + B4 and must result in HB polymer.

5. Synthesis of polyimides from "AB" monomers

The one-pot high-temperature catalytic polycyclocondensation in molten BA was successfully used to obtain homo- and co-PEIs from 4-(3-aminophenoxy)-phthalic acid (3-APPA) which has a structure of latent "AB" monomer [29]. In solid state, 3-APPA exists in a zwitterionic form. In molten BA at 140°C, it transfers to the "open" form which is able to dehydrate to give monomer with amino and anhydride groups. It is able for auto-polycondensation with moderate rate. Homo-PEI prepared from 3-APPA in molten BA is high molecular weight amorphous thermoplastic PEI with a glass temperature of 210–220°C soluble in chloroform and DMSO. The yield and degree of imidization are close to 100%. In contrast, homo-PEI prepared at the same conditions from isomeric 4-(4-aminophenoxy) phthalic acid (4-APPA) is insoluble and intractable till 400°C. Properties of copolyimides 3-APPA/4-APPA depend on composition. When the 3-APPA/4-APPA ratio in staring monomer mixture is 30:70 or less, the polymer is amorphous and soluble in organic solvents. At 3-APPA/4-APPA ratios of 40:60 or higher, CPIs lose solubility and tractability.

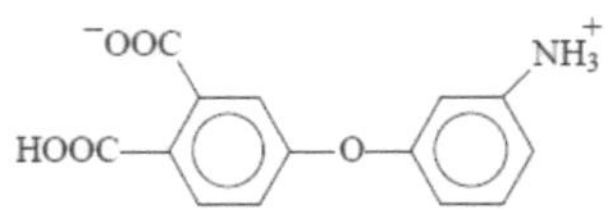

3-APPA

6. Synthesis of star-shaped polyimides with narrow molecular weight distribution

Star-shaped polymers (SSP) are of special interest among a variety of branching polymers. The state of the art in a field of SSP is considered in several excellent reviews [30]. In many papers, SSP with narrow MWR was synthesized using the methods of controlled chain polymerization. Yokozawa et al. [31]were the first to show that the "step growth" processes also can be applied to obtain SSP with narrow MWD. They synthesized SSP using polycondensation by the scheme (B_n + AB), where B_n is a multifunctional core initiator and AB is a low reactive monomer with two different functional groups A and B.

Narrow MWD can be achieved by using special AB' monomers, which are not able for auto-polycondensation, but able to react to active B groups of B_n core initiator. When AB' moiety attaches to B_n, terminal B′ group becomes much more reactive due to the changing mesomeric effect in the course of condensation, and selective arm growth occurs.

To synthesize star-shaped oligoimides (SOI), we have used known general reaction scheme $B_3(B_4)$ + AB, but another principle providing selective star arm growth not requiring special monomers AB [32, 33]. 1,3,5-(4-Aminophenoxy)toluene (TAPT) was used as B_3, bis(3,5-diaminophenoxy)benzene (BAPB) as B_4, and 3-APPA as latent AB monomer.

In the course of synthesis, APPA was introduced slowly to the reaction mixture containing TAPT and BAPB, providing that the current AB concentration in molten BA always being much less than that of TAPT or BAPB. The reaction scheme is given below (**Figure 18**). To check the possibility of obtaining stars with different average arm lengths, the overall 3-APPA/TAPT ratio was varied in a row: 10:1, 20:1, 40:1, and 100:1.

In IR spectra of SOI samples, characteristic absorption bands appear at imide cycle at 1720 and 1780 cm^{-1} (symmetric and asymmetric C=O vibrations). With increase in 3-APPA/TAPT ratio, the intensity of N-H vibrations in NMR ^{1}H of the model compound obtained by treatment TAPT with excess of phthalic anhydride (TAPT-PhA) of terminal NH_2 group at 3380–3480 cm^{-1} decreases.

Comparison of the NMR ^{1}H spectra of TAPT (**Figure 19**) and the model compound TAPT-PhA, obtained by treatment of TAPT with phthalic anhydride (PhA) (**Figure 19**), shows that the replacement of the amino group on phthalimide group results in disappearance of the signal of the amino group protons (5.0 ppm) and the signal shift (from 5.8 to 6.6 ppm) of c, d, e, and f protons (**Figure 20**) belonging to central aromatic ring. There are new signals in the region of 8.0 ppm, referred to the protons "u, z" of the phthalimide fragment. Similar changes are observed in the spectra of SOI: c, d, e, and f signals are shifted download. From the comparison of NMR spectra of SOI 10 and SOI 40 (**Figure 19**), it is seen that the intensity of proton signals of terminal amino groups "p" at 5.4 ppm and methyl group of the central fragment (2.08 ppm) decreases with an increase in the APPA/TAPT ratio. Weak signals at 6.2–6.5 ppm are referred to aromatic protons "k, l, n" of the terminal fragment containing amino group. The structure of SOI is confirmed by the results of integration of NMR ^{1}H signals.

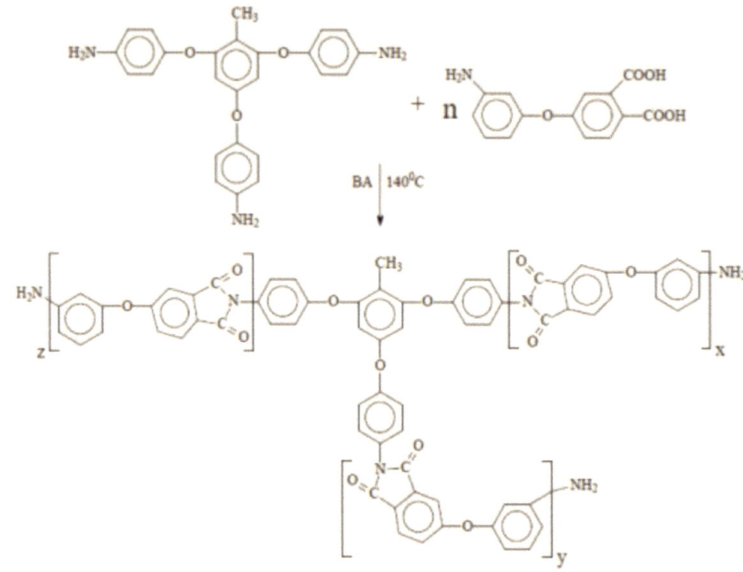

Figure 18.
Synthesis of 3-arm star-shaped polyimide.

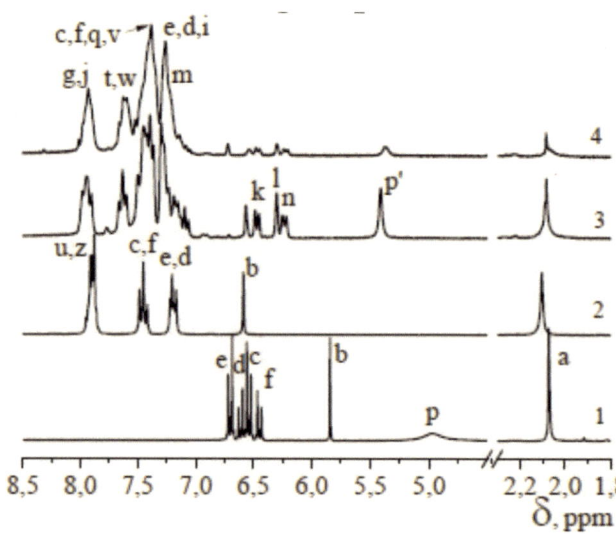

Figure 19.
NMR ^1H spectra of TAPT (1) and TAPT-PA (2), SOI prepared at different 3-APPA/TAPT mole ratios: 10/1 (3) and 40/1 (4).

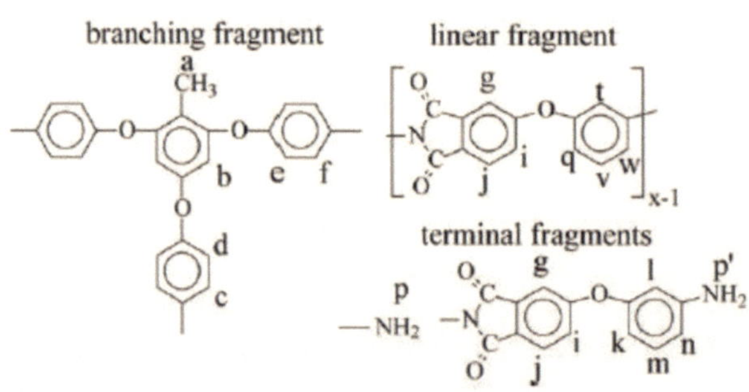

Figure 20.
Assignment of signals in NMR ^1H spectrum to corresponding hydrogen atoms in branching, linear and terminal.

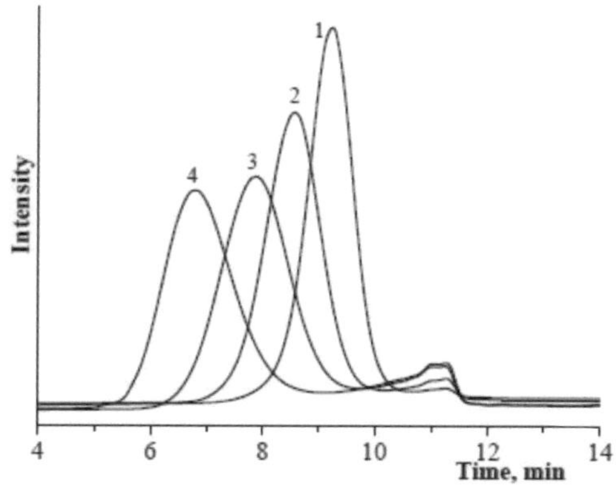

Figure 21.
GPC chromatogram of SOI prepared at different APPA/TAPT mole ratios: 10/1 (1), 20/1 (2), 40/1 (3), and 100/1 (4).

Sample APPA/TAPT	M_w	M_n	M_w/M_n	M_w/M_n (without low MW fraction)
SOI 10	13,100	11,600	1.13	1.08
SOI 20	19,800	15,200	1.3	1.08
SOI 40	29,200	16,860	1.73	1.16
SOI 100	53,350	20,700	2.57	1.17
Poly APPA	24,400	9000	2.71	—

Table 7.
Molecular weight characteristics of the oligoimides (GPC).

All SOI samples obtained are soluble in chloroform. For samples SOI40 and SOI 100, mechanically strong self-supporting films were obtained by casting from a solution in chloroform. According to DSC, increasing the APPA/TAPT ratio leads to an increase in the glass transition temperature.

In **Figure 21**, the GPC chromatograms of SOI samples are given. With an increase in the total ratio of 3-APPA/TAPT, the main peak on the GPC chromatogram shifts toward high molecular weights accomplished by small broadening of the peaks. All SOI samples have narrow MWD with M_w/M_n = 1.1–1.2 (**Table 7**), in which values are much less than the corresponding values for PI obtained by APPA auto-polycondensation in molten BA (M_w/M_n~3) in the absence of additives. Small polydispersity is very unusual for polycondensation processes. It indicates that there is a selective growth of arms on the initiator molecules, similar to the process of "chain polycondensation" [30].

There is a small peak at low molecular weights in all chromatograms (**Figure 21**). With increasing duration of synthesis, this peak does not change its position. Therefore, we refer it to low molecular cyclic oligomers. The M_w/M_n values calculated without this peak (**Table 7**, columns 5) do not differ much, in all cases M_w/M_n < 2. Formation of small quantity of cyclic oligomer can be the evidence of the low-rate by-reaction of AB auto-polycondensation to form linear oligomer with two complementary end groups A and B which can react to give cyclic oligomer.

The effective reactivity of AB in auto-polycondensation and in the reaction with terminal amino group of the growing star is thought to hardly differ from each other. The predominance of selective arm growth and narrow PDI is mainly due to the concentration of terminal arm's amino groups that is always greater than the current concentration of AB monomer in reaction system.

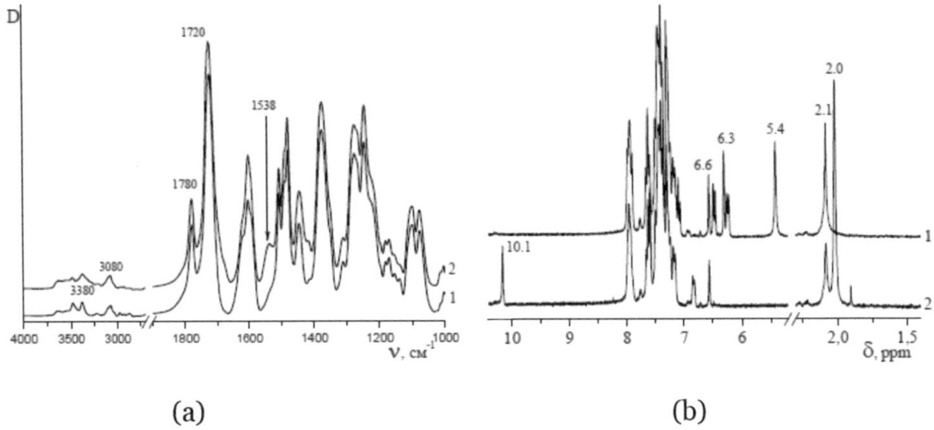

Figure 22.
IR spectra (a) and NMR ^1H spectra (b) of SOI 10 (1) and SOI-ac10 (2).

Thus, it is shown that star-shaped oligo- and polyimides with narrow MWD (Mw/Mn = 1.1–1.2) can be synthesized by polycondensation under these conditions.

To confirm the presence of reactive amino groups in SOI molecule, the following experiments were performed. In **Figure 22a**, the IR spectrum is shown of acetyl derivative SOI-ac10 obtained by the treatment of SOI 10 with acetic anhydride. The reaction of SOI with acetic anhydride leads to the disappearance absorption band of the amino group at 3480 cm^{-1} and the appearance of a new band at 1538 cm^{-1} (amide II). In the NMR ^1H spectrum of the SOI-ac10 (**Figure 22b**), the signals at 5.4 ppm (NH2) and 6.3–6.4 ppm (aromatic protons of terminal moiety) disappear, and new signals appear at 10.1 ppm (—NH—C=O)and 2.0 ppm (CH_3—C=O), in full correspondence with our expectations.

Presently, two novel tetrafunctional amines and corresponding tetra-arm star-shaped oligomers with $M_w/M_n < 1.6$ were also synthesized via B4 + AB scheme, where AB is 2-APPA and B4 is the product of direct condensation of di-N-BOC-protected 3,5-diaminobenzoic acid with aromatic diamines (ODA, AFL) [34] in the presence of triphenyl phosphate-pyridine.

7. Conclusions

1. The one-pot high-temperature catalytic polycondensation of diamines and tetracarboxylic acids dianhydrides in molten benzoic acid at 140–160°C is a new effective, technologically simple, and ecologically friendly method for obtaining polyimides and copolyimides of different chemical structures and topologies including linear polyimides from low reactive monomers, polyimides based on AB monomers, random and multiblock copolyimides, hyperbranched polyimides, star-shaped oligomers with narrow MMD and controlled arm length, etc.

2. Kinetics and mechanism of this method were investigated in detail. Due to catalysis of the first stage of the process—transient polyamic acid formation, combined with its low equilibrium constant—this stage is kinetically insignificant, and imidization becomes the limiting reaction.

3. Imidization reaction at the conditions of polyimide synthesis at 140–160° C in molten benzoic acid in open system with slow inert gas flow behaves like

nonreversible reaction. Elimination of both reversible stages in polyimide synthesis makes it possible to control chain microstructure of copolyimides by means of varying intermonomer loading.

4. Mathematical modeling of copolyimide chain microstructure formation in molten BA has been developed. Results of prediction of chain microstructure on the basis of independent kinetic data are in good consistence with experimental values obtained from NMR ^{13}C data of copolyimides.

Acknowledgements

The work is supported by the Russian Foundation of Basic Research, Grant #19-03-00820, and the Ministry of High Education and Science of Russian Federation.

Author details

Kuznetsov Alexander Alexeevich* and Tsegelskaya Anna Yurievna
Enikolopov Institute of Synthetic Polymer Materials RAS, Moscow, Russian Federation

*Address all correspondence to: kuznets24@yandex.ru

IntechOpen

© 2019 The Author(s). Licensee IntechOpen. This chapter is distributed under the terms of the Creative Commons Attribution License (http://creativecommons.org/licenses/by/3.0), which permits unrestricted use, distribution, and reproduction in any medium, provided the original work is properly cited.

References

[1] Ghosh M, Mittal K, editors. Polyimides: Fundamentals and Applications. New York: Marcel Dekker; 1996. p. 891

[2] Bessonov M, Koton M, Kudryavtsev V, Layus L. Polyimides—Thermally Stable Polymers. New York: Consultant Bureau; 1987. p. 318

[3] Sroog C. Polyimides. Progress in Polymer Science. 1991;**16**:561-694. DOI: 10.1016/0079-6700(91)90010-I

[4] Ardashnikov A, Kardash I, Pravednikov A. The equilibrium character of the reaction of aromatic anhydrides with aromatic amines and its role during the synthesis of polyimides. Polymer Science U.S.S.R. 1971;**8**:1863-1869. DOI: 10.1016/0032-3950(71)90411-4

[5] Kardash I, Ardashnikov A, Lavrov S, Pravednikov A, et al. The role of equilibrium state of the reaction of formation of polyamido acids in the process of their thermal cyclization in solution. Russian Chemical Bulletin. 1979;**28**:1983. DOI: 10.1007/BF00952499

[6] Korshak V, Vinogradova S, Vygodskii Ya, Pavlova S, Boiko L. Thermally stable soluble polyimides. Russian Chemical Bulletin. 1967;**16**: 2172-2178. DOI: 10.1007/BF00913301

[7] Vinogradova S, Vygodskii Ya, Korshak V. Some features of the synthesis of polyimides by single-stage high-temperature polycondensation. Polymer Science U.S.S.R. 1970;**12**: 2254-2262. DOI: 10.1007/BF01106305

[8] Vinogradova S, Vasnev V, Vygodskii Ya. Cardo polyheteroarylenes. Synthesis, properties, and characteristic features. Russian Chemical Reviews. 1996;**65**:249-277. DOI: 10.1070/RC1996v065n03ABEH000209

[9] Lozinskaya E, Shaplov A, Vygodskii Ya. Direct polycondensation in ionic liquids. European Polymer Journal. 2004;**40**:2065-2075. DOI: 10.1016/j.eurpolymj.2004.05.010

[10] Said-Galiyev E, Vygodskii Ya, Nikitin L, Vinokur R, Khokholov A, Pototskaya I, et al. Synthesis of polyimides in supercritical carbon dioxide. Polymer Science, Russia. 2004; **46**:634-638. DOI: 10.1016/S0896-8446(03)00146-3

[11] RF Patent No. 1809612. 1996

[12] Kuznetsov A, Tsegelskaya A, Belov M, Berendyaev V, Lavrov S, Semenova G, et al. Acid-catalyzed reactions in polyimide synthesis. Macromolecular Symposia. 1998;**128**:203-219. DOI: 10.1002/masy.201600202

[13] Kuznetsov A. One-pot polyimide synthesis in carboxylic acid medium. High Performance Polymers. 2000;**12**:445-460. DOI: 10.1134/S0965545X0711003X

[14] Vygodskii Ya, Spirina T, Nechayev P, Chudina L, Zaikov G, Korshak V, Vinogradova S. The study of the kinetics of formation of poly(amido)acids for carding. Polymer Science U.S.S.R. 1977; **19**:1738-1745. DOI: 10.1016/0032-3950(77)90186-1

[15] Kuznetsov A, Semenova G, Tsegel'skaya A, Yablokova M, Krasovskii V. Interaction of diamines with benzoic acid without a solvent: IR study and phase diagram analysis. Russian Journal of Applied Chemistry. 2008;**81**:78-81. DOI: 10.1134/S1070427208010187

[16] Kuznetsov A, Tsegelskaya A, Buzin P, Yablokova M, Semenova G. High temperature polyimide synthesis in "active" medium: Reactivity leveling of

the high and the low basic diamines. High Performance Polymers. 2007;**19**: 711-721. DOI: 10.1177/0954008307 081214

[17] Kuznetsov A, Tsegelskaya A, Buzin P. One-pot high-temperature synthesis of polyimides in molten benzoic acid: Kinetics of reactions modeling stages of polycondensation and cyclization. Polymer Science, Series A. 2007;**49**: 1157-1164. DOI: 10.1134/ S0965545X0711003X

[18] Vasnev V, Kuchanov S. Non-equilibrium copolycondensation in homogeneous systems. Russian Chemical Reviews. 1973;**42**: 1020-1033. DOI: 10.1070/ RC1973v042n12ABEH002782

[19] Kuznetsov A, Tsegelskaya A, Perov N. ^{13}C-NMR analysis of chain microstructure of copolyimides on the basis of 2,2-bis[(3,4-dicarboxyphenoxy) phenyl]propane dianhydride synthesized in molten benzoic acid. High Performance Polymers. 2012;**24**: 58-63. DOI: 10.1177/0954008311429501

[20] Batuashvili M, Tsegelskya A, Perov N, Semenova G, Abramov I, Kuznetsov A. Chain microstructure of soluble copolyimides containing moieties of aliphatic and aromatic diamines and aromatic dianhydrides prepared in molten benzoic acid. High Performance Polymers. 2014;**26**:470-476. DOI: 10.1177/0954008313518950

[21] Yamadera R, Murano M. The determination of randomness in copolyesters by high resolution nuclear magnetic resonance. Journal of Polymer Science Part A: Polymer Chemistry. 1967;**5**:2259-2268. DOI: 10.1002/ pol.1967.150050905

[22] Vygodskii Ya, Vinogradova S, Nagiev Z, et al. A study of copolyimide synthesis, structure and properties. Acta Polymerica. 1982;**33**:131. DOI: 10.1002/ actp.1982.010330206

[23] Batuashvili M, Tsegelskaya A, Perov N, Semenova G, Orlinson B, Kuznetsov A. Formation of the chain microstructure in the synthesis of adamantine containing copolyimides in molten benzoic acid. Russian Chemical Bulletin. 2015;**64**:930-935. DOI: 10.1007/s11172-015-0957-8

[24] Batuashvili M, Kaminskii V, Tsegelskaya A, Kuznetsov A. Formation of the chain microstructure in copolyimide synthesis by high-temperature polycondensation in molten benzoic acid. Russian Chemical Bulletin. 2014;**63**:2711-2718. DOI: 10.1007/s11172-014-0804-3

[25] Kuznetsov A, Batuashvili M, Kaminskii V. Modeling of copolyimide chain microstructure formation in a course of the one-pot copolycondensation in molten benzoic acid. High Performance Polymers. 2017;**29**:716-723. DOI: 10.1177/0954008317702953

[26] Kuznetsov A, Akimenko S, Tsegelskaya A, Perov N, Semenova G, Shakhnes A, et al. Synthesis of branched polyimides based on 9,9-bis (4-aminophenyl)fluorene and an oligomeric trianhydride, a 1,3,5-triaminotoluene derivative. Polymer Science, Series B. 2014;**56**:41-48. DOI: 10.1134/S1560090414010060

[27] Chukova S, Shakhnes A, Perov N, Krasovskii V, Shevelev S, Kuznetsov A. 2,4,6-Tris(4-aminophenoxy)toluene and hyperbranched polyimide derived from it. Russian Chemical Bulletin. 2015;**64**:473-474. DOI: 10.1007/ s11172-015-0890-x

[28] Tsegelskaya A, Dutov M, Serushkina O, Semenova G, Kuznetsov A. The one-stage synthesis of hyperbranched polyimides by (A2+B4) scheme in catalytic solvent.

Macromolecular Symposia. 2017;**375**: 1600202. DOI: 10.1002/masy.2016 00202

[29] Buzin P, Yablokova M, Kuznetsov A, Smirnov A, Abramov I. New AB polyetherimides obtained by direct polycyclocondensation of aminophenoxy phthalic acids. High Performance Polymers. 2004;**16**: 505-514. DOI: 10.1177/09540083040 39991

[30] Gao H. Development of star polymers as unimolecular containers for nanomaterials. Macromolecular Rapid Communications. 2012;**33**:722-734. DOI: 10.1002/marc.201200005

[31] Sugi R, Hitaka Y, Yokoyama A, Yokozawa T. Well-defined star-shaped aromatic polyamides from chain-growth polymerization of phenyl 4-(alkylamino)benzoate with multifunctional initiators. Macromolecules. 2005;**38**:5526-5531. DOI: 10.1021/ma0473420

[32] Kuznetsov A, Soldatova A, Tokmashev R, Tsegelskaya A, Semenova G, Shakhnes A, et al. Synthesis of reactive three-arm star-shaped oligoimides with narrow molecular weight distribution. Journal of Polymer Science, Part A: Polymer Chemistry. 2018;**56**:2004-2009. DOI: 10.1002/pola.29088

[33] Tsegelskaya A, Soldatova A, Semenova G, Dutov M, Abramov I, Kuznetsov A. One-stage high temperature catalytic synthesis of star-shaped oligoimides by (B4+AB) scheme. Polymer Science, Series B. 2019;**61**: 148-154. DOI: 10.1007/s11172-018-2345-7

[34] Soldatova A, Tsegelskaya A, Semenova G, Abramov I, Kuznetsov A. Synthesis of tetrafunctional aromatic amines and star-shaped oligoimides using the B4+AB scheme. Russian Chemical Bulletin. 2018;**67**:2152-2154. DOI: 10.1007/s11172-0182345-7

Section 2

Ionic Liquids

Chapter 4

Extraction of Aromatic Compounds from Their Mixtures with Alkanes: From Ternary to Quaternary (or Higher) Systems

Ángeles Domínguez, Begoña González, Patricia F. Requejo and Sandra Corderí

Abstract

Ionic liquids have been proposed as separation agents for liquid extraction of aromatic compounds from their mixtures with alkanes, with the aim of improving the separation process and replacing conventional organic solvents. A significant number of experimental liquid-liquid equilibrium data for ternary system alkane + aromatic compound + ionic liquid can be found in literature; however there are few data for quaternary or higher systems involving more than one aliphatic compound, several aromatic compounds or a mixture of ionic liquids as separation agent. These data are also necessary because molecular interactions between the compounds in the mixture can modify the affinity of the solvent for the aromatic compound of interest. In this chapter we review the published data involving more than three components, and we present new liquid-liquid equilibrium data for the quaternary systems heptane + cyclohexane + toluene +1-ethyl-3-methylimidazolium bis{(trifluoromethyl)sulfonyl}imide and heptane + cyclohexane + toluene + 1-hexyl-3-methylimidazolium bis{(trifluoromethyl)sulfonyl}imide.

Keywords: ionic liquids, aromatic compounds, liquid extraction

1. Introduction

Aromatic compounds, such as benzene, toluene, ethylbenzene and xylenes (BTEX), are raw materials for the production of polymers, resins paints and other products of industrial interest. They are mainly obtained from catalytic reforming and cracking processes in oil refineries, as a mixture of aromatic and aliphatic hydrocarbons. Aromatic content depends on the process characteristics, and it can range between 20 and 65 wt% for reformate gasoline and between 50 and more than 90% for pyrolysis gasoline [1].

The separation of aromatic from aliphatic compounds is difficult because they usually have close boiling points and many of their mixtures show azeotropic behaviour. One of the most used processes to separate these aromatic compounds is liquid-liquid extraction, due to the fact that it can be used for a wide range of aromatic concentration in the mixture. Sulfolane, among other solvents, is widely used in these processes because of its high selectivity (S) for aromatics although it is generally used when aromatic content is high. After the extraction process, further separation units are needed to recover sulfolane from raffinate and extract phases.

The increasing concern for the environment has induced the search of new solvents that reduce pollution and energy costs, making the extraction process more environmentally friendly. One of the alternatives proposed is the use of ionic liquids (ILs) as solvents for the extraction of aromatic compounds, and consequently a large number of papers reporting liquid-liquid equilibria (LLE) data of ternary system aliphatic + aromatic + IL can be found in the literature. From experimental data, selectivity and distribution coefficients (β) can be calculated and used to evaluate the capability of the solvent since high selectivity together with high distribution coefficients are suitable. Canales and Brennecke [1] compared ILs and conventional solvents for the extraction of aromatic from aliphatic compounds, finding that ILs are potentially alternatives to currently used solvents.

Even though the information obtained from ternary systems is crucial, we cannot be certain that selectivity and distribution coefficients obtained from ternary systems will remain unchanged when more components, aromatics or aliphatic are present in the mixture. Because of this, selectivity and solute distribution coefficients should be determined for complex mixtures, in order to obtain more actual data.

Some papers related to quaternary or higher systems using sulfolane to extract aromatics from aliphatic compounds can be found in the literature [2–8], while similar information using ILs as extractant agents is scarce, taking into account the huge number of ILs that can be used [9–22]. It also should be noted that mixtures of ILs or organic solvent + IL can be selected as separation agents [23], greatly increasing the number of systems that should be studied.

In this chapter, a literature review on the results obtained using ILs for the extraction of aromatic compounds in systems with more than three components is performed. Furthermore new liquid-liquid equilibrium experimental data for the quaternary systems heptane + cyclohexane + toluene + 1-ethyl-3-methylimidazolium bis{(trifluoromethyl)sulfonyl}imide, [EMim][NTf_2], and heptane + cyclohexane + toluene + 1-hexyl-3-methylimidazolium bis{(trifluoromethyl)sulfonyl}imide, [HMim][NTf_2], are presented.

2. Literature review

2.1 Quaternary systems with two aliphatic compounds

Up to now, a quite wide range of papers dealing with liquid-liquid equilibrium data of hydrocarbon + aromatic + IL ternary systems can be found in literature. These studies are essential to determine the capability of ILs to extract aromatic compounds; however, many other compounds are present in refinery streams, and these compounds can modify the extraction process.

With the aim of analysing the influence of more than one aliphatic compound on the extraction of aromatics, liquid-liquid equilibrium data of several quaternary systems have been published [9–14].

Requejo et al. [9–11] studied the extraction of benzene from its mixtures with octane and decane using tributylmethylammonium bis(trifluoromethylsulfonyl)imide, [N_{4441}][NTf_2]; 1-butyl-1-methylpyrrolidinium dicyanamide, [BMpyr][DCA]; and 1-butyl-1-methylpyrrolidinium bis(trifluoromethylsulfonyl)imide, [BMpyr][NTf_2], and the results of the quaternary system were compared with those obtained for the ternary systems octane + benzene + IL and decane + benzene + IL.

Using [N_{4441}] [NTf_2], the highest values of β were obtained for the ternary system octane + benzene + [N_{4441}] [NTf_2], while the β values calculated for the ternary system decane + benzene + $[N_{4441}][NTf_2]$ and for the quaternary system are similar. Selectivity values are quite similar for the ternary and quaternary systems; consequently the presence of two alkanes does not seem to affect them [9].

Solute distribution coefficient is similar for both ternary systems octane + benzene + [BMpyr][DCA] and decane + benzene + [BMpyr][DCA] and for the quaternary system octane + decane + benzene + [BMpyr][DCA], while the selectivity for the quaternary system is lower than for the ternary system with decane and higher than for the ternary system with octane [10].

A different influence was found comparing the ternary systems octane + benzene + $[BMpyr][NTf_2]$ and decane + benzene + $[BMpyr][NTf_2]$ with the quaternary system decane + octane (1) + benzene (2) + $[BMpyr][NTf_2]$ [11]. In that case, both solute distribution coefficients and selectivity are lower for the quaternary system.

The extraction of toluene from its mixtures with heptane and cyclohexane, using ethyl-methylimidazolium-based ILs, was studied by Corderí et al. [12–14]. The selected anions were methylsulphate $[MSO_4]$, acetate [OAc] and dicyanamide $[N(CN)_2]$. According to their results, solute distribution coefficient clearly follows the trend:

$[EMim][N(CN)_2]$ > [EMim][OAc] > $[EMim][MSO_4]$.

With regard to selectivity, higher values were obtained with $[EMim][N(CN)_2]$.

Comparing quaternary systems with the respective ternary ones, the following results were obtained:

i. Solute distribution coefficients for the ternary systems with $[EMim][MSO_4]$ are quite similar and higher than those obtained for the quaternary system at higher toluene composition in the hydrocarbon-rich phase. When the toluene concentration in the hydrocarbon-rich phase is lower than 0.2, similar values of β were obtained for the ternary and the quaternary systems.

ii. In the case of [EMim][OAc], β values for the quaternary system are slightly higher than those obtained for the ternary systems independent of the toluene concentration in the hydrocarbon-rich phase.

iii. Similar β values were obtained for the ternary and quaternary systems using $[EMim][N(CN)_2]$.

iv. In all cases, selectivity values for the quaternary system are between the values of the ternary systems.

As it can be seen, both solute distribution coefficients and selectivity values can be influenced by the presence of another hydrocarbon in different ways. The lack of more experimental data does not allow us to take general conclusions about the influence of other aliphatic compounds present in the mixture; consequently more experimental data would be necessary.

2.2 Quaternary systems with IL mixtures as extraction agents

Among all the ILs that have been studied as separation agents, only a small number of them have shown simultaneously solute distribution coefficients and selectivity higher than sulfolane since, generally, high selectivities are related to low solute distribution coefficients and vice versa. On the other hand, most ILs present quite high viscosity, which make their use difficult in extraction processes. The proper selection of two ILs could lead to a separation agent, whose physical properties and extraction performance overcome those of sulfolane. Therefore, some pairs of ILs have been tested to extract aromatics, mainly benzene or toluene, from their mixtures with alkanes.

Sakal et al. [15] carried out the extraction of benzene from benzene + cyclohexane mixtures using 1,3-dimethylimidazolium dimethylphosphate, [MMim][DMP], and 1-methylimidazolium tetrafluoroborate, [Mim][BF_4], and the mixtures [Mim][BF_4] wt. 50% + [MMim][DMP] wt. 50% and [Mim][BF_4] wt. 25% + [MMim][DMP] wt. 75%. Both selectivity and solute distribution coefficient follow the same trend: [MMim][DMP] > [Mim][BF_4] wt. 25% + [MMim][DMP] wt. 75% > [Mim][BF_4] wt. 50% + [MMim][DMP] wt. 50%. Additionally, IL mixtures of 1-methylimidazolium perchlorate, [Mim][ClO_4] and [MMim][DMP], at the same mass ratios were also checked, finding that the addition of [Mim][BF_4] or [Mim][ClO_4] does not improve the extraction capacity of [MMim][DMP].

Potdar et al. [16] determined LLE data for the quaternary system hexane + benzene + 1-ethyl-3-methylimidazolium ethylsulphate, [EMim][ESO_4], + 1-ethyl-3-methylimidazolium methylsulphate, [EMim][MSO_4]. The comparison between β and S obtained using the mixture of ILs and each of them separately shows that the mixture of ILs has a lower extraction capacity. These worse results can be due to the fact that there is less free volume in a combination of these two ILs, hindering the solution of benzene in the ILs mixture.

The mixtures of N-butylpyridinium tetrafluoroborate, [Bpy][BF_4], with N-butylpyridinium bis(trifluoromethylsulfonyl)imide, [Bpy][NTF_2], and with 1-butyl-4-methylpyridinium bis(trifluoromethylsulfonyl) imide, [4BMpy][NTf_2], were tested for the extraction of toluene from the mixture heptane-toluene. These ILs were selected because of the higher selectivity and the lower solute distribution coefficient showed by [Bpy][BF_4] and the opposite behaviour showed by [Bpy][NTf_2] and [4BMpy][NTf_2] [17–18]. In both cases, β and S values higher than those obtained using sulfolane were achieved with a [Bpy][BF_4] mole fraction of 0.7 in the mixture of ILs.

On the basis of the results obtained using 1-ethyl-3-methylimidazolium tricyanomethanide, [EMim][TCM], as well as 1-ethyl-3-methylimidazolium dicyanamide, [EMim][DCA], for the extraction of toluene from heptane, a mixture of [EMim][TCM] + [EMim][DCA] with a [EMim][TCM] mole fraction of 0.8 was selected to carry out the same extraction process [19]. Both solute distribution coefficient and selectivity are higher using this IL mixture than using sulfolane. Since selectivity values using the IL mixture were almost double than those using sulfolane, toluene extracted by the IL mixture would be significantly purer.

The capability of the binary mixture of 1-ethyl-4-methylpyridinium bis(trifluoromethylsulfonyl)imide, [4EMpy][NTf_2], and 1-ethyl-3-methylimidazolium dicyanamide, [EMim][DCA], as solvent for the extraction of toluene from its mixtures with heptane or 2,3-dimethylpentane or cyclohexane was evaluated by Larriba et al. [20]. The IL [4EMpy][NTf_2] was selected because it showed solute distribution coefficients for toluene higher than sulfolane. On the other hand, toluene selectivity of [EMim][DCA] is substantially higher than that of sulfolane. The extraction of toluene from other alkanes (hexane, octane or nonane) was also performed [21]. According to the authors, selectivity values increase when n-alkane chain length increases, while toluene distribution coefficients follow the opposite trend. Taking into account the results, the IL mixture can be an alternative to sulfolane in the extraction of toluene from mixtures toluene-n-alkane with low concentration of toluene.

The same mixture of ILs was selected for the extraction of benzene or ethylbenzene or xylenes from heptane [22]. The mixture with a [4EMpy][NTf_2] molar fraction of 0.3 was chosen in all cases because of its density and viscosity similar to sulfolane and good extraction performance obtained in preliminary tests. As well as using a single IL, best results were obtained in the separation of benzene, whereas the lowest values of β and S were obtained in the separation of ethylbenzene. Compared to sulfolane, the mixture of ILs showed higher extractive properties in the extraction of benzene and higher selectivity and slightly lower solute distribution coefficients for p-xylene/heptane separation.

As mentioned above, one of the main drawbacks of the use of ILs as extraction agents is their high viscosity that is difficult for the separation process. The selection of an appropriate solvent to mix with the IL could break the hydrogen bonds between the cation and the anion, lowering the viscosity but retaining the intrinsic properties of the IL. With the aim of lowering viscosity and decreasing costs, the mixtures of [EMim][OAc] + acetonitrile and [EMim][ESO_4] + acetonitrile were used to extract benzene from a benzene + hexane mixture [23]. It is worth noting that solute distribution coefficients are very low compared to the values obtained using a single IL; however, both hexane composition in extract phase and solvent composition in raffinate phase are practically zero. This fact would considerably simplify the solvent recovery.

2.3 Systems with more than four components

LLE data for systems with more than four components are very scarce. As a continuation of the extraction studies using a mixture of ILs, Larriba et al. [24] analysed the separation of BTEX fraction from a reformer gasoline model (n-hexane, n-heptane, n-octane). In that case, aromatic distribution coefficients using the IL mixture were significantly lower than using sulfolane, whereas selectivity values were higher. It is interesting to point out that in all pseudoternary system alkane + aromatic + IL mixture previously studied by the authors, they achieved better results using the mixture [4EMpy][NTf_2] + [EMim][DCA] than using sulfolane. In order to improve these results, [EMim][DCA] was replaced by [EMim][TCM], and a mixture of [4EMpy][NTf_2] + [EMim][TCM] with a [EMim][TCM] mole fraction of 0.6 was selected on the bases of extraction yield and thermophysical properties of the mixture [25]. The extraction results were used to perform the simulation of the extraction process including the extraction column and the recovery section of the solvent.

In order to study the extraction of BTEX fraction from naphtha reformate using 1-butyl-3-methylimidazolium hexafluorophosphate, [BMim][PF_6], Al-Rashed et al. [26] determined liquid-liquid equilibrium data for the system hexane + heptane + octane + benzene + toluene + o-xylene + [BMim][PF_6]. Comparing the aromatic extraction between this system and those systems with only one aromatic, they concluded that a combination of the three aromatic compounds affected negatively on the extractive capability of the IL. According to the authors, this can be because the aromatic compounds in one phase can strongly associate between them through π-π forces. Furthermore, the electrostatic association between the aromatic alkyl chains and alkanes present in the mixture also influences the extractive capability of the IL.

The gasoline or naphtha model was constituted by equal amounts of alkanes in all the above-mentioned studies.

3. Experimental part

In order to analyse the influence of the alkyl chain length of the IL cation on the extraction of toluene from its mixture with heptane and cyclohexane, liquid-liquid equilibrium data of the quaternary systems heptane + cyclohexane + toluene + 1-ethyl-3-methylimidazolium bis{(trifluoromethyl)sulfonyl}imide, [EMim][NTf_2], and heptane + cyclohexane + toluene + 1-hexyl-3-methylimidazolium bis{(trifluoromethyl)sulfonyl}imide, [HMim][NTf_2], were determined. Experimental data were compared to the results obtained for the ternary systems heptane + toluene + [EMim][NTf_2], cyclohexane + toluene + [EMim][NTf_2], heptane + toluene + [HMim][NTf_2] and cyclohexane + toluene + [HMim][NTf_2] [27, 28].

3.1 Chemicals and experimental procedure

The ILs [EMim][NTf_2] and [HMim][NTf_2] were supplied by IoLiTec GmbH. They were subjected to vacuum and moderate temperature (P = 0.2 Pa, T = 343 K) for several days to eliminate water and volatile compounds that could be present. Cyclohexane, heptane and toluene were supplied by VWR Prolabo, and they were used without further purification. Mass purity of all chemicals was higher than 99%. In order to prevent water absorption, all chemicals were kept and manipulated under inert gas atmosphere.

Figure 1 shows a representation of the quaternary systems studied in this work where grey surfaces represent the miscible areas on the top and the base of the tetrahedron. Due to the very negligible miscibility of these ILs in the aliphatic mixture (checked by ^{1}H-NRM), no IL is present in the aliphatic hydrocarbon phase. For the determination of liquid-liquid equilibrium data, samples of a mixture of the four components were prepared by weight using a Mettler AXE-205 Delta Range balance. In order to select the composition of the initial samples, a sectional plane (SP), in which the mole fraction of the ionic liquid is constant, was selected as it can be seen in **Figure 1**. This sectional plane is perpendicular to the tie lines, and the compositions of the other three compounds involved in the initial quaternary mixtures were selected in such a way that they cover the sectional plane.

The samples were vigorously stirred by using a magnetic stirrer for 6 hours and left to settle down overnight. Afterwards, a sample of each phase was withdrawn with a syringe, and it was analysed by gas chromatography.

The compositions of cyclohexane, heptane and toluene were analysed with a Hewlett-Packard 5890 Series II gas chromatograph with a Hewlett-Packard 5971 mass selective detector and a Hewlett-Packard-5MS capillary column (60 m × 0.250 mm × 0.25 μm). Because of the IL negligible vapour pressure, they cannot be analysed by gas chromatography; consequently, their composition was calculated through a mass balance. An empty precolumn was used to avoid the IL that could not be retained by the liner to go into the chromatograph. The temperature programme has initial temperature of 343.15 K for 10.30 min, ramp of 15 K min^{-1} and final temperature of 368.15 K for 4.30 min. The injector and detector were maintained at 553.15 K, and the helium carrier gas flow rate was kept constant at 1 mL min^{-1} in the column. Two analyses of each sample were performed to obtain a mean value.

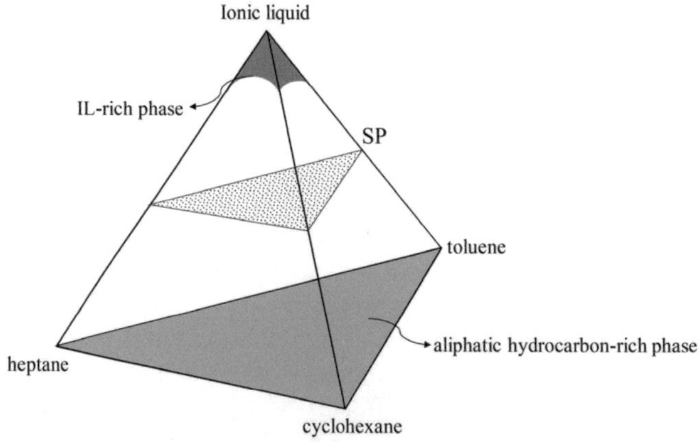

Figure 1.
Schematic representation of the quaternary system and the sectional plane (SP).

In order to determine the error in the composition, quaternary mixtures with known composition were analysed by chromatography, and their obtained compositions were compared with those obtained by weight. The error in the compositions was estimated to be ±0.004.

3.2 Results and discussion

From experimental values, solute distribution coefficient, β, and selectivity, S, were calculated by the following equations:

$$\beta = x_3^{II}/x_3^{I} \tag{1}$$

$$S = x_3^{II}x_{1+2}^{I}/x_3^{I}x_{1+2}^{II} \tag{2}$$

Hydrocarbon-rich phase		IL-rich phase			β	S
x_1^I	x_2^I	x_1^{II}	x_2^{II}	x_3^{II}		
0.778	0.151	0.018	0.010	0.062	0.87	28.97
0.717	0.123	0.020	0.010	0.133	0.83	23.28
0.639	0.182	0.021	0.015	0.147	0.82	18.73
0.522	0.167	0.019	0.014	0.242	0.78	16.25
0.445	0.149	0.018	0.014	0.294	0.72	13.44
0.367	0.153	0.017	0.015	0.329	0.69	11.14
0.196	0.180	0.010	0.022	0.405	0.65	7.63
0.561	0.241	0.03	0.024	0.153	0.77	11.48
0.408	0.264	0.014	0.024	0.249	0.76	13.42
0.351	0.226	0.014	0.022	0.301	0.71	11.41
0.255	0.254	0.010	0.026	0.341	0.69	9.82
0.143	0.271	0.007	0.031	0.392	0.67	7.29
0.604	0.326	0.029	0.029	0.057	0.81	13.06
0.423	0.352	0.013	0.031	0.183	0.81	14.33
0.317	0.357	0.011	0.033	0.248	0.76	11.65
0.264	0.328	0.010	0.033	0.295	0.72	9.95
0.523	0.427	0.014	0.031	0.041	0.82	17.31
0.340	0.427	0.011	0.037	0.191	0.82	13.10
0.229	0.444	0.004	0.039	0.246	0.75	11.77
0.165	0.403	0.005	0.041	0.313	0.72	8.95
0.436	0.501	0.009	0.036	0.052	0.83	17.19
0.322	0.510	0.012	0.049	0.138	0.82	11.20
0.247	0.511	0.007	0.045	0.190	0.79	11.44
0.350	0.589	0.024	0.072	0.052	0.85	8.34
0.241	0.587	0.004	0.066	0.138	0.80	9.49
0.150	0.623	0.006	0.071	0.182	0.80	8.05

Table 1.
Experimental LLE data, solute distribution coefficients, β, and selectivity, S, for the quaternary system heptane (1) + cyclohexane (2) + toluene (3) + [EMim][NTf$_2$] (4).

where x is the mole fraction; the superscripts I and II indicate the hydrocarbon-rich phase and the IL-rich phase, respectively; and the subscripts 1 + 2 refer to the mole fraction of heptane plus cyclohexane and 3 refers to toluene.

Tables 1 and **2** show the experimental data at 298.15 K, β and S for the system heptane + cyclohexane + toluene + $[EMim][NTf_2]$ or $[HMim][NTf_2]$, respectively.

As it can be seen from the tables, selectivity is higher using $[EMim][NTf_2]$, while higher solute distribution coefficients were obtained using $[HMim][NTf_2]$. Since S > 1 in both cases, $[HMim][NTf_2]$ would be the best choice due to β values higher than 1 which facilitate the extraction process, and less solvent and less separation stages would be necessary. **Figures 2** and **3** show the tie lines for the pseudoternary system (heptane + cyclohexane) (1) + toluene (2) + IL (3). The higher β values obtained for the extraction using $[HMim][NTf_2]$ are reflected in the positive slope of the tie lines for toluene molar fractions up to 0.45.

Hydrocarbon-rich phase		**IL-rich phase**			**β**	**S**
x_1^I	x_2^I	x_1^{II}	x_2^{II}	x_3^{II}		
0.776	0.149	0.095	0.03	0.112	1.49	11.05
0.688	0.168	0.099	0.038	0.193	1.35	8.43
0.631	0.142	0.097	0.034	0.287	1.26	7.46
0.544	0.170	0.099	0.045	0.334	1.17	5.79
0.458	0.164	0.076	0.041	0.416	1.10	5.85
0.368	0.175	0.071	0.052	0.459	1.00	4.43
0.264	0.177	0.062	0.061	0.513	0.92	3.29
0.182	0.183	0.049	0.069	0.554	0.87	2.70
0.675	0.245	0.087	0.053	0.115	1.44	9.45
0.590	0.258	0.090	0.065	0.198	1.30	7.13
0.527	0.240	0.078	0.059	0.286	1.23	6.87
0.431	0.265	0.070	0.070	0.347	1.14	5.67
0.352	0.254	0.065	0.074	0.417	1.06	4.63
0.164	0.279	0.039	0.102	0.510	0.92	2.88
0.593	0.327	0.081	0.076	0.116	1.45	8.50
0.478	0.362	0.070	0.090	0.206	1.29	6.76
0.351	0.369	0.056	0.100	0.332	1.18	5.45
0.158	0.376	0.031	0.124	0.456	0.98	3.37
0.507	0.409	0.066	0.095	0.121	1.44	8.20
0.398	0.445	0.059	0.115	0.204	1.30	6.30
0.259	0.464	0.040	0.131	0.319	1.15	4.87
0.420	0.495	0.056	0.121	0.120	1.41	7.30
0.159	0.548	0.025	0.159	0.326	1.11	4.28
0.330	0.585	0.045	0.150	0.118	1.39	6.51
0.229	0.615	0.031	0.158	0.199	1.28	5.41
0.161	0.741	0.019	0.194	0.135	1.38	5.83

Table 2.
Experimental LLE data, solute distribution coefficients, β, and selectivity, S, for the quaternary system heptane (1) + cyclohexane (2) + toluene (3) +$[HMim][NTf_2]$ (4).

Selectivity and solute distribution coefficients for the quaternary system heptane + cyclohexane + toluene + [EMim][NTf_2] can be compared to the results obtained for the ternary systems heptane + toluene + [EMim][NTf_2] and cyclohexane + toluene + [EMim][NTf_2] [27]. β values are similar for the ternary and quaternary systems, while S for the ternary system heptane + toluene + [EMim][NTf_2] (between 3 and 24) and the quaternary system are similar and higher than the ternary system with cyclohexane (between 8 and 4).

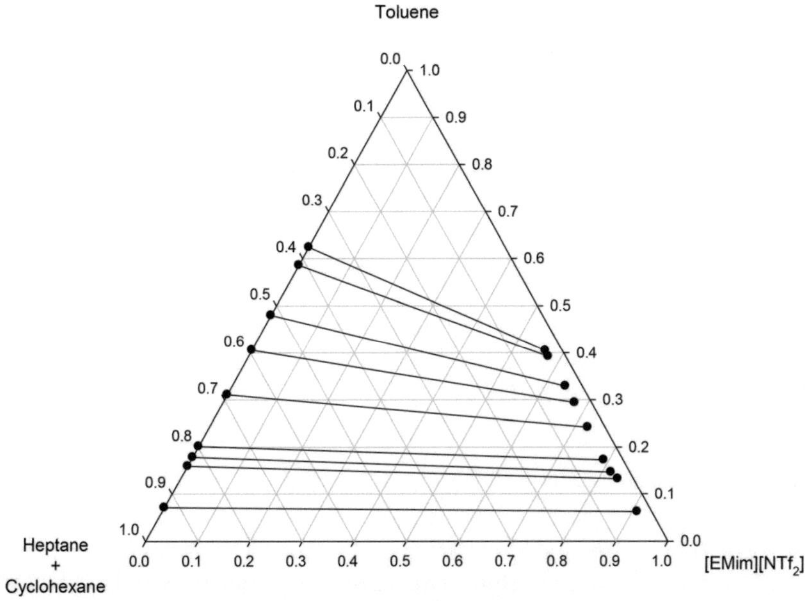

Figure 2.
Tie lines for the quaternary systems heptane + cyclohexane + toluene + [EMim][NTf_2].

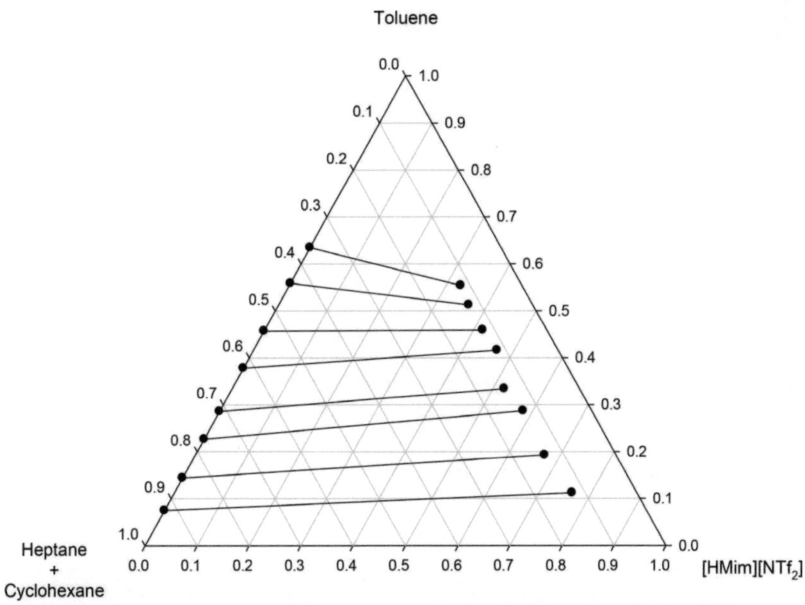

Figure 3.
Tie lines for the quaternary systems heptane + cyclohexane + toluene + [HMim][NTf_2].

Regarding the quaternary system heptane + cyclohexane + toluene + [HMim][NTf_2], the experimental data of the ternary systems heptane + toluene + [HMim][NTf_2] and cyclohexane + toluene + [HMim][NTf_2] can also been found in literature [28]. In that case, both β and S are higher for the quaternary system, which implies that, in some cases, the addition of another component can enhance the extraction of the aromatic compound.

Solute distribution coefficient and selectivity can be compared to those obtained using 1-ethyl-3-methylimidazolium methylsulphate, [EMim][MSO_4] [13]; 1-ethyl-3-methylimidazolium dicyanamide, [EMim][$N(CN)_2$] [14]; or 1-ethyl-3-methylimidazolium acetate ionic liquid, [EMim][OAc] [12]. Solute distribution coefficient follows the following trend:

[HMim][$HNTf_2$] > [EMim][$HNTf_2$] > [EMim][$N(CN)_2$] > [EMim][OAc] > [EMim][MSO_4].

β values obtained with ionic liquids with NTf_2 anion are significantly higher, reaching values greater than one in the case of [HMim][$HNTf_2$]. Regarding the selectivity, the trend is

[EMim][$N(CN)_2$] > [EMim][MSO_4] > [EMim][$HNTf_2$] > [EMim][OAc] > [HMim][$HNTf_2$].

Since selectivity and solute distribution coefficient do not follow the same trend, the selection of the most adequate IL should be a compromise between these two factors affecting extraction process.

4. Conclusions

Single ILs, IL mixtures and also mixtures of an IL and an organic solvent have been used to extract aromatics from its mixtures with aliphatic compounds. The obtained results show that, in general, ILs can be considered to replace conventional solvents such as sulfolane. However, before this replacement and taking into account the different results obtained for systems with more than three components, the study of the extraction of aromatic compounds from real naphtha reformate or pyrolysis gasoline, including the solvent recovery process, must be carried out.

Author details

Ángeles Domínguez*, Begoña González, Patricia F. Requejo and Sandra Corderí
Department of Chemical Engineering, University of Vigo, Vigo, Spain

*Address all correspondence to: admguez@uvigo.es

IntechOpen

© 2019 The Author(s). Licensee IntechOpen. This chapter is distributed under the terms of the Creative Commons Attribution License (http://creativecommons.org/licenses/by/3.0), which permits unrestricted use, distribution, and reproduction in any medium, provided the original work is properly cited.

References

[1] Canales RI, Brennecke JF. Comparison of ionic liquids to conventional solvents for extraction of aromatic from aliphatics. Journal of Chemical & Engineering Data. 2016;**61**:1685-1699

[2] Ashcroft SJ, Clayton AD, Shearn RB. Liquid-liquid equilibriums for three ternary and six quaternary systems containing sulfolane, n-heptane, toluene, 2-propanol, and water at 303.15 K. Journal of Chemical & Engineering Data. 1982;**27**:148-151

[3] Chen J, Duan L-P, Mi J, Fei W, Li Z-C. Liquid–liquid equilibria of multi-component systems including n-hexane, n-octane, benzene, toluene, xylene and sulfolane at 298.15 K and atmospheric pressure. Fluid Phase Equilibria. 2000;**173**:109-119

[4] Chen J, Li Z, Duan L. Liquid–liquid equilibria of ternary and quaternary systems including cyclohexane, 1-heptene, benzene, toluene, and sulfolane at 298.15 K. Journal of Chemical & Engineering Data. 2000;**45**:689-692

[5] Mohsen-Nia M, Paikar I. Liquid + liquid equilibria of ternary and quaternary systems containing n-hexane, toluene, m-xylene, propanol, sulfolane, and water at T = 303.15 K. The Journal of Chemical Thermodynamics. 2007;**39**:1085-1089

[6] Chen J, Mi J, Fei W, Li Z. Liquid–liquid equilibria of quaternary and quinary systems including sulfolane at 298.15 K. Journal of Chemical & Engineering Data. 2001;**46**:169-171

[7] Santiago RS, Aznar M. Liquid–liquid equilibria for quaternary mixtures of nonane + undecane + (benzene or toluene or m-xylene) + sulfolane at 298.15 and 313.15 K. Fluid Phase Equilibria. 2007;**253**:137-141

[8] Santiago RS, Aznar M. Quinary liquid–liquid equilibria for mixtures of nonane + undecane + two pairs of aromatics (benzene/toluene/m-xylene) + sulfolane at 298.15 and 313.15 K. Fluid Phase Equilibria. 2007;**259**:71-76

[9] Requejo PF, Calvar N, Domínguez A, Gómez E. Application of the ionic liquid tributylmethylammonium bis(trifluoromethylsulfonyl)imide as solvent for the extraction of benzene from octane and decane at T = 298.15 K and atmospheric pressure. Fluid Phase Equilibria. 2016;**417**:137-143

[10] Requejo PF, Calvar N, Domínguez A, Gómez E. Determination and correlation of (liquid + liquid) equilibria of ternary and quaternary systems with octane, decane, benzene and [BMpyr][DCA] at T = 298.15 K and atmospheric pressure. The Journal of Chemical Thermodynamics. 2016;**94**:197-203

[11] Requejo PF, Calvar N, Domínguez A, Gómez E. Comparative study of the LLE of the quaternary and ternary systems involving benzene, n-octane, n-decane and the ionic liquid [BMpyr][NTf_2]. The Journal of Chemical Thermodynamics. 2016;**98**:56-61

[12] Corderí S, Gómez E, Calvar N, Domínguez A. Measurement and correlation of liquid–liquid equilibria for ternary and quaternary systems of heptane, cyclohexane, toluene, and [EMim][OAc] at 298.15 K. Industrial and Engineering Chemistry Research. 2014;**53**:9471-9477

[13] Corderí S, Calvar N, Gómez E, Domínguez A. Quaternary (liquid + liquid) equilibrium data for the extraction of toluene from alkanes using the ionic liquid [EMim][MSO_4]. The Journal of Chemical Thermodynamics. 2014;**76**:79-86

[14] Corderí S, Gómez E, Domínguez A, Calvar N. (Liquid + liquid) equilibrium of ternary and quaternary systems containing heptane, cyclohexane, toluene and the ionic liquid [EMim][$N(CN)_2$]. Experimental data and correlation. The Journal of Chemical Thermodynamics. 2016;**94**:16-23

[15] Sakal SA, Shen C, Li C. (Liquid + liquid) equilibria of {benzene + cyclohexane + two ionic liquids} at different temperature and atmospheric pressure. The Journal of Chemical Thermodynamics. 2012;**49**:81-86

[16] Potdar S, Ramalingam A, Banerjee T. Aromatic extraction using mixed ionic liquids: Experiments and COSMO-RS predictions. Journal of Chemical & Engineering Data. 2012;**57**:1026-1035

[17] García S, Larriba M, García J, Torrecilla JS, Rodríguez F. Liquid–liquid extraction of toluene from n-heptane using binary mixtures of N-butylpyridinium tetrafluoroborate and N-butylpyridinium bis(trifluoromethylsulfonyl)imide ionic liquids. Chemical Engineering Journal. 2012;**180**:210-215

[18] García S, Larriba M, García J, Torrecilla JS, Rodríguez F. Separation of toluene from n-heptane by liquid–liquid extraction using binary mixtures of [bpy][BF_4] and [4bmpy][Tf_2N] ionic liquids as solvent. The Journal of Chemical Thermodynamics. 2012;**53**:119-124

[19] Larriba M, Navarro P, Garcia J, Rodriguez F. Liquid–liquid extraction of toluene from n-heptane by {[emim][TCM] + [emim][DCA]} binary ionic liquid mixtures. Fluid Phase Equilibria. 2014;**364**:48-54

[20] Larriba M, Navarro P, García J, Rodríguez F. Separation of toluene from n-heptane, 2,3-dimethylpentane, and cyclohexane using binary mixtures of [4empy][Tf_2N] and [emim][DCA] ionic liquids as extraction solvents. Separation and Purification Technology. 2013;**120**:392-401

[21] Larriba M, Navarro P, García J, Rodríguez F. Liquid–liquid extraction of toluene from n-alkanes using {[4empy][Tf_2N] + [emim][DCA]} ionic liquid mixtures. Journal of Chemical & Engineering Data. 2014;**59**:1692-1699

[22] Larriba M, Navarro P, García J, Rodríguez F. Extraction of benzene, ethylbenzene, and xylenes from n-heptane using binary mixtures of [4empy][Tf_2N] and [emim][DCA] ionic liquids. Fluid Phase Equilibria. 2014;**380**:1-10

[23] Manohar CV, Banerjee T, Mohanty K. Co-solvent effects for aromatic extraction with ionic liquids. Journal of Molecular Liquids. 2013;**180**:145-153

[24] Larriba M, Navarro P, García J, Rodríguez F. Liquid–liquid extraction of BTEX from reformer gasoline using binary mixtures of [4empy][Tf_2N] and [emim][DCA] ionic liquids. Energy & Fuels. 2014;**28**:6666-6676

[25] Larriba M, Navarro P, Delgado-Mellado N, González C, García J, Rodríguez F. Dearomatization of pyrolysis gasoline with an ionic liquid mixture: Experimental study and process simulation. AICHE Journal. 2017;**63**:4054-4065

[26] Al-Rashed OA, Fahima MA, Shaaban M. Prediction and measurement of phase equilibria for the extraction of BTX from naphtha reformate using BMIMPF6 ionic liquid. Fluid Phase Equilibria. 2014;**363**:248-262

[27] Corderí S, Calvar N, Gómez E, Domínguez A. Capacity of ionic liquids [EMim][NTf_2] and [EMpy][NTf_2] for extraction of toluene from mixtures with alkanes: Comparative study of

the effect of the cation. Fluid Phase Equilibria. 2012;**315**:46-52

[28] Corderí S, González EJ, Calvar N, Domínguez A. Application of [HMim][NTf_2], [HMim][TfO] and [BMim][TfO] ionic liquids on the extraction of toluene from alkanes: Effect of the anion and the alkyl chain length of the cation on the LLE. The Journal of Chemical Thermodynamics. 2012;**53**:60-66

Chapter 5

Imidazolium Ionic Liquid-Supported Schiff Base and Its Transition Metal Complexes: Synthesis, Physicochemical Characterization and Exploration of Antimicrobial Activities

Biswajit Sinha and Sanjoy Saha

Abstract

New Co(II), Ni(II) and Cu(II) metal complexes from an imidazolium ionic liquid supported Schiff base, 1-{2-(2-hydroxy-5-nitrobenzylideneamino)ethyl}-3-ethylimidazolium tetrafluoroborate were synthesized and characterized by different analytical and spectroscopic techniques such as elemental analysis (CHN analysis), UV-Visible, ^{1}H NMR, ^{13}C NMR, FT-IR, powder X-ray diffraction, mass-spectra, magnetic susceptibility measurements and molar conductance data. From these spectroscopic and analytical data, tetra coordinated 1:2 metal-ligand stoichiometry was suggested for the metal complexes. The molar conductance data of the complexes revealed their electrolytic nature (1:2). The synthesized complexes along with the ligand were screened for *in vitro* antibacterial applications against Gram-negative and Gram-positive bacteria to assess their inhibition potentials. The complexes were proved very effective against the tested organisms.

Keywords: ionic liquid-based Schiff base, Co(II) complex, Ni(II) complex, Cu(II) complex

1. Introduction

Ionic liquids (ILs) may be defined as "ionic materials," with low melting points (below 100°C) generally composed of inorganic or organic anions paired with large, usually asymmetric organic cations. Ionic liquids (ILs) pose a plethora of unique physicochemical and solvation characteristics that can be tuned for specific applications and often producing interesting results when employed instead of traditional molecular solvents [1, 2]. In addition, most ILs show negligible vapor pressure [3] as well as high thermal stability [4–6]. Due to these attractive features they are termed as neoteric solvents or green solvents. In recent years, ILs were extensively studied for their wide electrochemical window, high ionic conductivity [7] and a broad temperature range of the liquid state. Moreover, the physical properties of ILs including density, melting point, polarity, Lewis acidity, viscosity and enthalpy of vaporization can all be tuned

IntechOpen

by changing their cation and anion pairing [8]. IL-based solvent system typically exhibits enhanced reaction kinetics resulting in the efficient use of time and energy [1]. Due to these properties, ILs are treated as a new generation of solvents for catalysis, ecofriendly reaction media for organic synthesis and a successful replacement for conventional media in chemical processes [1, 9]. Recently, many researchers have focused on the synthesis of new ionic liquids called functionalized ionic liquids (FILs) with different functional groups in the cationic moiety [10–15]. Such functionalization of the cation can easily be done in a single reaction step and thus both the cationic and anionic moieties of the FILs can be altered as required for specific applications like increased catalytic stability and reduced catalyst leaching, etc. [16, 17].

Of note Schiff base being a salient class of multidentate ligand has played a key role in coordination chemistry. They exhibit varied denticities, chelating capability [18–20], functionalities [21] and diverse range of biological, pharmacological and antitumor activities. Schiff-bases containing hetero-atom such as N, O, and S are drawn special interest for their varied ways of coordination with different transition metal ions and having unusual configurations [22–24]. The present chapter describes the syntheses and physicochemical characterizations of an IL-supported Schiff base, 1-{2-(2-hydroxy-5-nitrobenzylideneamino)ethyl}-3-ethylimidazolium tetrafluoroborate and its Co(II), Ni(II) and Cu(II) complexes. The ligand and its metal complexes were screened for their *in vitro* antibacterial activities against Gram-negative bacteria *Escherichia coli*, *Pseudomonas aeruginosa*, *Proteus vulgaris*, *Enterobacter aerogenes* and Gram-positive bacteria *Staphylococcus aureus* and *Bacillus cereus*. The complexes and the ligand were found most effective against the tested Gram-negative/positive bacteria.

2. Materials and physical measurements

Analytical grade chemicals were used for synthesis without further purification. 1-ethyl imidazole, 2-bromoethylamine hydrobromide, 5-nitro-2-hydroxybenzaldehyde and $NaBF_4$ (sodium tetrafluoroborate) were purchased from Sigma Aldrich, Germany. Metal acetates and other reagents were used as obtained from SD Fine Chemicals, India. CH_3OH, petroleum ether, $CHCl_3$, DMF and DMSO were used after purification by standard methods described in the literature. FT-IR spectra were recorded by KBr pellets on a Perkin-Elmer Spectrum FT-IR spectrometer (RX-1). ^1H NMR and ^{13}C NMR spectra were recorded on a FT-NMR (Bruker Avance-II 400 MHz) spectrometer by using D_2O and DMSO-d_6 as solvents. Powder X-ray diffraction (XRD) data were obtained on INEL XRD Model Equinox 1000 using Cu Kα radiation (2θ = 0–90°). Elemental microanalysis (CHN analysis) was performed on Perkin-Elmer (Model 240C) analyzer. Metal content was obtained from AAS (Varian, SpectrAA 50B) by using standard metal solutions procured from Sigma-Aldrich, Germany. ESI-MS spectra were obtained on a JMS-T100LC spectrometer. The purity of the synthesized products was confirmed by thin layer chromatography (TLC) Merck 60 F254 silica gel plates (layer thickness 0.25 mm) and the spots were visualized using UV-light. The UV-visible spectra were obtained from JascoV-530 double beam spectrophotometer using CH_3OH as solvent. Specific conductance was measured at (298.15 ± 0.01) K with a Systronic conductivity TDS-308 metre. Magnetic susceptibility was measured with a Sherwood Scientific Ltd. magnetic susceptibility balance (Magway MSB Mk1) at ambient temperature. The melting point of synthesized compounds was determined by open capillary method. Antibacterial activity (*in vitro*) of the synthesized ligand and complexes were evaluated by well diffusion method against six bacterial strains (two Gram-positive and four Gram-negative). The bacterial strains were obtained from MTCC, Chandigarh, India.

2.1 Synthesis of 1-(2-aminoethyl)-3-ethylimidazolium tetrafluoroborate [2-aeeim]BF_4 (1a)

The FIL was synthesized by following a literature procedure [25]. [2-aeeim]BF_4 was obtained as yellow oil; (98 mg, 70%); ^1H NMR (400 MHz, D_2O, TMS): δ = ^1H NMR (400 MHz, D_2O): δ = 3.63 (m, 2H, NH_2—CH_2), 4.16 (s, 3H, CH_3), 4.49 (t, 1H, N—CH_2), 4.56 (t, 1H, N—CH_2), 7.40 (s, 1H, NCH), 7.50 (s, 1H, NCH), 8.61 (s, 2H, NH_2), 8.87 (s, 1H, N(H)CN); IR (KBr): υ = 3447, 3086, 2896, 1626, 1452, 1084. ESI-MS (*m/z*) calc for $[C_7H_{14}N_3]^+$: 140, found: 140 $[M\text{-}BF_4]^+$. Anal. calcd. for $C_7H_{14}F_4N_3B$: C 37.04, H 6.22, N 18.51, found: C 36.99, H 6.14, N 18.43.

2.2 Synthesis of imidazolium ionic liquid-supported Schiff base, LH (2a)

5-nitro-2-hydroxybenzaldehyde (1.67 g, 10 mmol) and [2-aeeim]BF_4 (2.27 g, 10 mmol) were taken in methanol and stirred at 25°C for 4 h. After completion of reaction, the product was diluted using ethanol. The precipitate was filtered, washed with cold EtOH and dried properly to collect the expected ligand as a yellowish brown solid; (282 mg, 75%). mp. 95–97°C. ^1H NMR: (400 MHz, DMSO-d_6, TMS): δ = 3.36 (q, 2H, N—CH_2), 3.60 (s, 3H, CH_3), 3.92 (t, 2H, N—CH_2), 4.60 (t, 2H, N—CH_2), 7.44 (s, 1H, NCH), 7.52 (s, 1H, NCH), 7.53 (s, 1H, N=CH), 7.61–7.59 (m, 3H, Ar-H), 8.65 (s, 1H, N(H)CN), 8.88 (s, 1H, OH). ^{13}C NMR: (400 MHz, DMSO-d_6, TMS): δ = 159.76, 138.43, 134.08, 130.47, 130.31, 123.89, 119.80, 118.65, 110.65, 39.86, 39.65, 39.24, 39.03 and 38.82. IR (KBr): υ = 3448 (O—H), 3071, 1664 (C=N), 1343 (N—O), 1293 (C—O), 1095 (B—F). UV/vis (methanol, λ_{max}): 206, 234, 306 nm; ESI-MS (CH_3OH, *m/z*) $[M\text{-}BF_4]^+$ calcd. for $[C_{14}H_{17}N_4O_3]^+$: 289, found; 289. Anal. calcd. for $C_{14}H_{17}N_4O_3BF_4$ (376): C 44.71, H 4.56, N 14.90. found: C 44.64, H 4.49, N 14.83.

2.3 Synthesis of the metal complexes (3a, 4a and 5a)

To an ethanolic solution of ligand, LH (2c) (0.376 g, 1 mmol) in round bottomed flask, metal acetate salt Co(II), Ni(II) and Cu(II), *viz.*, (0.5 mmol) dissolved in ethanol was added and the reaction mixture was refluxed for 12 h until the starting materials were completely consumed as monitored by TLC. On completion of the reaction, solvents were evaporated and the reaction mixture was cooled to room temperature. The precipitate was collected by filtration, washed successively with cold ethanol (10 mL × 3). Finally it was dried in vacuum desiccators to obtain the solid product. The complexes were soluble in *N,N*-dimethylformamide, dimethylsulfoxide, acetonitrile, methanol and water. A schematic representation of the syntheses is given in **Figure 1**.

*2.3.1 Co(II)complex (**3a**)*

Brown solid; (0.54 g, 67%), decomposes at ~293°C. IR (KBr): υ = 3386 (O—H), 1648 (C=N), 1332 (N—O), 1177 (C—O), 1106 (B—F), 651 (M—O), 510 (M—N). UV/vis (methanol, λ_{max}/nm): 227, 246, 358. ESI-MS (CH_3OH, *m/z*) $[M\text{-}2BF_4]^+$ calcd. for $[C_{28}H_{32}CoN_8O_6]^+$: 635, found: 635; anal. calcd. for $C_{28}H_{36}CoB_2F_8N_8O_8$ (809): C 41.56, H 3.99, N 13.85, Co 7.28, found: C 41.36, H 3.71, N 13.55, Co 7.12.

*2.3.2 Ni(II)complex (**4a**)*

Light green solid; (0.56 g, 69%), decomposes at ~293°C. IR (KBr): υ = 3396 (O—H), 1637 (C=N), 1330 (N—O), 1172 (C—O), 1102 (B—F), 646 (M—O), 526 (M—N).

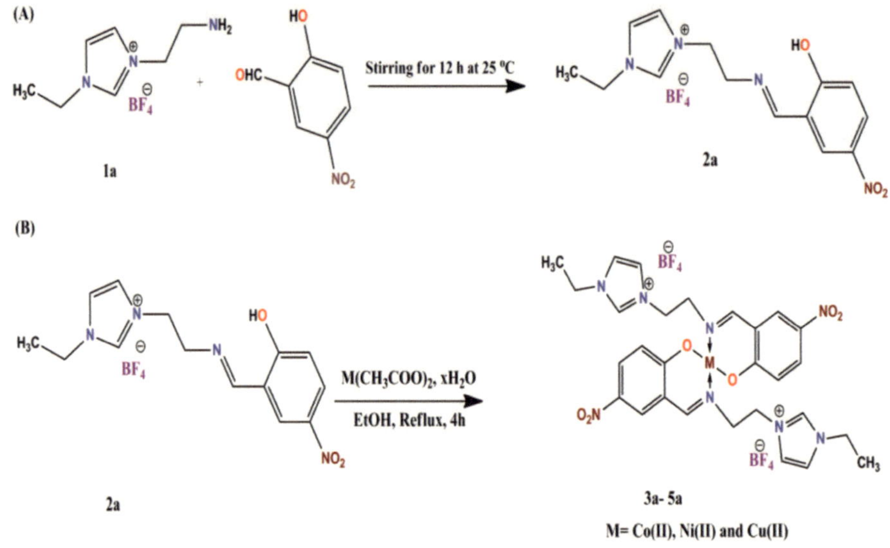

Figure 1.
*Synthesis of Ionic liquid supported Schiff base (**2a**) and its metal complexes **3a**, **4a** and **5a** from **2a**.*

UV/vis (methanol, λ_{max}/nm): 220, 340, 400. ESI-MS (CH_3OH, *m/z*) $[M\text{-}2BF_4]^+$ calcd. for $[C_{28}H_{32}NiN_8O_6]^+$: 634, found: 634; anal. calcd. for $C_{28}H_{36}NiB_2F_8N_8O_8$ (809): C 41.57, H 3.99, N 13.85, Ni 7.26, found: C 41.22, H 3.63, N 13.46, Ni 7.11.

*2.3.3 Cu(II) complex (**5a**)*

Dark green solid; (0.57 g, 70%), decomposes at ~295°C. IR (KBr): υ = 3429 (O—H), 1656 (C═N), 1334 (N—O), 1175 (C—O), 1103 (B—F), 633 (M—O), 471 (M—N). UV/vis (methanol, λ_{max}/nm): 226, 244, 354. ESI-MS (CH_3OH, *m/z*) $[M\text{-}2BF_4]^+$ calcd. for $[C_{28}H_{32}CuN_8O_6]^+$: 634, found: 634. Anal. calcd. for $C_{28}H_{36}CuB_2F_8N_8O_8$ (813.76): C 41.33, H 3.96, N 13.77, Cu 7.81, found: C 41.12, H 3.61, N 13.46, Cu 7.61.

2.4 Antibacterial assay

The synthesized ligand (**2a**) and complexes (**3a**, **4a** and **5a**) were screened against the Gram-negative bacteria (*E. coli*, *P. aeruginosa*, *P. vulgaris* and *E. aerogenes*) and Gram-positive bacteria (*S. aureus* and *B. cereus*) strains. The tests were performed using agar disc diffusion method [26]. The nutrient agar (Hi-Media Laboratories Limited, Mumbai, India) was put in an autoclave at 121°C and 1 atm for 15–20 min. The sterile nutrient medium was kept at 45–50°C and then 100 μL of bacterial suspension containing 10^8 colony-forming units (CFU)/mL was mixed with sterile liquid nutrient agar and poured into the sterile Petri dishes. All the stock solutions were made by dissolving the compounds in dimethylsulfoxide (DMSO). The concentrations of the tested compounds were 10, 20, 30, 40 and 50 μg/mL. The tested microorganisms were grown on nutrient agar medium in Petri dishes. The samples were soaked in a filter paper disc of 1 mm thickness and 5 mm diameter. The discs were kept on Petri plates and incubated for 24 h at 37°C. The diameter of the inhibition zone (including disc diameter of 5 mm) was measured. Each experiment was carried out three times to minimize the error and the mean values were accepted.

3. Results and discussion

All the isolated compounds were stable at room temperature to be characterized by different analytical and spectroscopic methods. The complexes are soluble in *N*,*N*-dimethylformamide, dimethylsulfoxide, acetonitrile, methanol and water.

3.1 FT-IR spectral studies

The assignments of the IR bands of the synthesized Co(II), Ni(II) and Cu(II) complexes had been made by comparing with the bands of ligand (LH) to determine the coordination sites involved in chelation. FT-IR spectra of LH (**2a**) showed a strong broad band at 3448–3071 cm^{-1}; which was due to the hydrogen bonded phenolic group (—OH) with H—C(=N) group in the ligand (OH...N=C) [27, 28]. The broad band appeared at 3386–3429 cm^{-1} for the metal complexes (**3a**, **4a** and **5a**) suggested the presence of the solvated water molecules (probably for the presence of $—NO_2$ group in the ligand and intrinsic property of the anion tetrafluoroborate) [29–31]. The band corresponding to the azomethine group (—C=N) of the ligand was found at 1664 cm^{-1}. This band gets shifted in the range 1637–1656 cm^{-1} because of coordination of N atom of azomethine linkage to the Co^{2+}, Ni^{2+} and Cu^{2+} ions respectively [32]. The band for phenolic C—O of free ligand was observed at 1293 cm^{-1} which moved to lower wave number 1172–1177 cm^{-1} for the complexes (**3a**, **4a** and **5a**) upon complexation. This fact established the bonding of ligand (**2a**) to the metal atoms through the N atom of azomethine and O atom of phenolic group [33]. The bands appeared in the region of 1102–1107 cm^{-1} for the metal complexes were assigned for B—F stretching frequency. FT-IR spectra of the LH (**2a**) and its complexes showed strong bands at 1330–1343 cm^{-1} which were assigned for the NO_2 group [34]. The spectra of the metal complexes exhibited bands at 633–651 and 471–526 cm^{-1} were attributed to M—O and M—N stretching vibrations, respectively [35]. IR spectra are given in **Figures 2–8**.

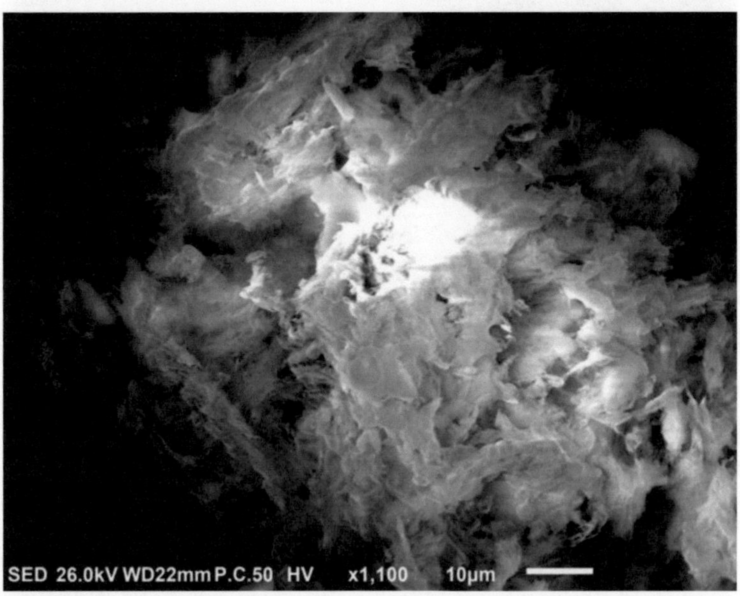

Figure 2.
SEM image of Co(II) complex (3a).

Figure 3.
*SEM image of Ni(II) complex (**4a**).*

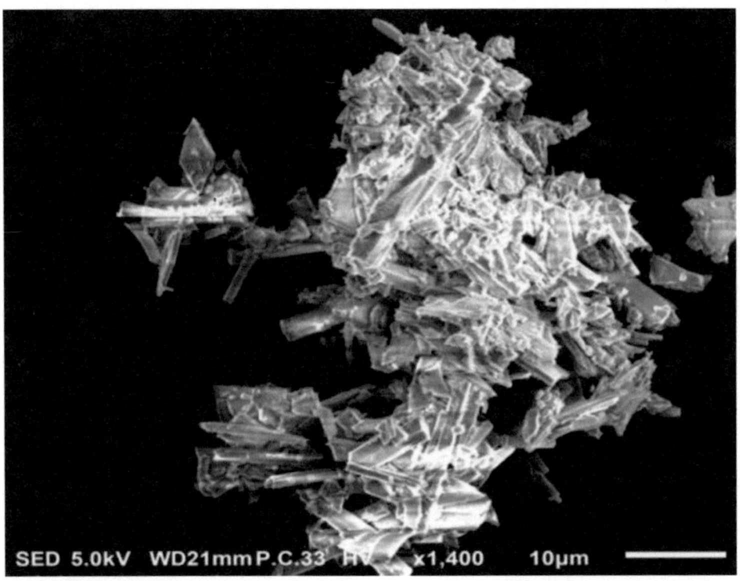

Figure 4.
*SEM image of Cu(II) complex (**5a**).*

3.2 ^{1}H NMR and ^{13}C NMR spectral studies

^{1}H NMR and ^{13}C NMR spectra of Schiff base were recorded in DMSO-d_6 (as shown in **Figures 9** and **10**). ^{1}H NMR spectra of the ligand showed singlet at 7.60 ppm which was assignable to proton of the azomethine linkage (—CH=N—) might be because of the effect of the *ortho*-hydroxyl group in the aromatic ring. A singlet at 8.88 ppm was assigned to hydroxyl proton (—OH). The downfield shift of the phenolic (—OH) proton was observed due to intramolecular (O—H...N) hydrogen bonding in the ligand [36]. ^{13}C NMR spectra of ligand exhibited peaks at δ 159.76 and 138.43 which were detected for the phenolic (C—O) and imino

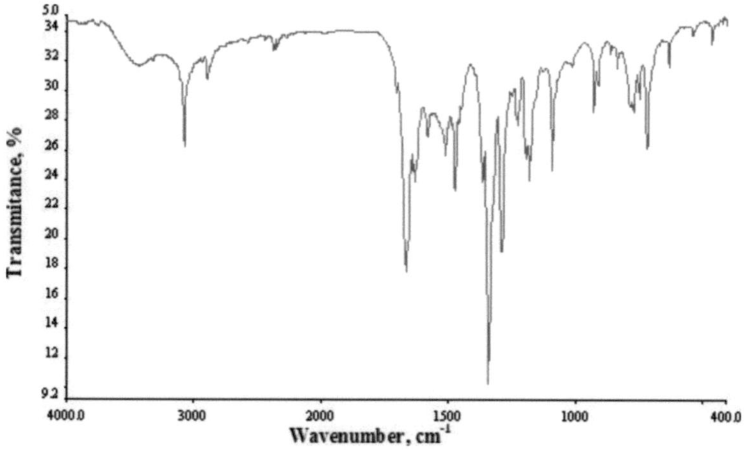

Figure 5.
*FT-IR spectra of ligand (**2a**).*

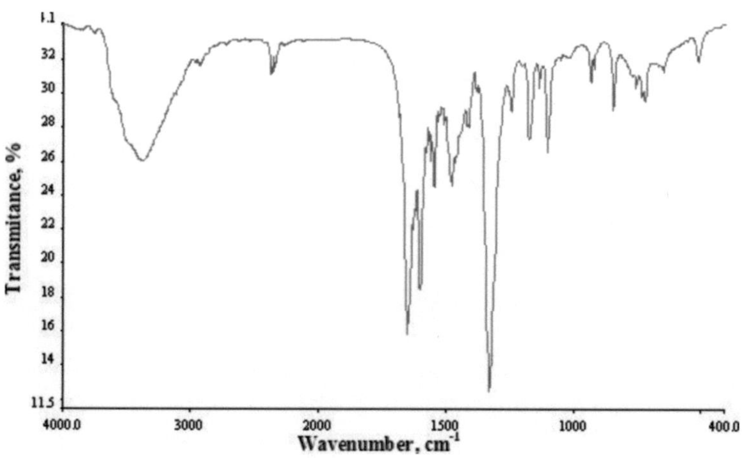

Figure 6.
*FT-IR spectra of Co(II) complex (**3a**).*

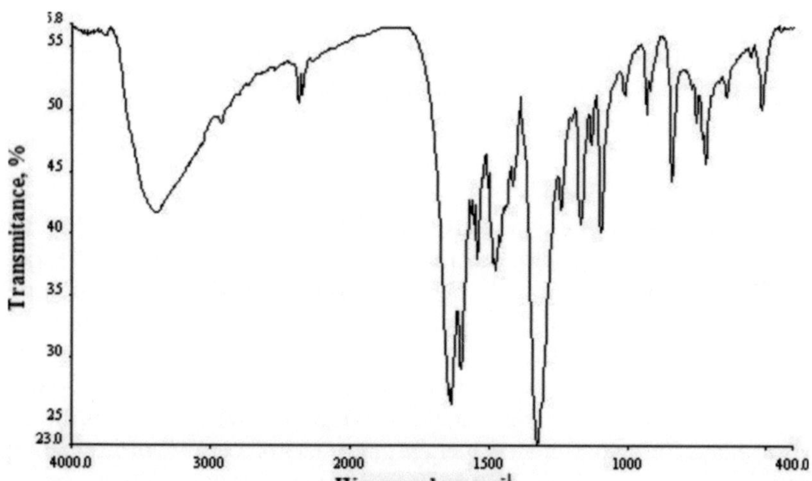

Figure 7.
*FT-IR spectra of Ni(II) complex (**4a**).*

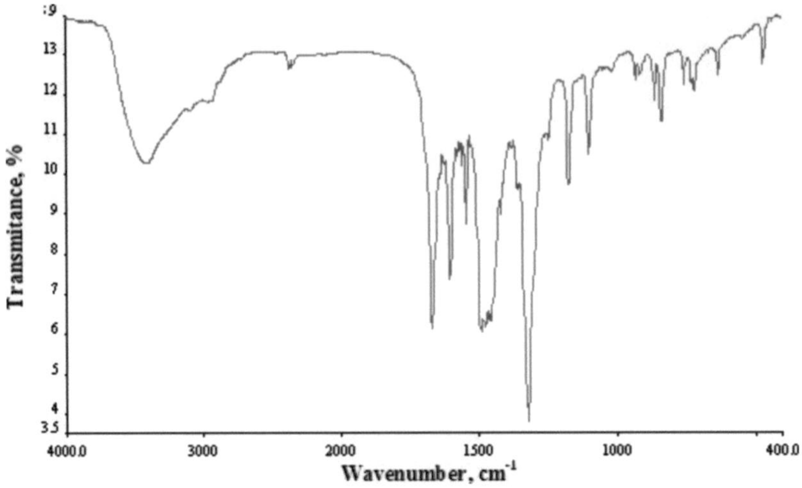

Figure 8.
FT-IR spectra of Cu(II) complex (5a).

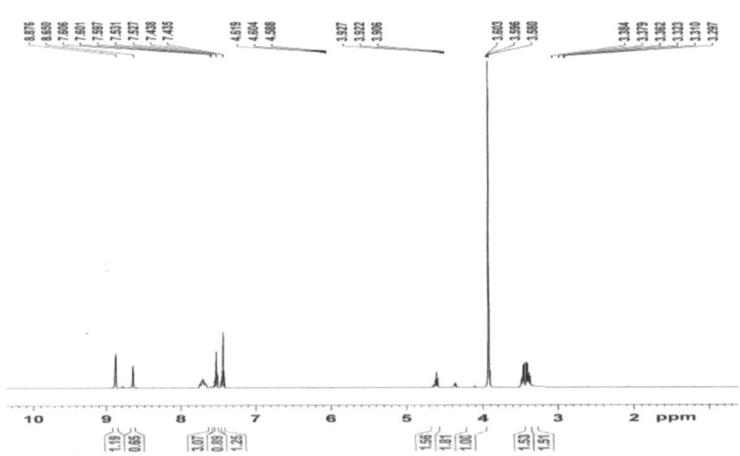

Figure 9.
^{1}H NMR spectra of LH (2a).

(—CH=N) carbon atoms (due to keto-imine tautomerism). The aromatic carbons showed pecks at δ 134.08, 130.47, 130.31, 123.89, 119.80 and 118.65.

3.3 PXRD analysis

The PXRD analysis of the synthesized compounds was carried out to find whether the particle nature of the samples was amorphous or crystalline. The PXRD spectrum of ligand (LH) exhibited sharp peaks because of their crystalline nature although the spectra of the two complexes did not show such peaks for their amorphous nature (as shown in **Figures 11–14**). The crystalline sizes were calculated using Debye Scherer's equation: $D = 0.9\ \lambda/\beta\cos\theta$, where constant 0.9 is the shape factor, λ is the X-ray wavelength (1.5406 Å), β is the full width at half maximum (FWHM) and θ is the Bragg diffraction angle. The experimental average grain sizes of LH and its metal complexes were found to be 31.71 nm (**2a**), 7.76 nm (**3a**), 3.26 nm (**4a**) and 4.52 nm (**5a**).

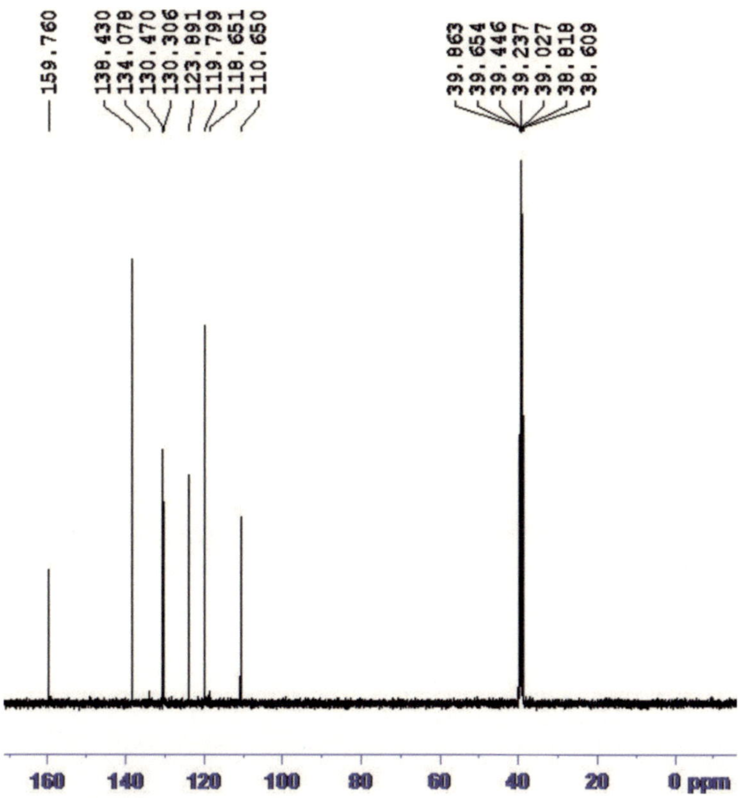

Figure 10.
*^{13}C NMR spectra of LH (**2a**).*

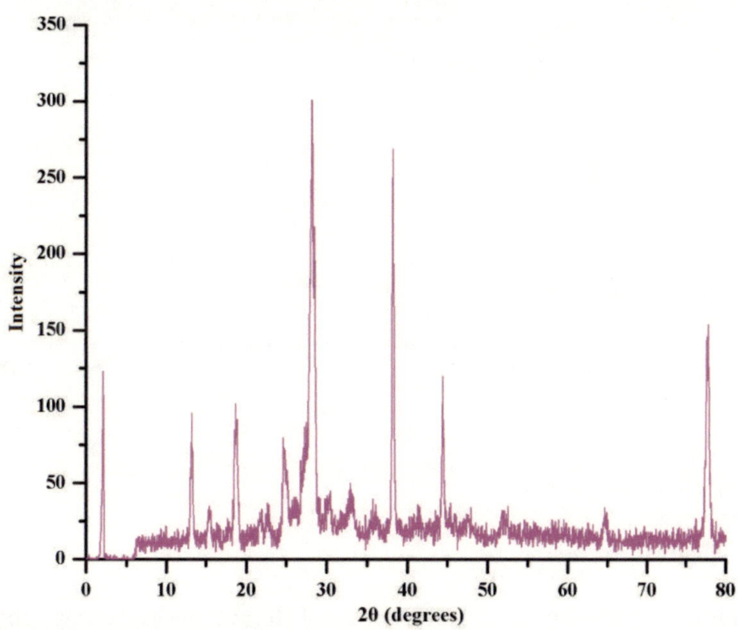

Figure 11.
*PXRD spectra of LH (**2a**).*

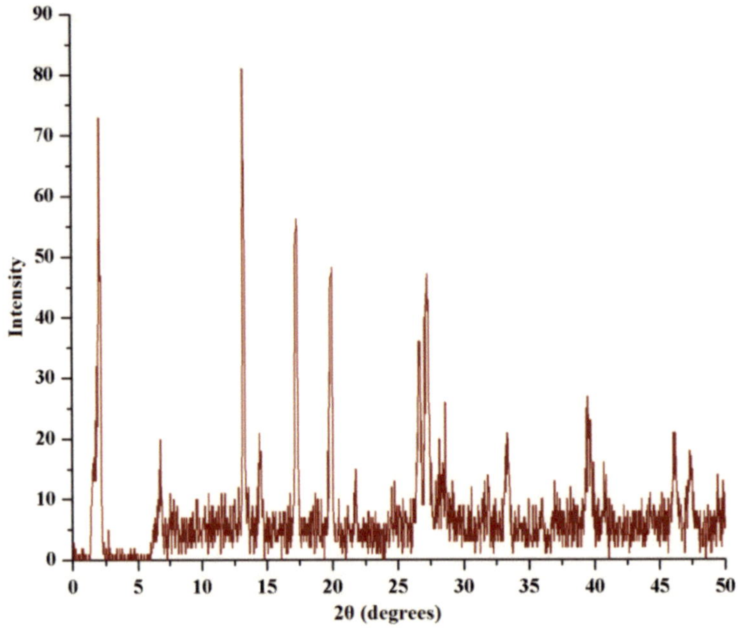

Figure 12.
PXRD spectra of Co(II) complex (3a).

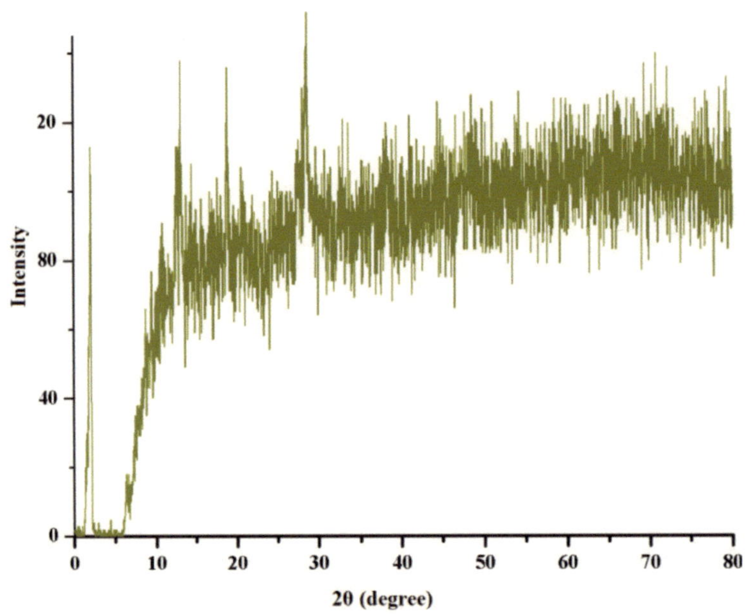

Figure 13.
PXRD spectra of Ni(II) complex (4a).

3.4 Mass spectral studies

To get information regarding the structure of the synthesized compounds at the molecular level, electrospray ionization (ESI) mass spectrometry was performed using methanol as solvent. Mass-spectra of the LH (**2a**) had a molecular ion peaks at m/z 289, that corresponds to $[M\text{-}BF_4]^+$, $[M = C_{14}H_{17}N_4O_2]^+$. The metal complexes (**3a, 4a** and **5a**) exhibited molecular ion peaks (*m/z*) at 635 ($M = [C_{28}H_{32}CoN_8O_6]^+$), at 634 ($M = [C_{28}H_{32}NiN_8O_6]^+$) and at 639 ($M = [C_{28}H_{32}CuN_8O_6]^+$) which confirmed their stoichiometry as $Co(L)_2$, $Ni(L)_2$ and $Cu(L)_2$ respectively. The mass spectra of the ligand and complexes were in good agreement with the respective structures as revealed by the elemental and other spectral analyses.

3.5 Electronic spectra and magnetic moment

The UV-visible spectra of the Schiff base and its metal complexes (as depicted in **Figure 15**) were recorded at room temperature using methanol as solvent.

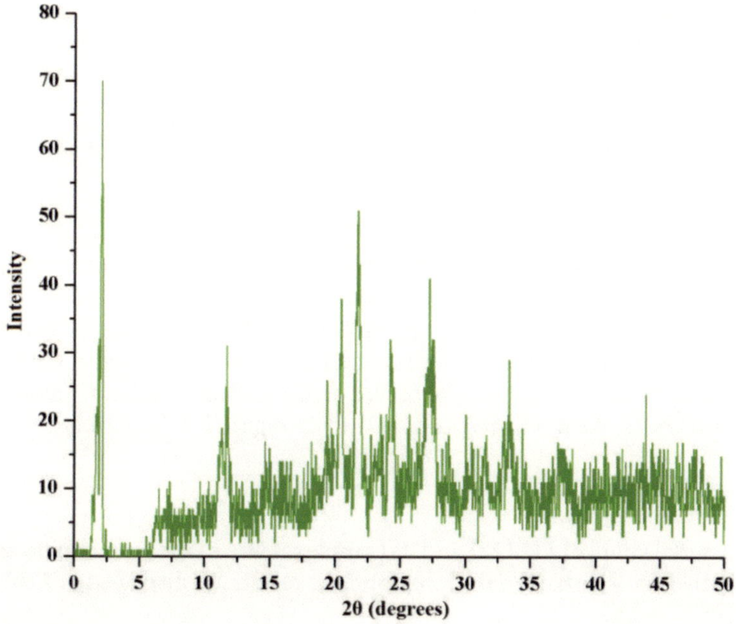

Figure 14.
PXRD spectra of Cu(II) complex (5a).

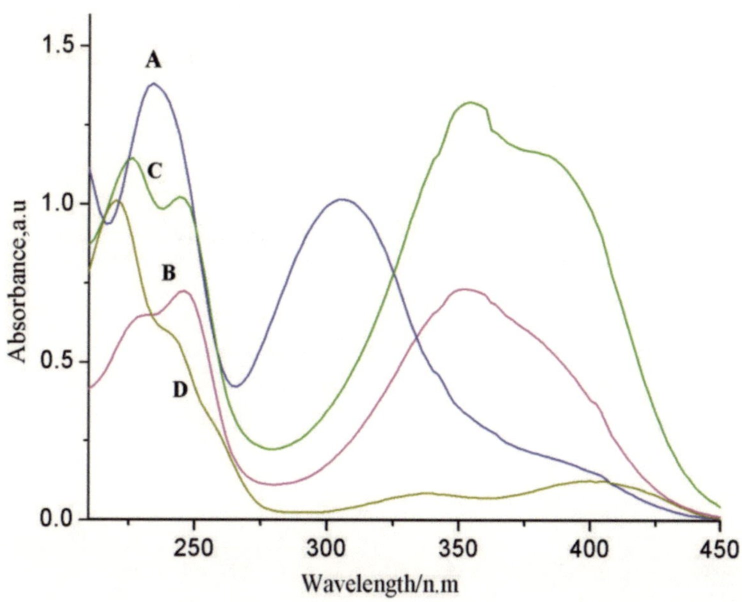

Figure 15.
*UV-visible spectra in methanol (concentration of the solutions 1×10^{-4} M): (A) LH (**2a**); (B) Co(II) complex (**3a**); (C) Ni(II) complex (**4a**) and (D) Cu(II) complex (**5a**).*

The LH (**2a**) exhibited three absorption bands at 306, 234 and 206 nm due to $n \rightarrow \pi^*$, $\pi \rightarrow \pi^*$ and transitions involved with the imidazolium moiety, respectively [37, 38]. For the complexes, the bands that appeared below 350 nm were ligand centered transitions ($n \rightarrow \pi^*$ and $\pi \rightarrow \pi^*$). The Co(II) complex (**3a**) displayed a band at 354 nm which could be attributed to the combination of $^2B_{1g} \rightarrow {}^1A_{1g}$ and $^1B_{1g} \rightarrow {}^2E_g$ transitions and supporting square planar geometry [39, 40]. The complex (**3a**) showed magnetic moment of 2.30 B.M. due to one unpaired electron. The Ni(II) complex (**4a**) was diamagnetic and the band appeared at around 400 nm due to $^1A_{1g} \rightarrow {}^1B_{1g}$ transition is consistent with low spin square planar environment [41]. UV-visible spectra of Cu(II) complex (**5a**) exhibited $d \rightarrow \pi^*$ metal-ligand charge transfer transition (MLCT) at the region 358 nm had been assigned to the combination of $^2B_{1g} \rightarrow {}^2E_g$ and $^2B_{1g} \rightarrow {}^2B_{2g}$ transitions in a distorted square planar geometry. The experimental magnetic moment value for **5a** was 1.84 B.M. consistent with the presence of an unpaired electron [42, 43].

3.6 Molar conductance

The molar conductance (Λ_m) of the metal complexes was determined by applying the relation $\Lambda_m = 1000 \times \kappa / c$, where κ and c stands for the specific conductance and molar concentration of metal complexes respectively. The complexes (1×10^{-3} M) were dissolved in DMF and their specific conductance was measured at (298.15 ± 0.01) K. The molar conductance data was observed as 123, 128 and 131S cm^{-1} mol^{-1} for the metal complexes **3a**, **4a** and **5a** respectively indicating their 1:2 electrolytic natures.

3.7 Antimicrobial activity

Antibacterial study of LH (**2a**) and its complexes was carried out *in vitro* against the Gram-negative/positive bacterial strains, and the results are displayed in **Tables 1** and **2**

Specimen	Concentration (μg/mL)														
	E. coli					*S. aureus*					*B. cereus*				
	10	20	30	40	50	10	20	30	40	50	10	20	30	40	50
LH	—	6	7	8	12	7	9	10	10	12	—	—	6	8	12
Co(II) complex	—	—	6	7	8	6	7	7	9	10	—	6	6	8	10
Ni(II) complex	6	7	8	9	9	—	—	7	8	10	—	—	6	8	10
Cu(II) complex	8	9	14	15	18	6	8	10	17	17	—	—	—	—	7

Table 1.
*Antibacterial activity data of Schiff base (**2a**) and its metal complexes (**3a**, **4a** and **5a**) against E. coli, S. aureus and B. cereus with their minimum zone of inhibition and MIC (μg/mL) mm values.*

Specimen	Concentration (μg/mL)														
	P. aeruginosa					*P. vulgaris*					*E. aerogenes*				
	10	20	30	40	50	10	20	30	40	50	10	20	30	40	50
LH	—	6	9	15	16	—	6	9	10	14	—	6	8	10	13
Co(II) complex	—	7	9	10	13	—	—	—	6	7	8	10	13	15	17
Ni(II) complex	—	—	6	7	9	—	—	6	7	8	8	10	12	12	16
Cu(II) complex	—	6	12	12	14	—	7	7	8	16	—	—	6	7	10

Table 2.
*Antibacterial activity data of Schiff base (**2a**) and its metal complexes (**3a**, **4a** and **5a**) against P. aeruginosa, P. vulgaris and E. aerogenes with their minimum zone of inhibition and MIC (μg/mL) mm values.*

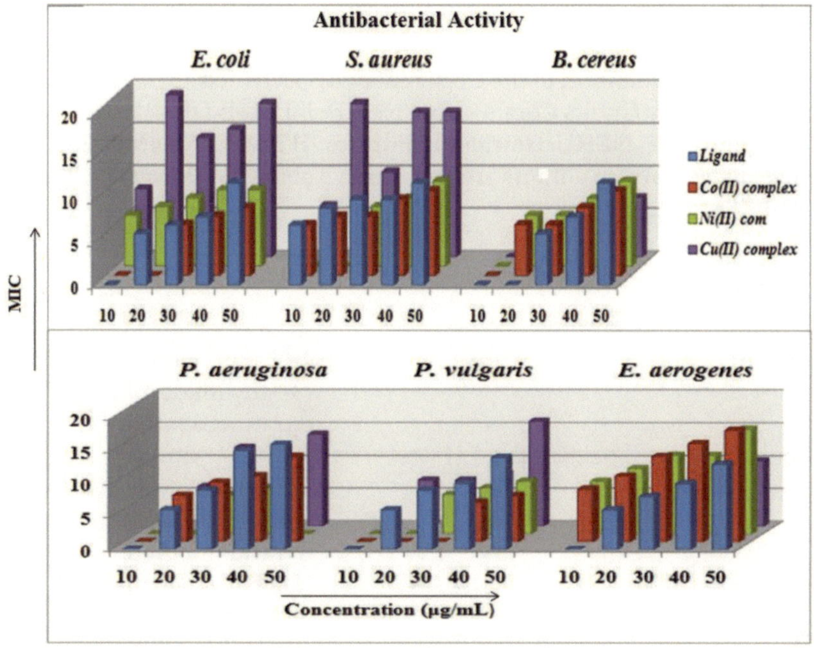

Figure 16.
Inhibition zones for the LH (2a), Co(II) complex (3a), Ni(II) complex (4a) and Cu(II) complex (5a).

and also in **Figure 16**. Minimum inhibitory concentration (MIC) was measured by broth micro dilution susceptibility method. No inhibition zone was found for the solvent control (DMSO) for each bacterial suspension. A serial dilution of sample extracts was made in nutrient broth medium. Then 1 mL of standard (0.5 Mc Farland) bacterial suspension was inoculated into each of these tubes. A similar nutrient broth tube without sample extract was also inoculated and used as control. The samples under investigation have shown promising results against the tested bacterial strains. The LH (**2a**) was most effective against *S. aureus* only. The Co(II) complex (**3a**) showed most effectiveness against *S. aureus*, *E. aerogenes*. The Ni(II) complex (**4a**) showed higher activity against *E. aerogenes*. Although in other cases it showed moderate activity. It was found that Cu(II) complex (**5a**) was most effective against the tested bacteria. The observation suggested that the chelation could facilitate the capability of the complexes to penetrate bacterial cell membrane [44]. Such a chelation could enhance the lipophilic property of the corresponding metal ions that favors permeation towards the lipid layer of cell membrane. The activity of both the complexes and ligand enhanced as the concentration was increased which were due to the growth of degree of inhibition.

4. Conclusion

Herein this chapter, new Co(II), Ni(II) and Cu(II) complexes of an ionic liquid-supported Schiff base, 1-{2-(2-hydroxy-5-nitrobenzylideneamino)ethyl}-3-ethylimidazolium tetrafluoroborate were synthesized and characterized by different spectral and analytical techniques. The Schiff base ligand played as a potential bidentate ligand coordinating through the N-atom of azomethine and O-atom of phenolic group to the metal ions and thus formed 1:2 (M:L) complexes. Spectral and magnetic susceptibility data revealed that the ligand was arranged in square planner geometry around the central metal ions. The antibacterial study of the synthesized compounds was performed and metal complexes have exhibited promising activity against the tested bacteria.

Acknowledgements

The authors are thankful to the Departmental Special Assistance Scheme, under the University Grants Commission, New Delhi (SAP-DRS-III, NO.540/12/DRS/2013) and SAIF, NEHU, Guwahati, India for ^1H NMR, ^{13}C NMR, ESI-MS and elemental analysis. Again authors are grateful to Annamalai University, Tamil Nadu, India for PXRD analysis.

Nomenclature

[2-aeeim][BF_4] (1a)	1-(2-aminoethyl)-3-ethylimidazolium tetrafluoroborate
Ionic liquid-supported Schiff base, LH (**2a**)	1-{2-(2-hydroxy-5-nitrobenzylmine) ethyl}-3-ethylimidazolium tetrafluoroborate
Co(II) complex (**3a**)	[Di(1-{2-(2-hydroxy-5-nitrobenzylidene amino) ethyl}-3-ethylimidazolium) Co(II)] tetrafluoroborate
Ni(II) complex (**4a**)	[Di(1-{2-(2-hydroxy-5-nitrobenzylidene amino) ethyl}-3-ethylimidazolium) Ni(II)] tetrafluoroborate
Cu(II) complex (**5a**)	[Di(1-{2-(2-hydroxy-5-nitrobenzylidene amino) ethyl}-3-ethylimidazolium) Cu(II)] tetrafluoroborate

Author details

Biswajit Sinha[1] and Sanjoy Saha[2*]

1 Department of Chemistry, University of North Bengal, Darjeeling, India

2 Department of Chemistry, Kalimpong College, Kalimpong India

*Address all correspondence to: sanjoychem83@yahoo.com

IntechOpen

© 2019 The Author(s). Licensee IntechOpen. This chapter is distributed under the terms of the Creative Commons Attribution License (http://creativecommons.org/licenses/by/3.0), which permits unrestricted use, distribution, and reproduction in any medium, provided the original work is properly cited.

References

[1] Welton T. Room-temperature ionic liquids. Solvents for synthesis and catalysis. Chemical Reviews. 1999;**99**:2071-2083

[2] Chiappe C, Pieraccini D. Kinetic study of the addition of trihalides to unsaturated compounds in ionic liquids. Evidence of a remarkable solvent effect in the reaction of ICl_2^-. The Journal of Organic Chemistry. 2004;**69**:6059-6064

[3] Earle MJ, Esperanc JMSS, Gilea MA, Lopes JNC, Rebelo LPN, Magee J, et al. The distillation and volatility of ionic liquids. Nature. 2006;**439**:831-834

[4] Rogers RD, Seddon KR. Ionic liquids—Solvents of the future? Science. 2003;**302**:792-793

[5] Sheldon R. Green solvents for sustainable organic synthesis: State of the art. Green Chemistry. 2005;**7**:267-278

[6] Wasserscheid P, Keim W. Ionic liquids—New "solutions" for transition metal catalysis. Angewandte Chemie, International Edition. 2000;**39**:3772-3789

[7] Sakaebe H, Matsumoto H. *N*-methyl-*N*-propylpiperidinium bis(trifluoromethane sulfonyl)imide (PP13–TFSI)—Novel electrolyte base for Li battery. Electrochemistry Communications. 2003;**5**:594-598

[8] Freire MG, Santos LMNBF, Fernandes AM, Coutinho JAP, Marrucho IM. An overview of the mutual solubilities of water-imidazolium-based ionic liquids systems. Fluid Phase Equilibria. 2007;**261**:449-454

[9] Sheldon R. Catalytic reactions in ionic liquids. Chemical Communications. 2001;**23**:2399-2407

[10] Yi F, Peng Y, Song G. Microwave-assisted liquid-phase synthesis of methyl 6-amino-5-cyano-4-aryl-2-methyl-4*H*-pyran-3-carboxylate using functional ionic liquid as soluble support. Tetrahedron Letters. 2005;**46**:3931-3933

[11] Bates ED, Mayton RD, Ntai I, Davis JH. CO_2 capture by a task-specific ionic liquid. Journal of the American Chemical Society. 2002;**124**:926-927

[12] Cole AC, Jensen JL, Ntai I, Tran KLT. Novel Brønsted acidic ionic liquids and their use as dual solvent-catalysts. Journal of the American Chemical Society. 2002;**124**:5962-5963

[13] Li J, Peng Y, Song G. Mannich reaction catalyzed by carboxyl-functionalized ionic liquid in aqueous media. Catalysis Letters. 2005;**102**:159-162

[14] Davis JH Jr, Forrester KJT, Merrigan J. Novel organic ionic liquids (OILs) incorporating cations derived from the antifungal drug miconazole. Tetrahedron Letters. 1998;**49**:8955-8958

[15] Jodry JJ, Mikami JK. New chiral imidazolium ionic liquids: 3D-network of hydrogen bonding. Tetrahedron Letters. 2004;**45**:4429-4431

[16] Fei Z, Geldbach TJ, Zhao D, Dyson PJ. From dysfunction to bis-function: On the design and applications of functionalised ionic liquids. European Journal of Chemistry. 2006;**12**:2122-2130

[17] Lee S. Functionalized imidazolium salts for task-specific ionic liquids and their applications. Chemical Communications. 2006;**14**:1049-1063

[18] Hadjikakou SK, Hadjiliadis N. Antiproliferative and anti-tumor activity of organotin compounds. Coordination Chemistry Reviews. 2009;**253**:235-249

[19] Garoufis A, Hadjikakou SK, Hadjiliadis N. Palladium coordination compounds as anti-viral, anti-fungal, anti-microbial and anti-tumor agents. Coordination Chemistry Reviews. 2009;**253**:1384-1397

[20] Liu CM, Xiong RG, You XZ, Liu YJ, Cheung KK. Crystal structure and some properties of a novel potent Cu_2Zn_2SOD model Schiff base copper(II) complex. Polyhedron. 1996;**15**:4565-4571

[21] Atkins AJ, Black D, Blake AJ, Marin-Bocerra A, Parsons S, Ruiz-Ramirez L, et al. Schiff-base compartmental macrocyclic complexes. Chemical Communications. 1996;**4**:457-464

[22] Goku A, Tumer M, Demirelli H, Wheatley RA. Cd(II) and Cu(II) complexes of polydentate Schiff base ligands: Synthesis, characterization, properties and biological activity. Inorganica Chimica Acta. 2005;**358**:1785-1797

[23] Mohindru A, Fisher JM, Rabinovitz M. Bathocuproine sulphonate: A tissue culture-compatible indicator of copper-mediated toxicity. Nature. 1983;**303**:64-65

[24] Palet PR, Thaker BT, Zele S. Preparation and characterisation of some lanthanide complexes involving a heterocyclic beta-diketone. Indian Journal of Chemistry Section A. 1999;**38**:563-567

[25] Song G, Cai Y, Peng Y. Amino-functionalized ionic liquid as a nucleophilic scavenger in solution phase combinatorial synthesis. Journal of Combinatorial Chemistry. 2005;**7**:561-566

[26] Ahmed I, Beg AJ. Antimicrobial and phytochemical studies on 45 Indian medicinal plants against multi-drug resistant human pathogens. Journal of Ethnopharmacology. 2001;**74**:113-123

[27] Yıldız M, Kılıc Z, Hökelek T. Intramolecular hydrogen bonding and tautomerism in Schiff bases, structure of 1,8-di[N-2-oxyphenyl-salicylidene]-3,6-dioxaoctane. Journal of Molecular Structure. 1998;**441**:1-10

[28] Yeap G-Y, Ha S-T, Ishizawa N, Suda K, Boey P-L, Mahmood WAK. Synthesis, crystal structure and spectroscopic study of para substituted 2-hydroxy-3-methoxybenzalideneanilines. Journal of Molecular Structure. 2003;**658**:87-99

[29] Abdel-Latif SA, Hassib HB, Issa YM. Studies on some salicylaldehyde Schiff base derivatives and their complexes with Cr(III), Mn(II), Fe(III), Ni(II) and Cu(II). Spectrochimica Acta Part A. 2007;**67**:950-957

[30] Wang J, Pei Y, Zhao Y, Hu Z. Recovery of amino acids by imidazolium based ionic liquids from aqueous media. Green Chemistry. 2005;**7**:196-202

[31] Han D, Row KH. Recent application of ionic liquids in separation technology. Molecules. 2010;**15**:2405-2426

[32] Kohawole GA, Patel KS. The stereochemistry of oxovanadium(IV) complexes derived from salicylaldehyde and polymethylenediamines. Journal of the Chemical Society Dalton Transactions. 1981;**6**:1241-1245

[33] Mahmoud MA, Zaitone SA, Ammar AM, Sallam SA. Synthesis, structure and antidiabetic activity of chromium(III) complexes of metformin Schiff-bases. Journal of Molecular Structure. 2016;**1108**:60-70

[34] Ulusoy N, Gürsoy A, Ötük G. Synthesis and antimicrobial activity of some 1,2,4-triazole-3-mercaptoacetic acid derivatives. II Farmaco. 2001;**56**:947

[35] Adams DM. Metal-Ligand and Related Vibrations: A Critical Survey of the Infrared and Raman Spectra

of Metallic and Organometallic Compounds. England: Edward Arnold (Publishers) Ltd London; 1967

[36] Li B, Li YQ, Zheng WJ, Zhou MY. Synthesis of ionic liquid supported Schiff base. ARKIVOC. 2009;**11**:165-171

[37] Peral F, Gallego E. Self-association of imidazole and its methyl derivatives in aqueous solution. A study by ultraviolet spectroscopy. Journal of Molecular Structure. 1997;**415**:187-196

[38] Shakir M, Nasam OSM, Mohamed AK, Varkey SP. Transition metal complexes of 13-14-membered tetraazamacrocycles: Synthesis and characterization. Polyhedron. 1996;**15**:1283-1287

[39] Chem LS, Cummings SC. Synthesis and characterization of cobalt(II) and some nickel(II) complexes with N,N′-ethylenebis(p-X-benzoylacetone iminato) and N,N′-ethylenebis (p-X-benzoylmonothioacetone iminato) ligands. Inorganic Chemistry. 1978;**17**:2358-2361

[40] Silverstein RM. Spectrometric Identification of Organic Compounds. 7th ed. United States of America: John Wiley & Sons; 2005

[41] Natarajan C, Tharmaraj P, Murugesan R. In situ synthesis and spectroscopic studies of copper(II) and nickel(II) complexes of 1-hydroxy-2-naphthylstyrylketoneimines. Journal of Coordination Chemistry. 1992;**26**:205-213

[42] Dehghanpour S, Bouslimani N, Welter R, Mojahed F. Synthesis, spectral characterization, properties and structures of copper(I) complexes containing novel bidentate iminopyridine ligands. Polyhedron. 2007;**26**:154-162

[43] Lever ABP. Inorganic Electronic Spectroscopy. 2nd ed. Amsterdam: Elsevier; 1984

[44] Tweedy BG. Plant extracts with metal ions as potential antimicrobial agents. Phytopathology. 1964;**55**:910-915

Chapter 6

Progress in Ionic Liquids as Reaction Media, Monomers and Additives in High-Performance Polymers

Dan He, Zhengping Liu and Liyan Huang

Abstract

In this chapter, we will review the progress in ionic liquids (ILs) widely used as reaction media, monomers and additives in the synthesis, chemical modification and physical processing of high-performance polymers (HPPs). Using ILs as reaction media in the syntheses of HPPs, the high-molecular-weight polymers were obtained in good yields and the shortened dehydration time compared to the conventional methods, the separation efficiency of products was improved. It is particularly noteworthy that the number of novel copolymers of HPPs with polymerisable ILs has steadily increased in recent years. In addition, ILs have been used as various types of additives such as the components of polymer materials, plasticizers and porogenic agents in the physical processing of HPPs, and the materials prepared include membranes, microcapsules, nanocomposites (NCs), electrolytes and grease.

Keywords: ionic liquids (ILs), high-performance polymers (HPPs), reaction media, monomers, additives, polyimides (PIs), polysulphones (PSFs)

1. Introduction

1.1 Ionic liquids (ILs) and their properties

The melting points of ILs, also known as low-melting-point organic salts, are usually below 100°C; ILs are composed of organic cations and inorganic/organic anions, as shown in **Figure 1** [1, 2]. The number of possible cation-anion combination has been estimated to be $>10^6$ [3]. ILs are most commonly abbreviated by writing the abbreviation/formula of the cation and anion in square brackets (without charges); e.g., [bmim][PF_6], [bmim][Tf_2N] and [emim][Cl] are the abbreviations for 1-butyl-3-methyl-imidazolium hexafluorophosphate, 1-butyl-3-methylimidazolium bis(trifluoromethanesulphonyl)imide and 1-ethyl-3-methylimidazolium chloride, respectively. Owing to their high chemical and thermal stability, low volatility, and low toxicity, ILs have attracted much attention for applications in chemistry and industry. In addition, the properties of ILs include high conductivity, wide electrochemical window, low flammability, ability to dissolve organic and inorganic solutes and gases, and recyclability. As far as vapour pressures are concerned, several

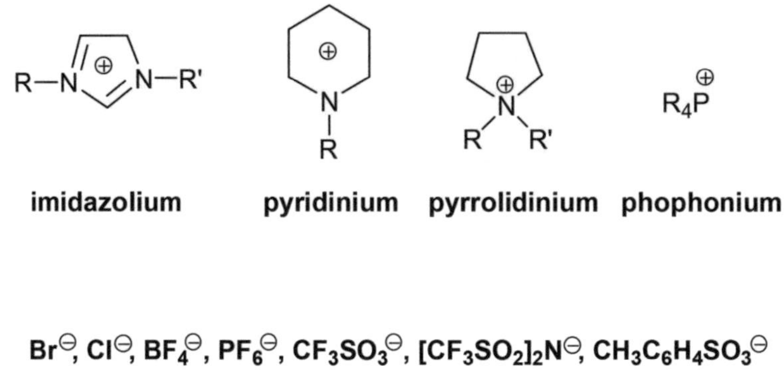

Figure 1.
Typical cationic and anionic components of ILs.

ILs can be vaporised under a high vacuum at 200–300°C and then recondensed [4]; however, ILs indeed have negligible vapour pressures at near ambient conditions. Thus, for general reactivities or processes, they may be considered as low-volatile reaction media. ILs are generally chemically stable reaction media, but this cannot be taken as granted. The proton at the C(2)-position of imidazolium cation is acidic, and under basic conditions, deprotonation leading to carbene is possible [5]. ILs are considered as rational designable solvents that can be easily tuned by using various combinations of cations and anions to achieve ILs exhibiting appropriate properties and achieving practical applications for a desired task (so-called task-specific ILs), making it possible to introduce ILs into specific synthesis processes [6].

Because of these special properties, ILs have emerged as novel and exciting reaction media in their own right. Every year, an increasing number of papers are being published on the applications of ILs for enhancing reactivities or processes in both chemical research and industry. So far, free-radical polymerisation, polycondensation and ionic polymerisation have been successfully carried out using ILs as the reaction media. In the step polymerisation field, there is a huge interest for high-performance polymers (HPPs). Despite improved synthesis methods and commercial availability of various ILs for replacing typical organic reaction media, they are still more expensive than typical organic reaction media. Therefore, the application of ILs as reaction media for enhancing reactivities or processes is limited.

1.2 High-performance polymers (HPPs) and their categories

HPPs are also known as high-temperature polymers, special engineering plastics, advanced engineering materials, and heat-resistant polymers [7]. They are defined as polymers that can retain the desirable properties when exposed to very harsh conditions, including, but not limited to, a high-temperature, a high-pressure, and corrosive environment. They are well known for outstanding thermal stability and service temperatures, good mechanical properties, dimensional and environmental stability, high resistance to most chemicals, gas barrier and electrical properties, etc., under extreme conditions, even at elevated temperatures [8, 9]. To better understand the reason for their strength, one must consider the bond strength that can be quantified by bond dissociation energy. First, the higher the bond dissociation energy, the harder it is to break the polymer chain, and thus the better the resistance of the polymer to harsh environment. The bond energies of C–C and C═C bonds are 83 and 145 kcal mol^{-1}, respectively; thus, it is harder to break a C═C bond than a C–C bond. Most HPPs contain more C═C bonds than C–C bonds. Similarly, the bond energies of C–H and C–F bonds are 99 and 123 kcal mol^{-1}, respectively; some of the C–H groups are also replaced with C–F groups. The resonance stabilisation is enhanced by adding aromatic components along the backbone, and it is estimated that the incorporation of resonance-stabilised units can add 40–70 kcal mol^{-1} to the bond strength. Such a molecular structure of HPPs improves their resistance and stability; thus, they can retain the desirable properties under very harsh conditions.

HPPs include polysulphones (PSFs), polyimides (PIs), polyaryletherketones (PAEKs), poly(arylene sulphide)s (PASs), polyarylates (PARs), liquid crystalline polymers (LCPs), fluoroplastics (PVDFs), *p*-hydroxybenzoic acid polymers, poly(naphthalene), poly(oxadiazole), and high-temperature nylon (HTN). HPPs can be divided into amorphous polymers, semi-crystalline polymers, and LCPs; e.g., PSFs are described as amorphous polymers and polyetheretherketones are semi-crystalline polymers. The applications of HPPs span across aerospace, defence, weaponry, energy, electronics, automotive, construction, nuke industry, membrane technologies, etc. In recent years, new HPPs and materials containing HPPs with enhanced application potential in more fields have been reported, including materials obtained by the chemical modification and blending of HPPs containing ILs.

1.3 Overview

The aim of this article is to review the recent progress in the field of ILs as reaction media, monomers and additives in the synthesis, chemical modification and physical processing of HPPs based on recent literatures, with the main emphasis on possible advantages, limitations and importance of the work. The article is structured as follows: section 2 focuses on progress in IL application in HPPs. Section 2.1 focuses on ILs as reaction media in the synthesis of HPPs, including PIs, PSFs, and PAEKs, and synthesis of HPPs in ILs under microwave (MW) irradiation. Section 2.2 focuses on ILs as monomers for the chemical modification of HPPs. The last part of Section 2 focuses on ILs as additives for the physical processing of HPPs, including membranes, microcapsules, electrolytes, nanocomposites (NCs) and grease.

2. Progress in IL application in HPPs

2.1 ILs as reaction media for synthesis of HPPs

Most HPPs are synthesised by step polymerisation reactions. Step polymerisation is one of the main polymerisation reactions for preparing polymers, usually requiring elevated temperatures, high-boiling-point reaction solvents, high vacuum and the removal of small molecules to reach the equilibrium. Therefore, it seems to be suitable to introduce ILs into step polymerisation owing to their intrinsic properties as described above. In 2002, high-molecular-weight aromatic PIs and polyamides were synthesised for the first time, obtaining polymers with inherent viscosities from 0.52 to 1.35 dL/g in ILs 1,3-dialkylimidazolium bromides [10]. The use of ILs as novel solvents for the synthesis of other HPPs has been reported, such as PAEK [3, 11] and PSF [12, 13].

2.1.1 Synthesis of PIs in ILs

In 1908, Jones et al. first synthesised PIs, but it was difficult to process and fabricate them [14]. Until the early 1960s, Du Pont, USA, made a substantial progress in the processing of PIs; thus, PIs were developed and widely utilised in various applications [15]. These polymers are known as HPPs and possess outstanding thermal stability, excellent electrical properties, improved mechanical properties and good resistance to organic solvents. They are widely applied in various modern industries such as gas separation membranes, insulator films for electrical/electronics, semi-conductor devices, coatings and composites, high-temperature adhesives, cell processing, and biochip design [16]. In general, PIs are produced in two steps via the formation of polyamic acids from diamines and dianhydrides. Other reported synthetic routes utilised tetracarboxylic acids, half-esters, a combination of bis(*o*-diiodoaromatics) with carbon monoxide, etc., [17]. Co-PIs with flexible linkages, such as ether and ester linkages between the aromatic rings of the main chain, have been synthesised, such as poly(amide-imide)s (PAIs), poly(ether amide-imide)s (PEAIs), poly(ester-amide-imide)s, poly(ether-imidazole-imide)s, and poly(amine-amide-imide)s [18].

In 2002, Vygodskii et al. first reported a novel one-step strategy for the synthesis of high-molecular-weight aromatic PIs by the polycyclisation reaction of 1,4,5,8-naphthalene tetracarboxylic acid dianhydride (DANTCA) with 3,3-bis(4′-aminophenyl)phthalide (Aph) in 1,3-dialkylimidazolium-based ILs without using catalyst at 180–200°C, as shown in **Figure 2** [10]. These ILs seem to be suitable reaction and activating media for the synthesis of aromatic PIs and polyamides. The polymerisation process and molecular weights of PIs are significantly affected by the structure and nature of ILs. The effects of IL structure include the following: (1) the best results were obtained in ILs with a symmetrical structure, obtaining polymers with a maximum inherent viscosity of 1.35 dL/g. (2) When using ILs with a symmetrical structure bearing alkyl chains of carbon atoms n = 2–6 and 12, polycyclisation occurred in a homogeneous solution for ILs with $n \leq 4$, but rapid precipitation of the PI occurred for ILs with $n > 4$. (3) High-molecular-weight polymers could be obtained in ILs with Br^-. (4) PIs insoluble in organic solvents do not precipitate from reaction solutions in ILs with anions SiF_6^-, HSO_4^-, NO_3^-, I^- and CH_3COO^-. (5) As far as cations were concerned, the molecular weights of PIs are lower in ILs based on quinoline and pyridine bearing the same alkyl chains longer than imidazole. In summary, it is possible to tailor ILs as active solvents for the step polymerisation reactions of PIs by varying the structures of the cations and anions of ILs. Studies on using ILs as reaction media for synthesising other step polymerisation polymers are in progress. Later, Vygodskii and co-workers reported studies on using different ILs as reaction media for synthesising other step polymerisation polymers with high molecular weights, such as poly(amide imide) s (PAIs) [19–21]. Although relatively high-molecular-weight PIs have been obtained in the absence of any added catalysts and lower reaction temperatures than the conventional synthetic method [17, 22, 23], the limited solubility of some aromatic substrates in ILs was still the main problem. In 2006, Ohno et al. reported that the solubility of starting materials was improved by adding imidazolium-type zwitterion (ZI), 1-(1-butyl-3-imidazolio)butane-4-sulphonate in ILs, leading to higher molecular weights of the resulting PIs [24]. On the other hand, in some studies of the step polymerisation of PIs, catalytic effect of ILs was observed. A type of PI was synthesised by the step polymerisation reactions of 1,4-bis(3-aminopropyl)

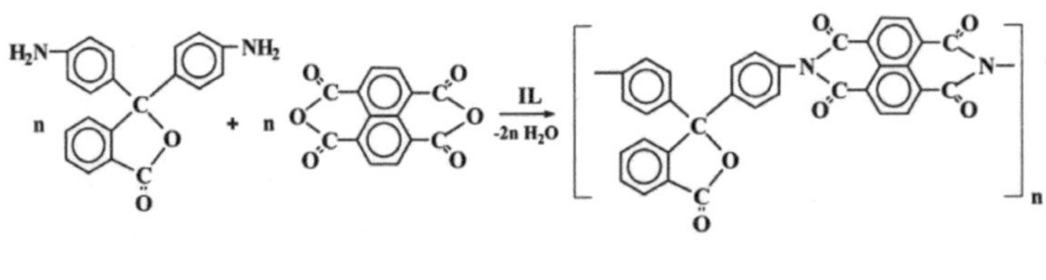

Figure 2.
Synthetic route for PI in IL.

piperazine with 3,3′,4,4′-benzophenonetetracarboxylic dianhydride (BTDA) in the presence of ILs as the catalyst and N-methylpyrrolidone (NMP) as the solvent. The degrees of polymerisation are not only high, but the IL also exerts a detectable effect on polymer solubility. The PI with a higher degree of polydispersity was obtained in ILs based on imidazole than that obtained with pyridine and alkylamine [25]. PI nanoparticles were obtained in an IL, namely, 1-ethyl-3-methylimidazolium bis(trifluoromethyl-sulphonyl)imide ([emim][Tf_2N]), as a continuous phase without the addition of any further activating or stabilising agents by the heterophase step polymerisation of different aromatic tetracarboxylic acids and diamines. These PI particles with a range of 100 nm showed a high thermal stability by TGA and a decomposition temperature of ~520°C [26]. A sulphonated co-PI (SPI) was prepared in an IL, 1-ethyl-3-methyl imidazolium bromide [emim][Br], without any catalyst. These co-PIs prepared showed superior properties compared to those prepared in a common solvent, indicating promising properties for applications in proton-exchange membrane fuel cells [27, 28]. The trifluoromethylated poly(ether-imidazole-imide)s (PEII)s based on an unsymmetrical-diamine-bearing carbazole and imidazole chromophores were obtained with 80–96% yields in imidazolium-based ILs. They were amorphous with good thermal and thermo-oxidative stability, excellent solubility, and ability to form tough and flexible films [29].

Optically active PAIs were successfully synthesised in an IL, namely, 1,3-dipropylimidazolium bromide ([*d*pim][Br]), using triphenyl phosphite (TPP) (a condensing agent) without any additional extra components such as LiCl and pyridine. Therefore, it was concluded that ILs not only act as solvents, but also act as catalysts in this step polymerisation [30]. At the same time, various types of ILs were investigated as solvents and catalysts for the polymerisation of PAIs [31]. A PAI based on 2-[5-(3,5-dinitrophenyl)-1,3,4-oxadiazole-2-yl]pyridine was synthesised in [bmim][Br]. Heterocyclic and optically active PAIs incorporating L-amino acids were synthesised in [pmim][Br] [18, 32]. Shadpour et al. later reported several relative articles in succession. For example, organosoluble and optically active PAIs bearing an *S*-valine moiety were synthesised by the step polymerisation of different aliphatic and aromatic diisocyanates with a chiral diacid monomer in tetrabutylammonium bromide (TBAB) IL. These polymers exhibited good thermal properties and were soluble in amide-type solvents [33]. The poly(amide-ether-imide-urea) s (PAEIU)s were synthesised by the step polymerisation of a chiral diamine with several diisocyanates in [*d*pim][Br] IL [34]. Heat-stable and optically active processable PAI nanostructures bearing a hydroxyl pendant group were synthesised by step polymerisation in the presence of IL and TPP [35]. In recent years, using TBAB IL/TPP and ultrasonic irradiation, optically active PAI/TiO_2 bio-NCs containing *N*-trimellitylimido-L-isoleucine linkages, poly(vinyl alcohol) (PVA) with chiral PI nanoparticles containing S-valine, and optically active PAI/zinc oxide bio-NCs (PAI/ZnO BNCs) containing L-valine were obtained one after the other [36–38].

2.1.2 Synthesis of PSFs in ILs

In 1965, PSF was first successfully developed and commercialised by Union Carbide, USA, currently known as Udel PSF. Usually, the number-average molecular weights of commercial products are in the range of 16,000–35,000, and the weight-average molecular weights are in the range of 35,000–80,000 [6]. They are well known for their outstanding thermal stability, good mechanical properties, electrical properties, transparency and resistance to most chemicals. They are widely used in various modern industries such as electrical/electronics, machineries, medical equipment, transportation and aerospace, and membrane separation technologies [39–42]. In recent studies, PSFs were mainly studied in the development of membrane technologies such as hemodialysis, micro-/ultra-filtration membrane and gas separation. PSFs are usually synthesised via nucleophilic aromatic substitution polymerisation (S_NAR) reactions [43]. For example, bisphenol-A PSF is synthesised from 2,2-bis(4-hydroxyphenyl) propane (bisphenol A) and 4,4′-dichlorodiphenylsulphone (DCDPS) or 4,4′-difluorodiphenylsulphone (DFDPS). Poly(ether sulphone)s (PESs) are usually synthesised from 4,4′-dihydroxydiphenylsulphone (bisphenol-S) and DCDPS/DFDPS [13].

Liu's research group is one of the most active research groups in this field. In 2012, high-molecular-weight PSFs were synthesised in high yields by step polymerisation of bisphenol A with DFDPS for the first time in various ILs/zwitterions (ZIs) as the reaction media in the presence of potassium carbonate (K_2CO_3) using a one-pot green protocol shown in **Figure 3**. In this work, the dehydration time was shortened to 80% (2.5–8 to 0.5 h) compared to the conventional methods, and the weight-average molecular weights ranged from 30,000 to 130,000, with great potential for commercial applications. The polarity of ILs strongly affected the molecular weight of PSF, and ILs containing PF_6^- were better [6]. Recently, PESs were also successfully synthesised by the step polymerisation of bisphenol S with DFDPS in ILs/ZIs as the reaction media in the presence of K_2CO_3 using a one-pot green protocol. The dehydration time was shortened to 0.5 h compared to the conventional methods, and a high solubility of bisphenol-s dipotassium salt in IL/ZI significantly lowered the reaction temperature (150°C) than the conventional temperature (220 – 300°C). The proposed method has clear advantages for synthesising PSF and PES compared to volatile organic solvents and, in principle, can be applied to the synthesis of other HPPs via nucleophilic aromatic substitution polymerisation reactions [13]. In 2017, the synthesis of poly(phenylene sulphide sulphone) (PPSS) in ILs was presented [44].

2.1.3 Synthesis of PAEKs in ILs

PAEK was reported by Bonner of Du Pont, USA, in 1962 and Goodman of Imperial Chemical Industries (ICI), UK, but the molecular weight of products synthesised was the lower. Until 1979, Rose et al. of ICI reported that PEK with a high molecular weight was synthesised, laying the foundation for the synthesis of PAEK. Commercially available as VICTREX® PEEK™, it was introduced into the market by ICI. Poly(ether ether ketone) (PEEK) is one of the most recently developed materials. PAEKs exhibit many outstanding characteristics including high thermal stability, excellent mechanical, thermo-oxidative, and chemical resistance

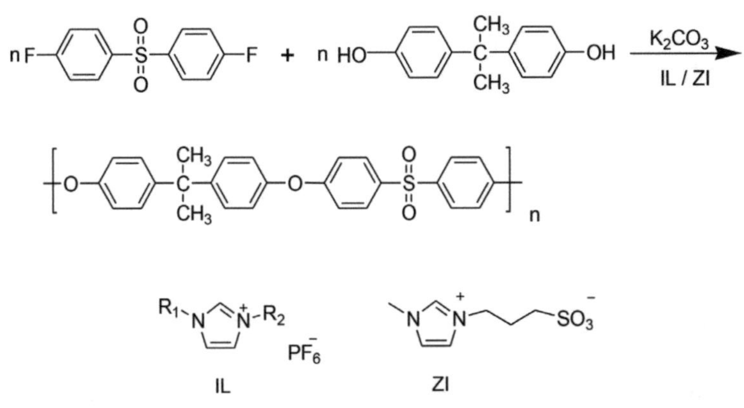

Figure 3.
Synthetic route for PSF in IL/ZI and structures of ILs and ZI.

properties under diverse conditions, and good electrical insulation [45, 46]. The PAEK applications span across automotive, aerospace or chemical industries, orthopaedics and surgery, cable insulation, and membrane technologies. In addition, high temperatures, intense mechanical stress, and/or exposure to harsh chemical environments are required; thus, PAEK can be a lightweight replacement for metals.

In 2013, PAEKs were successfully synthesised via S_NAR mechanism using ILs [*i*-pmim][PF_6] as the green reaction media. The optimised step polymerisation conditions were 50 wt% monomer concentration, dehydration at 150°C for 0.5 h and polymerisation at 180°C for further 1.5 h. The number-average molecular weights of PAEKs ranged from 10,000 to 18,000 g mol^{-1} with high yields. In addition, the interactions of bisphenol A with ILs were investigated, exhibiting a strong influence on the PAEK synthesis [11]. In the same year, PEEKs were also synthesised in IL [bmim][Tf_2N], by polycondensation reactions of hydroquinone with 4,4′-dihalobenzophenones in the presence of K_2CO_3 at 320°C. The materials produced in IL were similar to those produced in the industrial solvent of choice, but the molecular weights were lower, possibly due to the lower solubility of the polymer. The advantage of using IL [bmim][Tf_2N] over diphenyl sulphone as the solvent is that the separation efficiency significantly improved [3]. It is expected that more custom-designed ILs would be used for PAEK production, potentially widening the scope of the choice of solvents currently used.

2.1.4 Synthesis of HPPs in ILs under microwave irradiation

Microwave (MW) is a type of electromagnetic wave with frequency ranging from 300 MHz to 300 GHz, usually 2450 MHz for radiation chemical reaction; the temperature of the system depends on the power of MW and electrical properties of the medium. MW radiation can accelerate the reaction rate for a specific system and complete these reactions within a short period. Thus, as a non-conventional energy source, MW radiation has become an increasingly more practical and popular technology in organic chemistry, including polymerisation. Owing to their high ionic conductivity and polarisability, ILs act as excellent MW-absorbing agents and are, therefore, used in polymerisation. Mallakpour's research group is one of the most active research groups in this field. In 2007, poly(urea-urethane)s (PUU)s were prepared using IL 1,3-diallylimidazolium bromide and tetrabutyl-ammonium bromide (TBAB) under MW irradiation as well as conventional heating. The polymerisation reactions occurred rapidly, producing a series of PUUs in good yields and with moderate inherent viscosities of 0.21–0.46 dL/g. These PUUs showed a good solubility and could be readily dissolved in traditional organic solvents [47].

MW-assisted synthesis can provide higher yields and purer products than traditional heating; therefore, recently, MW radiation has been used for the synthesis of HPPs, especially PAIs. In 2010, poly(urethane-imide)s (PUIs) were prepared in the presence of IL TBAB under MW irradiation conditions. All the PUIs showed thermal stability and good solubility in various organic solvents. In in vitro toxicity studies, the prepared materials showed biological activity and non-toxicity to microbial growth and were classified as bioactive and biodegradable compounds [48]. In 2012, chiral-nanostructured PAIs were synthesised in TBAB IL under MW irradiation by the polymerisation reactions of several amino-acid-based chiral diacids with an aromatic diamine, 2-(3,5-diaminophenyl)-benzimidazole. The PAIs obtained were organo soluble, and the HPPs were obtained in high yields and with inherent viscosities in the range 0.40–0.52 dL/g. The materials synthesised were amorphous polymers with nanostructures containing nanosized particles in the range from 40 to 80 nm [49]. Chiral-PAI-nanostructures-bearing hydroxyphenyl pendant units in the side chain were also prepared using TBAB IL and TPP as the condensing agent under MW irradiation. The obtained PAIs had inherent viscosities in the range 0.32–0.49 dL/g; they were amorphous polymers with nanostructures in which the nanosized particles are in the range from 66 to 78 nm [50]. Soluble, thermally stable PAIs modified with siloxane linkages with a reduced dielectric constant were synthesised via the isocyanate method in TBAB, tetrabutyl-phosphonium bromide (TBPB), and 1-buthyl-3-methyl imidazolium chloride ([bmim][Cl]) under MW irradiation [51].

2.2 ILs as monomers for chemical modification of HPPs

In 2010, Li et al. reported that block co-PIs based on aromatic dianhydrides and diamines copolymerised with diamino IL monomers, specifically 1,3-di(3-amino-propyl) imidazolium bis[(trifluoromethyl)sulphonyl]imide ([DAPIM]-[Tf_2N]) and 1,12-di[3-(3-aminopropyl)imidazolium]dodecane bis[(trifluoromethyl)sulphonyl] imide ([C_{12}(DAPIM)$_2$][Tf_2N]$_2$), as shown in **Figure 4**, were synthesised by the Boc protection method and using diverse compositions. These two ILs were first reacted with 2,2-bis(3,4-carboxylphenyl)hexafluoropropane dianhydride (6FDA) to produce 6FDA-IL oligomers as the IL component for block co-PIs. Later, the oligomers were reacted with 6FDA and *m*-phenylenediamine (MDA) at an oligomer concentration from 6.5 to 25.8 mol% to form block co-PIs. As the concentration of 6FDA-IL oligomer increased in the block co-PIs, the thermal degradation temperature and glass transition temperature of the produced co-PIs decreased, but their density increased. Compared to pure 6FDA-MDA, the gas permeability of the IL-based block co-PI decreased, but the ideal permeability selectivity for CO_2/CH_4 gas pair increased [52]. The co-PIs were mainly used in the separation of gases such as O_2, N_2, CH_4 and CO_2 [53, 54].

Later, a series of poly(arylene ether sulphone)s containing bulky imidazole groups (PSf-Im-x) based on a monomer 2,2′-bis-(2-ethyl-4-methylimidazole-1-ylmethyl)-biphenyl-4,4′-diol (EMIPO) were successfully synthesised. After the quaternisation by *n*-bromobutane, these polymers were evaluated as alkaline anion exchange membranes (AEMs) as shown in **Figure 5**. 2-Ethyl-3-butyl-4-methylimidazolium was introduced as the functional group in these polymers; the bulky groups present around the imidazolium ring reduced the access of OH^- to imidazolium, thus increasing the alkaline stability of the membranes. The membrane showed an IEC value of 2.07 and ionic (OH^-) conductivity of 0.014 S cm^{-1} at 30°C, in which 80% of the ionic conductivity was maintained even for the treatment in 1 M KOH at 60°C for 144 h [12].

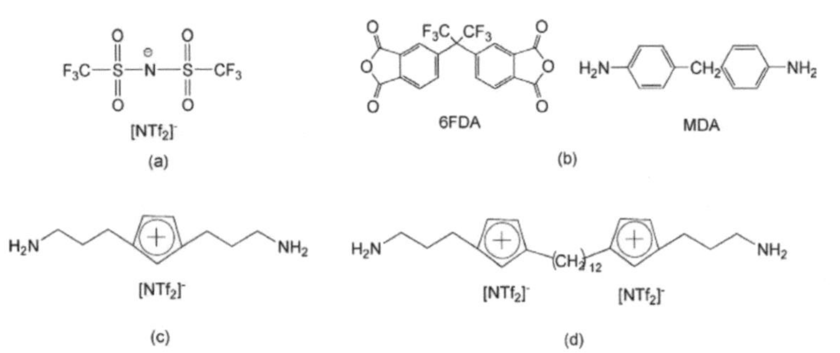

Figure 4.
Structures of monomers used in polyimide synthesis: (a) bis[(trifluoromethyl)sulphonyl]imide ([Tf_2N]), (b) 2,2-bis(3,4-carboxylphenyl)hexafluoropropane dianhydride (6FDA), m-phenylenediamine (MDA), (c) 1,3-di(3-aminopropyl)imidazolium bis[(trifluoromethyl)sulphonyl] imide ([DAPIM] [Tf_2N]), and (d) 1,12-di[3-(3-aminopropyl)imidazolium] dodecane bis[(trifluoromethyl)sulphonyl] imide ([C_{12}(DAPIM)$_2$][Tf_2N]$_2$).

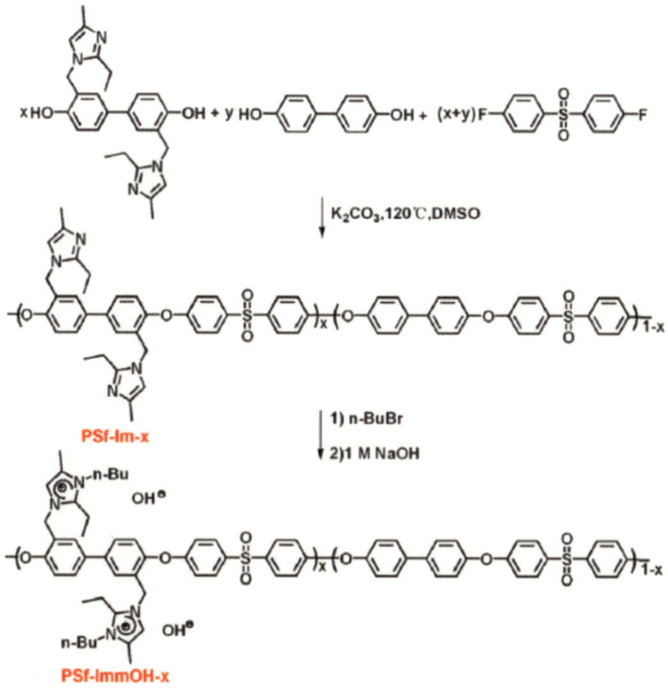

Figure 5.
Synthetic routes of PSf-Im-x and PSf-ImmOm-x.

Recently, a new synthetic method was reported for the modification of PIs; the PIs were first transformed to their ionic forms via the subsequent *N*-alkylation and quaternisation of benzimidazole or quinuclidine moieties; then, an ion exchange reaction was carried out to prepare polymers bearing the bis(trifluoro-methyl-sulphonyl)imide anion. High-molecular-weight (Mn = 22,000–97,000) cationic polyelectrolytes with the degree of quaternisation as high as 96% were obtained under the optimal conditions. These new materials showed excellent mechanical and thermal properties, adjustable surface wettability, and improved gas transport properties [55]. Several recent articles reported that the incorporation of IL moieties into HPP by copolymerisation is a promising strategy to prepare novel copolymers of ILs with HPPs with improved properties. It is presumed that further related work will be reported in the future.

2.3 ILs as additives for physical processing of HPPs

The application of ILs for HPPs is not limited to their use as reaction media in polymerisations for preparing HPPs, and ILs are miscible with some HPPs and used as additives in the materials such as the components of polymer materials, plasticizers, and porogenic agents. By blending ILs with HPPs, the properties of the obtained mixtures can be considerably affected [56]. Thus, applications of ILs are being explored in the fields of membranes, microcapsules, electrolytes, NCs and grease.

2.3.1 Membrane

Supported IL materials have two main processes. First, ILs are covalently linked to polymers, inorganic surfaces or particles, thereby supporting the IL materials. In such systems, the properties of the ILs are modified to some extent, but generally, the main features are retained. Second, ILs are dissolved and imbibed in a polymeric membrane, porous matrix, particle or bulk material as the components of the mixture, and the properties of the IL are retained [57]. In recent years, supported IL membranes (SILMs) have received considerable attention for their applications in gas separation, electrolyte, proton exchange, etc.

Early on, it was reported that ILs based on 1-*n*-alkyl-3-methylimidazolium cation (*n*-butyl, *n*-octyl, and *n*-decyl) can be used together with the anions PF_6^- or BF_4^-. Immobilisation of these ILs on a polyvinylidene fluoride (PVDF) membrane provides an extremely highly selective transport for secondary amines over tertiary amines [58]. Later, the PVDF/ILs composite membranes were prepared. A membrane using [emim][Tf_2N] and PVDF hollow fibre was prepared as a support for CO_2/N_2 separation [59].

A quasi-solid-state dye-sensitised solar cell based on poly(vinylidenefluoride-co-hexafluoro-propylene) P(VDF-HFP)/SBA-15 nanocomposite membranes was obtained using dimethyl-propylimidazolium iodide (DMPII) IL [60]. The SILM was prepared using a hydrophilic PVDF support immobilised in the IL 1-butyl-2,3-dimethylimidazolium hexafluorophosphate ([b*d*mim][PF_6]) [61]. The preparation of PVDF-blended membranes with dominating β-phase crystals was studied in ILs 1-butyl-3-methylimidazolium tetrafluoroborate ([bmim]BF_4), [bmim][PF_6] and 1-methylimidazolium trifluoromethanesulphonic ([mim]CF_3SO_3) as the co-solvents for zwitterionic copolymers [62]. A PVDF membrane with piezoelectric β-form was prepared by immersion precipitation in mixed solvents containing an IL [bmim][BF_4] [63]. Composite membranes were prepared as electroactive actuators using PVDF and [emim][Tf_2N] as the plasticiser [64]. A 1-butyl-3-methylimidazolium triflate ([bmim]OTf)/PVDF composite membrane was prepared by the impregnation method and used for the separation of C_6H_6/H_2 and C_6H_{12}/H_2, as shown in **Figure 6** [65]. Thin films containing 1-ethyl-3-methylimidazolium nitrate ([emim][NO_3]) IL and PVDF were investigated [66]. Another HPP-containing fluorine, polytetrafluoroethylene (PTFE), was often prepared using an SILM. For example, gelled SILMs were synthesised by the gelation of [bmim][PF_6] in the pores of PTFE hollow fibres and used in the separation of butanol from acetone-butanol-ethanol mixtures (ABE) by sweep gas pervaporation [67]. Amino-acid-IL-based-facilitated transport membranes containing PTFE were prepared via impregnation [68].

The SILMs prepared with PSF supports are often used for CO_2 separation such as CO_2/He, CO_2/CH_4 and CO_2/N_2 separation, even at elevated temperatures [69–71]. In addition, ion-conductive membranes were prepared by casting a solution of Udel-type PSF and IL 1-butyl-3-methylimidazolium trifluoromethane-sulphonate ([bmim][TfO]) or 1-ethylimidazolium trifluoromethanesulphonate ([eim][TfO]) [72]. The SILMs prepared with PES supports are frequently used in the separation of gases, especially SO_2 [73, 74] and CO_2 [75–78]. In addition, PES membranes with ion exchange and anti-biofouling properties were prepared by the surface immobilisation of Brønsted acidic ILs via double-click reactions [79]. A sulphonated PES (SPES) film containing ILs was obtained by solution casting and prepared using double-side, self-cleaning polymeric materials, as shown in **Figure 7** [80]. Hydrophilic porous PES membranes and microcapsules were prepared via non-solvent-induced phase separation (NIPS) using IL [bmim][PF_6] as the structure control agent [81].

The surface wettability of negatively charged PI films was tuned by the electrostatic self-assembly of ILs and formation of spherical nanoparticles, indicating the assembly of longer-substituent cations [82]. The membranes containing ILs prepared with PI supports were often used in gas separation [83–86] and fuel cells [87–89]. In addition, available PAIs [90], copolymers of poly(ethylene glycol) (PEG) and aromatic PI [91, 92], and SPI [93–95] were also used in the preparation

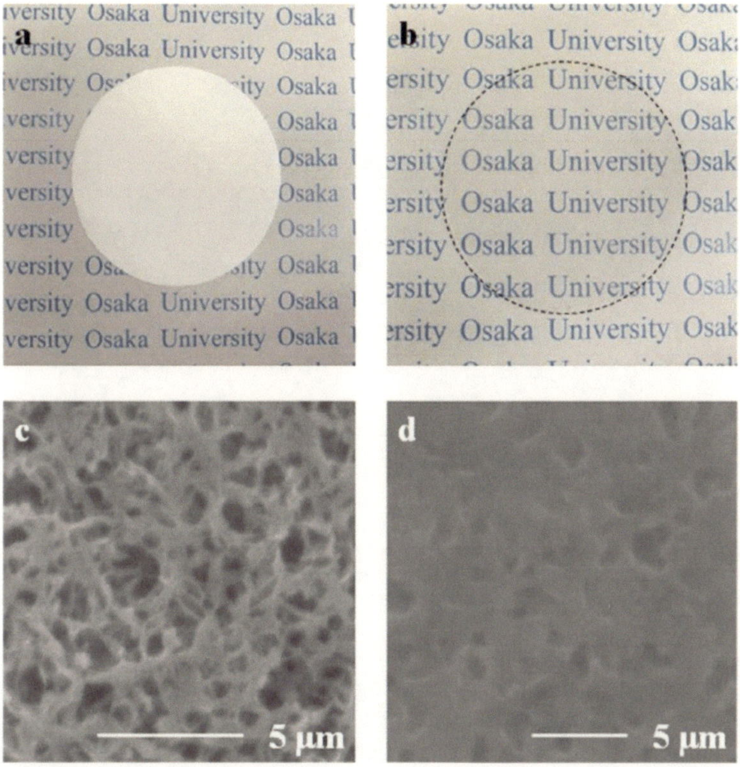

Figure 6.
Photographs (a and b) and SEM images (c and d) of the PVDF membrane before (a and c) and after impregnation (b and d) with [bmim] OTf.

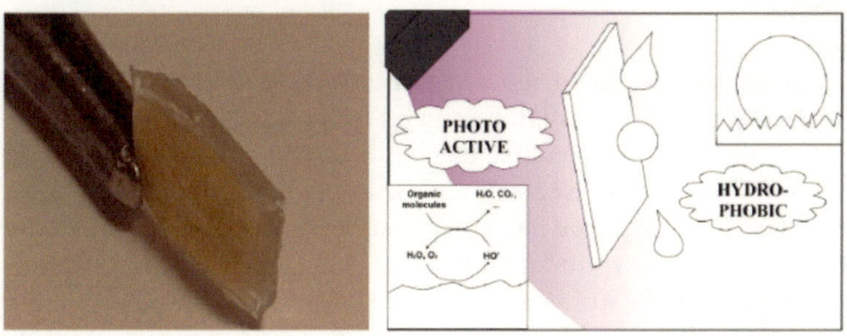

Figure 7.
Picture and schematic of the final material.

of membranes containing ILs. For example, SPI/IL composite membranes as proton-exchange membranes have been reported in recent years [96, 97].

Composite membranes based on sulphonated poly(ether ether)ketone (SPEEK) with ILs [$CH_3CH_2CH_2NH_3$][CF_3COO] (TFAPA), [bmim][Cl] and [bmim][PF_6] have been prepared [98, 99]. Composite membranes have been prepared using SPEEK ILs in the presence of Y_2O_3 [100].

2.3.2 Microcapsule

In 2007, monodispersed microcapsules enclosing [bmim][PF_6] were prepared via a two-stage microfluidic approach, as shown in **Figure 8**; the hollow PSF microcapsules showed an encapsulation capacity of 30.8% [101]. PSF microcapsules containing

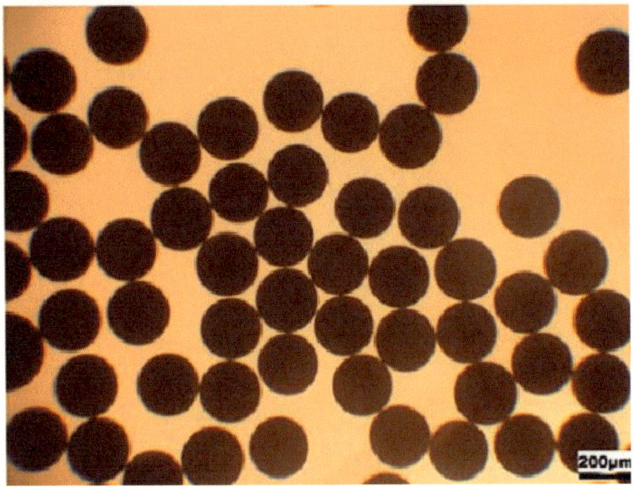

Figure 8.
Optical microscopic images and composition of organic phase is PSF: dichloromethane (DCM): [bmim] PF_6 = 5 g: 80 ml: 3 g; continuous phase is 0.1 wt% gelatin solution; inner size of nozzle: 0.6 mm, IL microcapsules, flow rate of continuous phases (CP): 30 ml/min; and flow rate of droplet phases (DP): 125 μl/min.

[bmim][PF_6] were also obtained by spraying a suspension dispersion with an encapsulation capacity of 29% [102]. PSF microcapsules have practical use such as the removal of caprolactam from water [103]. PEEK microcapsules containing trihexyl(tetradecyl) phosphonium chloride IL was obtained in *N,N*-dimethylformamide (DMF) as the dispersing phase and dodecane [104]. PTFE microcapsules containing [hmim][Tf_2N] IL lubricant with small sizes (below 10 μm) have been reported [105].

2.3.3 Electrolyte

ILs have also been used as solvents in extraction processes or as electrolytes. Polymer electrolytes comprising IL [emim][Tf_2N] and soluble SPI showed a high ionic conductivity and reliable mechanical strength, suitable for HPP actuators [106]. Gel polymer electrolytes (GPEs) [107–109] and solid polymer electrolytes [110–114] have been widely reported. A series of GPEs were prepared using the electrospun membranes of poly(vinylidene fluoride-co-hexafluoropropylene) [P(VdF-co-HFP)] incorporating ILs, 1-alkyl-3-methylimidazolium bis(trifluoromethylsulphonyl)imide in the presence of lithium bis(trifluoromethylsulphonyl)imide ($LiTf_2N$) [115]. An IL-GPE containing semi-crystal PVDF, amorphous polyvinyl acetate (PVAc) and ionic conductive [bmim] [BF_4] was prepared via the solution-casting method for solid supercapacitors [116]. A PVDF-HFP/PMMA-blended microporous gel polymer electrolyte incorporating [bmim][BF_4] was fabricated for lithium-ion batteries [117]. Solid polymer electrolytes using poly(vinylidene-fluoridetrifluoroethylene) and N,N,N-trimethyl-N-(2-hydroxyethyl) ammonium bis(trifluoromethylsulphonyl)imide ([$N_{1112(OH)}$] [Tf_2N]) IL were fabricated for energy storage applications [118]. PI/IL composite membranes for fuel cells operating at high temperatures were prepared by impregnating a porous Matrimid® membrane with protic ILs: 1-*n*-methyl-imidazolium dibutylphosphate ([C_1im][DBP]), 1-*n*-butylimidazolium dibutyl-phosphate ([C_4im][DBP]) and 1-*n*-butylimidazolium bis(2-ethylhexyl)phosphate ([C_4im] [BEHP]). The electrolyte membranes were used as a proton-exchange membrane fuel cell (PEMFC) [119]. An IL-polymer electrolyte film based on a low-viscosity IL (1-ethyl-3-methylimidazolium dicyanamide) incorporated into a polymer matrix

(PVDF-HFP) was prepared, exhibiting liquid-like conductivity, and the maximum conductivity of PVDF-HFP + 25 wt% IL was as high as 10^{-3} S/cm, as shown in **Figure 9** [120]. Quasi-solid-state electrolytes (QSEs) consisting of IL-$LiTf_2N$-fumed silica nanoparticles were prepared for use in bulk-type all-solid-state cell configuration lithium-sulphur rechargeable batteries [121].

2.3.4 Nanocomposite

Carbon nanotubes (CNTs) consist of rolled-up graphene sheets and can be used as electronic, conductive and reinforcing fillers for polymer composites. MWCNTs possess a nanoscale diameter, a high aspect ratio, excellent mechanical properties, and good electrical conductivity. In recent years, MWCNTs have gained considerable interests of scientists and engineers, especially in polymer composites containing HPPs and ILs. A PI composite film comprising finely IL-dispersed MWCNTs in IL was obtained with a high shielding effectiveness (SE) for use in the packaging of a 2.5-Gbps plastic transceiver module with numerous applications in fibre to the home lightwave transmission systems [122]. PEI NCs consisting of bucky gels of industrial-grade MWCNTs and [bmim][PF_6] were prepared; they are suitable for the aerospace and electronics industries, as shown in **Figure 10** [123]. Some PI and PEI NCs consisting of MWCNTs and polymerised ILs (PILs) were prepared, exhibiting differential function [124, 125].

The crystal structure of PVDF was modified by utilising the long alkyl chains of [C_{16}mim][Br] and IL-modified MWCNTs, and the crystallisation kinetics of the composites was investigated [126, 127]. A series of PVDF composites with 'bucky gels' of MWNTs and ILs were obtained by simple melt compounding. According to the DSC and XRD results, the addition of ILs in the composites changed the crystallinity and crystal form of the PVDF [128]. PTFT and PVDF as the components of NCs containing HPP and ILs have received increasingly more attention. The nanomaterials with a nanoscale structure were prepared using pyridinium, imidazolium and phosphonium ILs as new synthetic building blocks in a PTFT. The cation-anion combination and functionalisation of ILs affect the ionic networks and nanostructures of materials [129, 130]. These nanomaterials show optimised thermal and mechanical properties and have numerous potential applications such as supercritical CO_2 [131]. PVDF/IL/GraNCs were fabricated via the solution casting of PVDF with graphene (Gra) modified with a long alkyl chain IL [C_{16}mim][Br], exhibiting a low loss tangent and low conductivity in the PVDF/ionic liquid-modified graphene

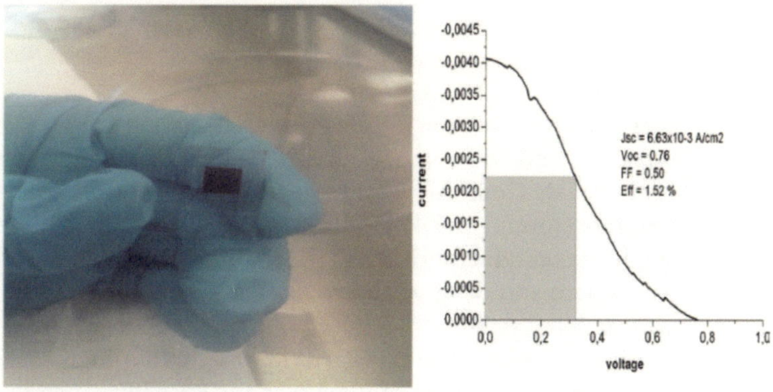

(a) (b)

Figure 9.
Photograph of (a) assemble DSSC and (b) I-V characteristics of DSSC comprising IL-incorporated PVDF-HFP polymer electrolyte film (maximum conductivity).

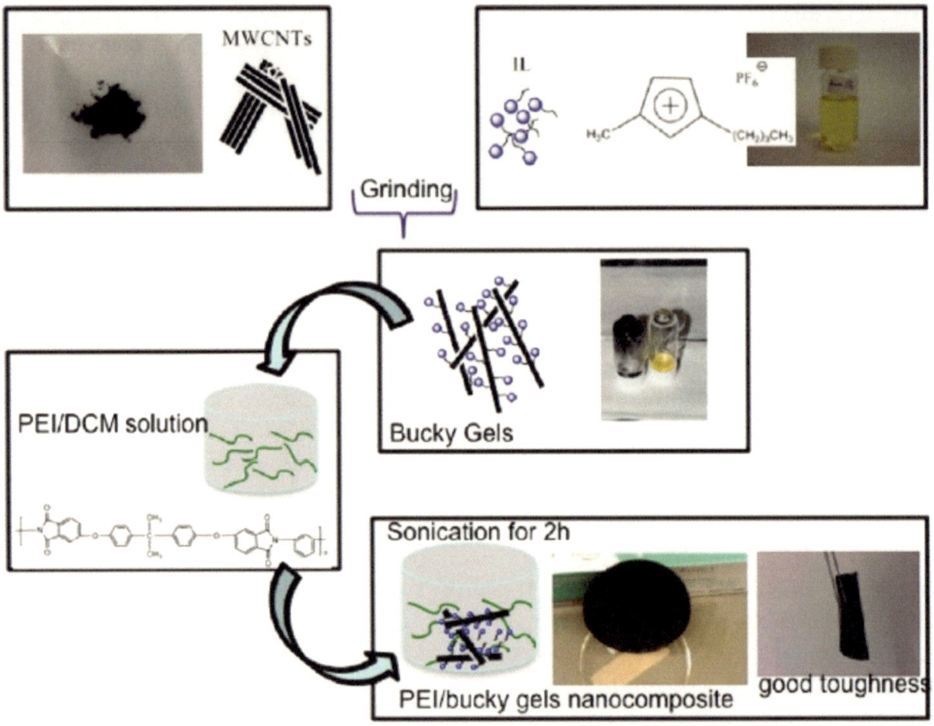

Figure 10.
Processing of PEI/bucky gel nanocomposites; the obtained composites had good toughness as the film could be curved adequately without any damage.

(GIL) composites [132, 133]. NCs, based on a homopolymer PVDF and IL, were fabricated, and the preparation process opens up a new synthesis route for nano-structured polymer composites. Dielectric NCs based on PVDF, conductive carbon black (CB) and IL 1-vinyl-3-ethylimidazolium tetrafluoroborate [VEIM][BF_4] were prepared via melt blending and electron beam irradiation (EBI) methods [134].

A nanostructured PAI was prepared in TBAB as a green medium by the step polymerisation reaction of 4,4′-methylenebis(3-chloro-2,6-diethyl trimellitimidobenzene) with 3,5-diamino-*N*-(4-hydroxy-phenyl)benzamide. Then, amino acid-functionalised multiwalled carbon nanotubes (f-MWCNTs)/PAI NCs were prepared by a solution mixing method [30].

2.3.5 Grease

In 2012, two types of conductive lubricating greases consisting of ILs (abbreviated as 'ILs lubricating greases') and HPPs were synthesised using 1-octyl-3-methylimidazolium hexafluorophosphate ([omim][PF_6]) (L-P108) and 1-octyl-3-methylimidazolium tetrafluoroborate ([omim][BF_4]) (LB108) as the base oil and PTFE as the thickener, exhibiting higher conductivities than the traditional conductive lubricating greases containing conductive stuffing, as shown in **Figure 11** [135]. The conductive lubricating greases using 1-hexyl-3-methylimidazolium tetrafluoro-borate ([hmim][BF_4]) and 1-hexyl-3-methylimida-zolium bis(trifluoromethylsulphonyl)amide ([hmim][Tf_2N]) were prepared, and the conductivity and tribological performance of these greases were studied [136, 137]. By changing the type of ILs, different types of conductive lubricating greases were obtained [138].

In addition, the materials obtained by the blending of ILs with HPPs also include IL marbles containing PTFE and IL [139], PAI hollow fibre containing [bmim]

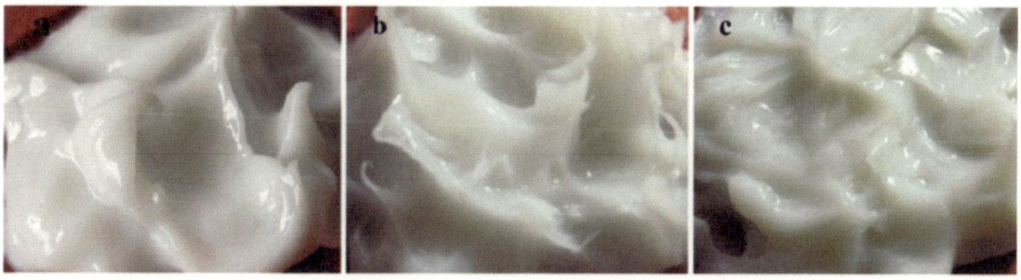

Figure 11.
The photo images of the IL lubricating greases and poly-a-olefin (PAO) grease: PAO10 (abbreviate as 'PAO lubricating grease') (a) L-P108 (b) and LB108 (c).

[Tf_2N] [140], IL-coated PTFE tube [141], PVDF/IL functionalised graphene oxide (GO-IL) composite (PGL) films [142] and electro-active electrospun fibre containing PVDF and [bmim][Tf_2N] [143]. For the materials obtained by the blending of ILs with HPPs, ILs based on imidazole and PF_6^- and Tf_2N^- are most often used [144, 145]. PVDF, PSF and PI are the most popular [44, 146–148].

3. Conclusions

Studies on ILs as the reaction media, monomers and additives in the synthesis, chemical modification and physical processing of HPPs are in progress. In 2017, the synthesis of poly(phenylene sulphide sulphone) (PPSS) in ILs was also presented. ILs are not only interesting as a replacement for traditional volatile and flammable organic solvents, but also have the potential to reduce energy consumption and increase chemical reactivity, thus leading to more efficient processes for the synthesis of HPPs. Using ILs in the synthesis of HPPs, promising and novel established approaches have been developed under mild conditions. Owing to their high ionic conductivity and polarisability, ILs, as excellent MW-absorbing agents, were introduced to polymerisation reactions to achieve higher yields and purer products than the traditional heating method. Notably, the number of novel copolymers of HPPs with polymerisable ILs has steadily increased in recent years, and they were mainly used in separating gases such as O_2, N_2, CH_4 and CO_2. ILs are not only used as reaction media in polymerisations for preparing HPPs and as monomers in the chemical modifications of HPPs, but also ILs are miscible with some HPPs and used as various types of additives (such as the components of polymer materials, plasticizers and porogenic agents) in the physical processing of HPPs. The materials prepared include membranes, microcapsules, electrolytes, NCs and grease. HPPs and inorganic substrates have been used to support IL materials by the covalent bonding of ILs, where the properties of ILs may be changed to some extent, and HPP membranes and NC-absorbed ILs exhibit concomitant changes in ionic conductivity and mobility. These novel green chemical approaches provide diverse possible new materials, and it is expected that more modified materials of HPPs and ILs with special properties and applications would be obtained. It is believed that more studies on ILs containing HPPs will be reported in the future.

Acknowledgements

The authors would like to acknowledge the National Science Foundation of China (grant no. 21274014), Beijing Municipal Commission of Education, the

Fundamental Research Funds for the Central Universities, and the Measuring Fund of Large Apparatus of Beijing Normal University for their financial support.

Author details

Dan He, Zhengping Liu* and Liyan Huang
Beijing Key Laboratory of Materials for Energy Conversion and Storage,
BNU Key Lab of Environmentally Friendly and Functional Polymer Materials,
College of Chemistry, Beijing Normal University, Beijing, PR China

*Address all correspondence to: lzp@bnu.edu.cn

IntechOpen

© 2019 The Author(s). Licensee IntechOpen. This chapter is distributed under the terms of the Creative Commons Attribution License (http://creativecommons.org/licenses/by/3.0), which permits unrestricted use, distribution, and reproduction in any medium, provided the original work is properly cited.

References

[1] Weber CC, Masters AF, Maschmeyer T. Structural features of ionic liquids: Consequences for material preparation and organic reactivity. Green Chemistry. 2013;**15**(10):2655-2679

[2] Kubisa P. Ionic liquids as solvents for polymerisation processes—Progress and challenges. Progress in Polymer Science. 2009;**34**(12):1333-1347

[3] Gunaratne HQN, Langrick CR, Puga AV, et al. Production of polyetheretherketone in ionic liquid media. Green Chemistry. 2013;**15**(5):1166-1172

[4] Earle MJ, Esperan AJSS, Gilea MA, et al. The distillation and volatility of ionic liquids. Nature. 2006;**439**(7078):831-834

[5] Chowdhury S, Mohan RS, Scott JL. Reactivity of ionic liquids. Tetrahedron. 2007;**63**(11):2363-2389

[6] Wang J, Liu Z. An efficient synthetic strategy for high performance polysulfone: Ionic liquid/zwitterion as reaction medium. Green Chemistry. 2012;**14**(11):3204-3210

[7] De Leon AC, Chen Q, Palaganas NB, et al. High performance polymer nanocomposites for additive manufacturing applications. Reactive and Functional Polymers. 2016;**103**:141-155

[8] Mittal V. High Performance Polymers and Engineering Plastics. John Wiley & Sons, Inc. Front Matter [M]; 2011. p. i-xix

[9] Nakamura M, Ishida H. Synthesis and properties of new crosslinkable telechelics with benzoxazine moiety at the chain end. Polymer. 2009;**50**(12):2688-2695

[10] Vygodskii YS, Lozinskaya EI, Shaplov AS. Ionic liquids as novel reaction media for the synthesis of condensation polymers. Macromolecular Rapid Communications. 2002;**23**(12):676-680

[11] Wang J, Liu Z. Ionic liquids as green reaction media for synthesis of poly(aryl ether ketone) s. Chinese Science Bulletin. 2013;**58**(11):1262-1266

[12] Yang Y, Wang J, Zheng J, et al. A stable anion exchange membrane based on imidazolium salt for alkaline fuel cell. Journal of Membrane Science. 2014;**467**:48-55

[13] Wang J, Wu Y, Liu Z-P. A facile and highly efficient protocol for synthesis of poly(ether sulfone)s in ionic liquid/zwitterion. Chinese Journal of Polymer Science. 2016;**34**(8):981-990

[14] Wilson D, Stenzenberger H-D, Hergenrother P-M. Polyimides. New York: Blackie; 1990

[15] Liou GS, Maruyama M, Kakimoto MA, et al. Preparation and properties of new soluble aromatic polyimides from 2,2′-bis(3,4-dicarboxyphenoxy) biphenyl dianhydride and aromatic diamines. Journal of Polymer Science Part A: Polymer Chemistry. 1998;**36**(12):2021-2027

[16] Liaw D-J, Liaw B-Y. Synthesis and properties of polyimides derived from 3,3′,5,5′-tetramethyl-bis[4-(4-aminophenoxy)phenyl] sulfone. European Polymer Journal. 1997;**33**(9):1423-1431

[17] Yoneyama M, Matsui Y. Direct polycondensation of aromatic tetracarboxylic acids with aromatic diamines in ionic liquids. High Performance Polymers. 2006;**18**(5):817-823

[18] Mansoori Y, Atghia SV, Sanaei SS, et al. New, organo-soluble, thermally stable aromatic polyimides and poly(amide-imide) based on 2-[5-(3,5-dinitrophenyl)-1,3,4-oxadiazole-2-yl]pyridine. Polymer International. 2012;**61**(7):1213-1220

[19] Kricheldorf HR, Schwarz G, Fan S-C. Cyclic polyimides—A comparison of synthetic methods. High Performance Polymers. 2004;**16**(4):543-555

[20] Vygodskii YS, Lozinskaya EI, Shaplov AS, et al. Implementation of ionic liquids as activating media for polycondensation processes. Polymer. 2004;**45**(15):5031-5045

[21] Lozinskaya EI, Shaplov AS, Vygodskii YS. Direct polycondensation in ionic liquids. European Polymer Journal. 2004;**40**(9):2065-2075

[22] Tsuda Y, Yoshida T, Kakoi T. Synthesis of soluble polyimides based on alicyclic dianhydride in ionic liquids. Polymer Journal. 2006;**38**(1):88-90

[23] Chen M, Wang S. Non-thermal polyimidization reaction using base-ionic liquid medium as a dual catalyst-solvent. RSC Advances. 2016;**6**(99):96914-96917

[24] Tamada M, Hayashi T, Ohno H. Improved solubilization of pyromellitic dianhydride and 4,4′-oxydianiline in ionic liquid by the addition of zwitterion and their polycondensation. Tetrahedron Letters. 2007;**48**(9):1553-1557

[25] Alici BK, Koytepe S, Seckın T. Synthesis of piperazine based polyimide in the presence of ionic liquids. Turkish Journal of Chemistry. 2007;**31**(6):569-578

[26] Frank H, Ziener U, Landfester K. Formation of polyimide nanoparticles in Heterophase with an ionic liquid as continuous phase. Macromolecules. 2009;**42**(20):7846-7853

[27] Akbarian-Feizi L, Mehdipour-Ataei S, Yeganeh H. Synthesis of new sulfonated copolyimides in organic and ionic liquid media for fuel cell application. Journal of Applied Polymer Science. 2012;**124**(3):1981-1992

[28] Akbarian-Feizi L, Mehdipour-Ataei S, Yeganeh H. Investigation on the preparation of new Sulfonated polyimide fuel cell membranes in organic and ionic liquid media. International Journal of Polymeric Materials. 2014;**63**(3):149-160

[29] Ghaemy M, Hassanzadeh M, Taghavi M, et al. Synthesis and characterization of trifluoromethylated poly(ether-imidazole-imide)s based on unsymmetrical diamine bearing carbazole and imidazole chromophores in ionic liquids: Study of electrochemical properties by using nanocomposite electrode. Journal of Fluorine Chemistry. 2012;**142**:29-40

[30] Mallakpour S, Kowsari E. Ionic liquids as novel solvents and catalysts for the direct polycondensation of N,N′-(4,4′-oxydiphthaloyl)-bis-L-phenylalanine diacid with various aromatic diamines. Journal of Polymer Science Part A: Polymer Chemistry. 2005;**43**(24):6545-6553

[31] Mallakpour S, Abdolmaleki A, Rostami M. Hybrid S-valine functionalized multi-walled carbon nanotubes/poly(amid-imide) nanocomposites containing trimellitimidobenzene and 4-hydroxyphenyl benzamide moieties: Preparation, processing, and thermal properties. Journal of Materials Science. 2014;**49**(21):7445-7453

[32] Zahmatkesh S. Ionic liquid catalyzed synthesis and characterization of heterocyclic and optically active poly

(amide-imide)s incorporating l-amino acids. Amino Acids. 2011;**40**(2):533-542

[33] Mallakpour S, Khania M. Synthesis and characterization of poly(amide-imide)s bearing a S-Valine moiety in molten ionic liquid. Designed Monomers & Polymers. 2011;**14**(3):221-232

[34] Mallakpour S. Synthesis of soluble poly(amide-ether-imide-urea)s bearing amino acid moieties in the main chain under green media (ionic liquid). Amino Acids. 2011;**40**(2):487-492

[35] Mallakpour S, Zadehnazari A. Synthesis and characterization of novel heat stable and Processable optically active poly(amide-imide) nanostructures bearing hydroxyl pendant group in an ionic green medium. Journal of Polymers and the Environment. 2013;**21**(1):132-140

[36] Mallakpour S, Barati A. Preparation and characterization of optically active poly(amide-imide)/TiO bionanocomposites containing N-trimellitylimido-L-isoleucine linkages: Using ionic liquid and ultrasonic irradiation. Journal of Polymer Research. 2012;**19**(2):1-8

[37] Mallakpour S, Dinari M. Reinforcement of poly(vinyl alcohol) with chiral poly(amide-imide) s nanoparticles containing S-valine under simple ultrasonic irradiation method. Colloid and Polymer Science. 2013;**291**(10):2487-2494

[38] Mallakpour S, Ahmadizadegan H. Manufacture of zinc oxide/chiral poly(amide-imide)-functionalized amino acid and thiazole bionanocomposites: Using ionic liquid and ultrasonic irradiation. Journal of Thermoplastic Composite Materials. 2015;**28**(5):672-685

[39] Xie W, Ju H, Geise GM, et al. Effect of free volume on water and salt transport properties in directly copolymerized disulfonated poly(arylene ether sulfone) random copolymers. Macromolecules. 2011;**44**(11):4428-4438

[40] Chen Y, Lee CH, Rowlett JR, et al. Synthesis and characterization of multiblock semi-crystalline hydrophobic poly(ether ether ketone)-hydrophilic disulfonated poly(arylene ether sulfone) copolymers for proton exchange membranes. Polymer. 2012;**53**(15):3143-3153

[41] Wang J, Xu Y, Zhu L, et al. Amphiphilic ABA copolymers used for surface modification of polysulfone membranes. Part 1: Molecular design, synthesis, and characterization. Polymer. 2008;**49**(15):3256-3264

[42] Temtem M, Pompeu D, Barroso T, et al. Development and characterization of a thermoresponsive polysulfone membrane using an environmental friendly technology. Green Chemistry. 2009;**11**(5):638-645

[43] Percec V, Clough RS, Rinaldi PL, et al. Termination by reductive elimination in the polyetherification of bis(aryl chlorides) activated by carbonyl groups, with bisphenolates. Macromolecules. 1991;**24**(21):5889-5892

[44] He D, Wu Y, Liu Z, et al. The synthesis of poly(phenylene sulfide sulfone) in ionic liquids at atmospheric pressure. RSC Advances. 2017;**7**(63):39604-39610

[45] Perng LH, Tsai CJ, Ling YC. Mechanism and kinetic modelling of PEEK pyrolysis by TG/MS. Polymer. 1999;**40**(26):7321-7329

[46] Roelofs KS, Hirth T, Schiestel T. Sulfonated poly(ether ether ketone)-based silica nanocomposite membranes for direct ethanol fuel cells. Journal of Membrane Science. 2010;**346**(1):215-226

[47] Mallakpour S, Rafiee Z. Efficient combination of ionic liquids and microwave irradiation as a green protocol for polycondensation of 4-(3-hydroxynaphthalene)-1,2,4-triazolidine-3,5-dione with diisocyanates. Polymer. 2007;**48**(19):5530-5540

[48] Mallakpour S, Tirgir F, Sabzalian MR. Novel biobased polyurethanes synthesized from nontoxic phenolic diol containing l-tyrosine moiety under green media. Journal of Polymers and the Environment. 2010;**18**(4):685-695

[49] Mallakpour S, Dinari M. Straightforward and green method for the synthesis of nanostructure poly(amide-imide)s-containing benzimidazole and amino acid moieties by microwave irradiation. Polymer Bulletin. 2013;**70**(3):1049-1064

[50] Mallakpour S, Zadehnazari A. Microwave-assisted synthesis and morphological characterization of chiral poly(amide–imide) nanostructures in molten ionic liquid salt. Advances in Polymer Technology. 2013;**32**(1):21333

[51] Rafiemanzelat F, Khoshfetrat SM, Kolahdoozan M. Fast and eco-friendly synthesis of novel soluble thermally stable poly(amide-imide)s modified with siloxane linkage with reduced dielectric constant under microwave irradiation in TBAB, TBPB and MeBuImCl via isocyanate method. Journal of Applied Polymer Science. 2013;**127**(4):2371-2379

[52] Li P, Zhao Q, Anderson JL, et al. Synthesis of copolyimides based on room temperature ionic liquid diamines. Journal of Polymer Science Part A: Polymer Chemistry. 2010;**48**(18):4036-4046

[53] Li P, Coleman MR. Synthesis of room temperature ionic liquids based random copolyimides for gas separation applications. European Polymer Journal. 2013;**49**(2):482-491

[54] Zhang C, Cao B, Coleman MR, et al. Gas transport properties in (6FDA-RTIL)-(6FDA-MDA) block copolyimides. Journal of Applied Polymer Science. 2016;**133**:43077

[55] Shaplov AS, Morozova SM, Lozinskaya EI, et al. Turning into poly(ionic liquid)s as a tool for polyimide modification: Synthesis, characterization and CO_2 separation properties. Polymer Chemistry. 2016;**7**(3):580-591

[56] Kubisa P. Ionic liquids in the synthesis and modification of polymers. Journal of Polymer Science Part A: Polymer Chemistry. 2005;**43**(20):4675-4683

[57] Lu J, Yan F, Texter J. Advanced applications of ionic liquids in polymer science. Progress in Polymer Science. 2009;**34**(5):431-448

[58] Branco LC, Crespo JG, Afonso CAM. Studies on the selective transport of organic compounds by using ionic liquids as novel supported liquid membranes. Chemistry—A European Journal. 2002;**8**(17):3865-3871

[59] Kim D-H, Baek I-H, Hong S-U, et al. Study on immobilized liquid membrane using ionic liquid and PVDF hollow fiber as a support for CO_2/N_2 separation. Journal of Membrane Science. 2011;**372**(1-2):346-354

[60] Yang C-C, Wey J-Y, Liou T-H, et al. A quasi solid state dye-sensitized solar cell based on poly(vinylidenefluoride-co-hexafluoropropylene)/SBA-15 nanocomposite membrane. Materials Chemistry and Physics. 2012;**132**(2-3):431-437

[61] Pilli SR, Banerjee T, Mohanty K. 1-Butyl-2,3-dimethylimidazolium hexafluorophosphate as a green solvent

for the extraction of endosulfan from aqueous solution using supported liquid membrane. Chemical Engineering Journal. 2014;**257**:56-65

[62] Yi Z, Zhu L-P, Zhang H, et al. Ionic liquids as co-solvents for zwitterionic copolymers and the preparation of poly(vinylidene fluoride) blend membranes with dominated β-phase crystals. Polymer. 2014;**55**(11):2688-2696

[63] Li C, Zhu Y, Lv R, et al. Poly(vinylidene fluoride) membrane with piezoelectric β-form prepared by immersion precipitation from mixed solvents containing an ionic liquid. Journal of Applied Polymer Science. 2014;**131**:40505

[64] Dias JC, Lopes AC, Magalh ESB, et al. High performance electromechanical actuators based on ionic liquid/poly(vinylidene fluoride). Polymer Testing. 2015;**48**:199-205

[65] Hirota Y, Maeda Y, Nishiyama N, et al. Separation of C_6H_6 and C_6H_{12} from H_2 using ionic liquid/PVDF composite membrane. AICHE Journal. 2016;**62**(3):624-628

[66] Frübing P, Wang F-P, Kühle T-F, Gerhard R. AC and DC conductivity of ionic liquid containing polyvinylidene fluoride thin films. Applied Physics A. 2016;**122**(1):1-10

[67] Plaza A, Merlet G, Hasanoglu A, et al. Separation of butanol from ABE mixtures by sweep gas pervaporation using a supported gelled ionic liquid membrane: Analysis of transport phenomena and selectivity. Journal of Membrane Science. 2013;**444**:201-212

[68] Kasahara S, Kamio E, Matsuyama H. Improvements in the CO_2 permeation selectivities of amino acid ionic liquid-based facilitated transport membranes by controlling their gas absorption properties. Journal of Membrane Science. 2014;**454**:155-162

[69] Ilconich J, Myers C, Pennline H, et al. Experimental investigation of the permeability and selectivity of supported ionic liquid membranes for CO_2/He separation at temperatures up to 125°C. Journal of Membrane Science. 2007;**298**(1-2):41-47

[70] Yoo S, Won J, Kang SW, et al. CO_2 separation membranes using ionic liquids in a Nafion matrix. Journal of Membrane Science. 2010;**363**(1-2):72-79

[71] Dai Z, Bai L, Hval KN, et al. Pebax®/TSIL blend thin film composite membranes for CO_2 separation. Science China Chemistry. 2016;**59**(5):538-546

[72] Schauer J, Sikora A, Plıškov M, et al. Ion-conductive polymer membranes containing 1-butyl-3-methylimidazolium trifluoromethanesulfonate and 1-ethylimidazolium trifluoromethanesulfonate. Journal of Membrane Science. 2011;**367**(1-2):332-339

[73] Jiang Y-Y, Zhou Z, Jiao Z, et al. SO_2 gas separation using supported ionic liquid membranes. The Journal of Physical Chemistry B. 2007;**111**(19):5058-5061

[74] Jiang Y, Wu Y, Wang W, et al. Permeability and selectivity of sulfur dioxide and carbon dioxide in supported ionic liquid membranes. Chinese Journal of Chemical Engineering. 2009;**17**(4):594-601

[75] Zhao W, He G, Zhang L, et al. Effect of water in ionic liquid on the separation performance of supported ionic liquid membrane for CO_2/N_2. Journal of Membrane Science. 2010;**350**(1-2):279-285

[76] Zhao W, He G, Nie F, et al. Membrane liquid loss mechanism of supported ionic liquid membrane for gas separation. Journal of Membrane Science. 2012;**411-412**:73-80

[77] Lan W, Li S, Xu J, et al. Preparation and carbon dioxide separation performance of a hollow fiber supported ionic liquid membrane. Industrial & Engineering Chemistry Research. 2013;**52**(20):6770-6777

[78] Mohshim DF, Mukhtar H, Man Z. The effect of incorporating ionic liquid into polyethersulfone-SAPO34 based mixed matrix membrane on CO_2 gas separation performance. Separation and Purification Technology. 2014;**135**:252-258

[79] Yi Z, Liu C-J, Zhu L-P, et al. Ion exchange and antibiofouling properties of poly(ether sulfone) membranes prepared by the surface immobilization of brønsted acidic ionic liquids via double-click reactions. Langmuir. 2015;**31**(29):7970-7979

[80] Soliveri G, Sabatini V, Farina H, et al. Double side self-cleaning polymeric materials: The hydrophobic and photoactive approach. Colloids and Surfaces A: Physicochemical and Engineering Aspects. 2015;**483**:285-291

[81] Lakshmi DS, Cundari T, Furia E, et al. Preparation of polymeric membranes and microcapsules using an ionic liquid as morphology control additive. Macromolecular Symposia. 2015;**357**(1):159-167

[82] Zhao Y, Li M, Lu Q. Tunable wettability of polyimide films based on electrostatic self-assembly of ionic liquids. Langmuir. 2008;**24**(8):3937-3943

[83] Shindo R, Kishida M, Sawa H, et al. Characterization and gas permeation properties of polyimide/ZSM-5 zeolite composite membranes containing ionic liquid. Journal of Membrane Science. 2014;**454**:330-338

[84] Nancarrow P, Liang L, Gan Q. Composite ionic liquid-polymer-catalyst membranes for reactive separation of hydrogen from carbon monoxide. Journal of Membrane Science. 2014;**472**:222-231

[85] Nagase Y, Suleimenova B, Umeda C, et al. Syntheses of aromatic polymers containing imidazolium moiety and the surface modification of a highly gas permeable membrane using the nanosheets. Polymer. 2018;**135**:142-153

[86] Abedini A, Crabtree E, Bara JE, et al. Molecular simulation of ionic polyimides and composites with ionic liquids as gas-separation membranes. Langmuir. 2017;**33**(42):11377-11389

[87] Deligöz H, Yılmazoğlu M, Yılmaztürk S, Şahin Y, Ulutaş K. Synthesis and characterization of anhydrous conducting polyimide/ionic liquid complex membranes via a new route for high-temperature fuel cells. Polymers for Advanced Technologies. 2012;**23**(8):1156-1165

[88] Kiatkittikul P, Nohira T, Hagiwara R. Advantages of a polyimide membrane support in nonhumidified fluorohydrogenate-polymer composite membrane fuel cells. Fuel Cells. 2015;**15**(4):604-609

[89] Pang H-W, Yu H-F, Huang Y-J, et al. Electrospun membranes of imidazole-grafted PVDF-HFP polymeric ionic liquids for highly efficient quasi-solid-state dye-sensitized solar cells. Journal of Materials Chemistry A. 2018;**6**(29):14215-14223

[90] Xing DY, Chan SY, Chung T-S. Fabrication of porous and interconnected PBI/P84 ultrafiltration membranes using [EMIM]OAc as the green solvent. Chemical Engineering Science. 2013;**87**:194-203

[91] Coletta E, Toney MF, Frank CW. Impacts of polymer-polymer interactions and interfaces on the structure and conductivity

of PEG-containing polyimides doped with ionic liquid. Polymer. 2014;**55**(26):6883-6895

[92] Coletta E, Toney MF, Frank CW. Effects of aromatic regularity on the structure and conductivity of polyimide-poly(ethylene glycol) materials doped with ionic liquid. Journal of Polymer Science Part B: Polymer Physics. 2015;**53**(7):509-521

[93] Lee S-Y, Yasuda T, Watanabe M. Fabrication of protic ionic liquid/sulfonated polyimide composite membranes for non-humidified fuel cells. Journal of Power Sources. 2010;**195**(18):5909-5914

[94] Delig ZH, Yılmazoğlu M. Development of a new highly conductive and thermomechanically stable complex membrane based on sulfonated polyimide/ionic liquid for high temperature anhydrous fuel cells. Journal of Power Sources. 2011;**196**(7):3496-3502

[95] Yasuda T, Nakamura S-I, Honda Y, et al. Effects of polymer structure on properties of sulfonated polyimide/protic ionic liquid composite membranes for nonhumidified fuel cell applications. ACS Applied Materials & Interfaces. 2012;**4**(3):1783-1790

[96] Chen B-K, Wu T-Y, Kuo C-W, et al. 4,4′-Oxydianiline (ODA) containing sulfonated polyimide/protic ionic liquid composite membranes for anhydrous proton conduction. International Journal of Hydrogen Energy. 2013;**38**(26):11321-11330

[97] Kowsari E, Zare A, Ansari V. Phosphoric acid-doped ionic liquid-functionalized graphene oxide/sulfonated polyimide composites as proton exchange membrane. International Journal of Hydrogen Energy. 2015;**40**(40):13964-13978

[98] Che Q, Sun B, He R. Preparation and characterization of new anhydrous, conducting membranes based on composites of ionic liquid trifluoroacetic propylamine and polymers of sulfonated poly (ether ether) ketone or polyvinylidenefluoride. Electrochimica Acta. 2008;**53**(13):4428-4434

[99] Che Q, He R, Yang J, et al. Phosphoric acid doped high temperature proton exchange membranes based on sulfonated polyetheretherketone incorporated with ionic liquids. Electrochemistry Communications. 2010;**12**(5):647-649

[100] Li D, Guo Q, Zhai W, et al. Research on properties of SPEEK based proton exchange membranes doped with ionic liquids and Y_2O_3 in different humidity. Procedia Engineering. 2012;**36**:34-40

[101] Yang WW, Lu YC, Xiang ZY, et al. Monodispersed microcapsules enclosing ionic liquid of 1-butyl-3-methylimidazolium hexafluorophosphate. Reactive and Functional Polymers. 2007;**67**(1):81-86

[102] Gao H, Xing J, Xiong X, et al. Immobilization of ionic liquid [BMIM][PF_6] by spraying suspension dispersion method. Industrial & Engineering Chemistry Research. 2008;**47**(13):4414-4417

[103] Chen D-X, Ouyang X-K, Wang Y-G, et al. Adsorption of caprolactam from aqueous solution by novel polysulfone microcapsules containing [Bmim][PF_6]. Colloids and Surfaces A: Physicochemical and Engineering Aspects. 2014;**441**:72-76

[104] Shanthana Lakshmi D, Figoli A, Fiorani G, et al. Preparation and characterization of ionic liquid polymer microspheres [PEEKWC/DMF/CYPHOS IL 101] using the phase-inversion technique. Separation and Purification Technology. 2012;**97**:179-185

[105] Bandeira P, Monteiro J, Baptista AM, et al. Tribological performance of PTFE-based coating modified with microencapsulated [HMIM] [NTf2] ionic liquid. Tribology Letters. 2015;**59**(1):1-15

[106] Imaizumi S, Ohtsuki Y, Yasuda T, et al. Printable polymer actuators from ionic liquid, soluble polyimide, and ubiquitous carbon materials. ACS Applied Materials & Interfaces. 2013;**5**(13):6307-6315

[107] Tigelaar DM, Palker AE, Meador MAB, et al. Synthesis and compatibility of ionic liquid containing rod-coil polyimide gel electrolytes with lithium metal electrodes. Journal of the Electrochemical Society. 2008;**155**(10):A768-A774

[108] Chaurasiask S, Singh RK, et al. Electrical, mechanical, structural, and thermal behaviors of polymeric gel electrolyte membranes of poly(vinylidene fluoride-co-hexafluoropropylene) with the ionic liquid 1-butyl-3-methylimidazolium tetrafluoroborate plus lithium tetrafluoroborate. Journal of Applied Polymer Science. 2015;**132**(7):41456

[109] Dong Z, Zhang Q, Yu C, et al. Effect of ionic liquid on the properties of poly(vinylidene fluoride)-based gel polymer electrolytes. Ionics. 2013;**19**(11):1587-1593

[110] Kuo P-L, Tsao C-H, Hsu C-H, et al. A new strategy for preparing oligomeric ionic liquid gel polymer electrolytes for high-performance and nonflammable lithium ion batteries. Journal of Membrane Science. 2016;**499**:462-469

[111] Wen X, Dong T, Liu A, et al. A new solid-state electrolyte based on polymeric ionic liquid for high-performance supercapacitor. Ionics. 2018;**25**(1):241-251

[112] Chen N, Xing Y, Wang L, et al. “Tai Chi” philosophy driven rigid-flexible hybrid ionogel electrolyte for high-performance lithium battery. Nano Energy. 2018;**47**:35-42

[113] Yang HM, Kwon YK, Lee SB, et al. Physically cross-linked homopolymer ion gels for high performance electrolyte-gated transistors. ACS Applied Materials & Interfaces. 2017;**9**(10):8813-8818

[114] Karuppasamy K, Reddy PA, Srinivas G, et al. An efficient way to achieve high ionic conductivity and electrochemical stability of safer nonaflate anion-based ionic liquid gel polymer electrolytes (ILGPEs) for rechargeable lithium ion batteries. Journal of Solid State Electrochemistry. 2016;**21**(4):1145-1155

[115] Raghavan P, Zhao X, Choi H, et al. Electrochemical characterization of poly(vinylidene fluoride-co-hexafluoro propylene) based electrospun gel polymer electrolytes incorporating room temperature ionic liquids as green electrolytes for lithium batteries. Solid State Ionics. 2014;**262**:77-82

[116] Yang L, Hu J, Lei G, et al. Ionic liquid-gelled polyvinylidene fluoride/polyvinyl acetate polymer electrolyte for solid supercapacitor. Chemical Engineering Journal. 2014;**258**:320-326

[117] Zhai W, Zhu H-J, Wang L, et al. Study of PVDF-HFP/PMMA blended micro-porous gel polymer electrolyte incorporating ionic liquid [BMIM]BF_4 for Lithium ion batteries. Electrochimica Acta. 2014;**133**:623-630

[118] Leones R, Costa CM, Machado AV, et al. Development of solid polymer electrolytes based on poly(vinylidene fluoride-trifluoroethylene) and the [N1 1 1 2(OH)][NTf2] ionic liquid for energy storage applications. Solid State Ionics. 2013;**253**:143-150

[119] Dahi A, Fatyeyeva K, Langevin D, et al. Polyimide/ionic liquid composite

membranes for fuel cells operating at high temperatures. Electrochimica Acta. 2014;**130**:830-840

[120] Nawaz A, Sharif R, Rhee H-W, et al. Efficient dye sensitized solar cell and supercapacitor using 1-ethyl 3-methyl imidazolium dicyanamide incorporated PVDF-HFP polymer matrix. Journal of Industrial and Engineering Chemistry. 2016;**33**:381-384

[121] Unemoto A, Ogawa H, Gambe Y, et al. Development of lithium-sulfur batteries using room temperature ionic liquid-based quasi-solid-state electrolytes. Electrochimica Acta. 2014;**125**:386-394

[122] Chang CM, Chiu JC, Lan YF, et al. High electromagnetic shielding of a 2.5-Gbps plastic transceiver module using dispersive multiwall carbon nanotubes. Journal of Lightwave Technology. 2008;**26**(10):1256-1262

[123] Chen Y, Tao J, Deng L, et al. Polyetherimide/Bucky gels nanocomposites with superior conductivity and thermal stability. ACS Applied Materials & Interfaces. 2013;**5**(15):7478-7484

[124] Cardiano P, Fazio E, Lazzara G, et al. Highly untangled multiwalled carbon nanotube@polyhedral oligomeric silsesquioxane ionic hybrids: Synthesis, characterization and nonlinear optical properties. Carbon. 2015;**86**:325-337

[125] Tunckol M, Hernandez EZ, Sarasua J-R, et al. Polymerized ionic liquid functionalized multi-walled carbon nanotubes/polyetherimide composites. European Polymer Journal. 2013;**49**(12):3770-3777

[126] Bahader A, Gui H, Li Y, et al. Crystallization kinetics of PVDF filled with multi wall carbon nanotubes modified by amphiphilic ionic liquid. Macromolecular Research. 2015;**23**(3):273-283

[127] Xing C, Zhao L, You J, et al. Impact of ionic liquid-modified multiwalled carbon nanotubes on the crystallization behavior of poly(vinylidene fluoride). The Journal of Physical Chemistry B. 2012;**116**(28):8312-8320

[128] Chen GX, Zhang S, Zhou Z, et al. Dielectric properties of poly(vinylidene fluoride) composites based on Bucky gels of carbon nanotubes with ionic liquids. Polymer Composites. 2015;**36**(1):94-101

[129] Livi S, Duchet-Rumeau J, Gerard J-F. Nanostructuration of ionic liquids in fluorinated matrix: Influence on the mechanical properties. Polymer. 2011;**52**(7):1523-1531

[130] Livi S, Duchet-Rumeau J, Gerard J-F. Application of supercritical CO_2 and ionic liquids for the preparation of fluorinated nanocomposites. Journal of Colloid and Interface Science. 2012;**369**(1):111-116

[131] Tiago G, Restolho J, Forte A, et al. Novel ionic liquids for interfacial and tribological applications. Colloids and Surfaces A: Physicochemical and Engineering Aspects. 2015;**472**:1-8

[132] Hu Y, Xu P, Gui H, et al. Effect of graphene modified by a long alkyl chain ionic liquid on crystallization kinetics behavior of poly(vinylidene fluoride). RSC Advances. 2015;**5**(112):92418-92427

[133] Xu P, Gui H, Wang X, et al. Improved dielectric properties of nanocomposites based on polyvinylidene fluoride and ionic liquid-functionalized graphene. Composites Science and Technology. 2015;**117**:282-288

[134] Xing C, Wang Y, Huang X, et al. Poly(vinylidene fluoride) nanocomposites with simultaneous

organic nanodomains and inorganic nanoparticles. Macromolecules. 2016;**49**(3):1026-1035

[135] Wang Z, Xia Y, Liu Z, et al. Conductive lubricating grease synthesized using the ionic liquid. Tribology Letters. 2012;**46**(1):33-42

[136] Fan X, Wang L. Highly conductive ionic liquids toward high-performance space-lubricating greases. ACS Applied Materials & Interfaces. 2014;**6**(16):14660-14671

[137] Fan X, Xia Y, Wang L. Tribological properties of conductive lubricating greases. Friction. 2014;**2**(4):343-353

[138] Fan X, Xia Y, Wang L, et al. Study of the conductivity and tribological performance of ionic liquid and lithium greases. Tribology Letters. 2014;**53**(1):281-291

[139] Gao L. Ionic liquid marbles. Langmuir. 2007;**23**(21):10445-10447

[140] Lee JS, Hillesheim PC, Huang D, et al. Hollow fiber-supported designer ionic liquid sponges for post-combustion CO_2 scrubbing. Polymer. 2012;**53**(25):5806-5815

[141] Stanisz E, Werner J, Matusiewicz H. Task specific ionic liquid-coated PTFE tube for solid-phase microextraction prior to chemical and photo-induced mercury cold vapour generation. Microchemical Journal. 2014;**114**:229-237

[142] Maity N, Mandal A, Nandi AK. Interface engineering of ionic liquid integrated graphene in poly(vinylidene fluoride) matrix yielding magnificent improvement in mechanical, electrical and dielectric properties. Polymer. 2015;**65**:154-167

[143] Dias JC, Correia DC, Lopes AC, et al. Development of poly(vinylidene fluoride)/ionic liquid electrospun fibers for tissue engineering applications. Journal of Materials Science. 2016;**51**(9):4442-4450

[144] Tu X, Zhang Y, Zhao T, et al. Rheological behavior of polyarylsulfone 1-Butyl-3-methylimidazolium chloride solutions. Journal of Macromolecular Science: Physics. 2006;**45**(4):665-669

[145] Xing C, Zhao M, Zhao L, et al. Ionic liquid modified poly(vinylidene fluoride): Crystalline structures, miscibility, and physical properties. Polymer Chemistry. 2013;**4**(24):5726-5734

[146] Zhu Y, Li C, Na B, et al. Polar phase formation and competition in the melt crystallization of poly(vinylidene fluoride) containing an ionic liquid. Materials Chemistry and Physics. 2014;**144**(1-2):194-198

[147] Soares BG, Pontes K, Marins JA, et al. Poly(vinylidene fluoride-co-hexafluoropropylene)/polyaniline blends assisted by phosphonium-based ionic liquid: Dielectric properties and β-phase formation. European Polymer Journal. 2015;**73**:65-74

[148] Qi H, Guo Y, Zhang L, et al. Covalently attached mesoporous silica-ionic liquid hybrid nanomaterial as water lubrication additives for polymer-metal tribopair. Tribology International. 2018;**119**:721-730

Chapter 7

On the Technology of Heterogenization of Transition Metal Catalysts towards the Synthetic Applications in Ionic Liquid Matrix

Alwar Ramani, Suresh Iyer and Murugesan Muthu

Abstract

With the invention of ionic liquids, synthetic chemistry reached a new arena towards the transition metal catalyzed reactions in the syntheses of fine, specialty, agricultural, commodity, fragrant chemicals and building blocks. Inside the ionic liquid matrix, the transition metal catalysts, when immobilized, offer a valuable solution in terms of heterogenization. This technology offers high level of recyclability without loss of activity and improves the turnover number with high selectivity. Synthetic chemists, chemical engineers and technologists continue their efforts to recover and recycle the transition metal catalysts through various methodologies to convert the processes cost effective. The processes that are reported in the literature reveals that the ionic liquids by virtue of their inertness coupled with ability to retain the catalytic materials provide an excellent solution in terms of high levels of recovery and recyclability. This chapter presents a short account on the recent development in the transition metal catalyzed reactions in ionic liquid systems where both the solvent and the catalyst were recycled and reused without any emission of volatile materials.

Keywords: immobilization, heterogenized ionic liquid, transition metal catalysts, synthetic organic chemistry, recyclability

1. Introduction

Transition metal catalysis plays a pivotal role in the production of bulk chemicals, active pharmaceutical ingredients, synthetic scaffolds and aesthetically interesting petrochemical-derived materials towards the interior decoration of our built environment; consequently, manufacturers globally started employing transition metals [1]. Transition metal catalysis is ever expanding in the field of organic synthesis owing to the way to achieve higher molecular complexity by limiting the catalyst and process cost, while the potentiality and efficiency of such methodologies cannot be attained by conventional organic synthetic pathways [2]. The process of catalysis is being exploited in the manufacture of 85% of all chemical products. Apart from the design and invention of new reactions in this field, process development is also a motivating factor since it renders highly selective organic functional group transformations in attaining structurally complex organic compounds with biological and pharmaceutical importance. Transition metals being situated at the middle part of the periodic table, either in their ionic or neutral form in solution, or *in situ*, are capable of accommodating up to nine valence shell orbitals accounting for the variable oxidation states and forms various transition states to generate hybrid molecular orbitals yielding the desired product with high chemo-, regio- and stereo-selectivity with high turnover number and turnover frequency [3]. With the advancement in the analytical techniques to understand the mechanistic pathway, transition states, and kinetics, chemists and technologists in this epoch were able to tune (through the selection of appropriate ligands and/or external additives) the catalyst according to the needs. However, the major disadvantage of the homogeneous catalytic system is that the catalyst cannot be separated easily and recycled, and also separation of product from the reaction mixture besides, in some cases, poses plethora of potential problems. In the process development of pharmaceutical drugs employing homogenous systems, the separation of the transition metallic species is very important, since if it remains in the drug molecules (product) even in the ppm level, inside our biological system, it can create undesirable side effects owing to its multivalence ability and can interfere in the biological and enzymatic process [4] and attack the immune system. Eventually, separation of the catalytic material, recovering it without losing it in the work up procedure, and subsequent reuse are the major challenges arising from the transition metal catalysis. In most of the cases, the transition metal catalyst loses its original form while quenching the reaction mixture with the treatment of water; eventually, the metal ion should be treated with new chemical procedure to acquire the original catalyst; in practice, the challenges are multifold as far as the catalyst's reuse and recyclability are concerned. On the other hand, though the heterogeneous catalytic system, separation of the catalytic material/s through a simple filtration technique is allowed, thereby the recovery and reuse are solved with ease; heterogeneous system does not offer most of the benefits that arise from that of homogeneous system; thus, the rich chemistry of transition metal catalysis remains at the top rank which is admired by academicians, chemists, technologists and entrepreneurs [5]. In the industrial processes, though the heterogeneous processes get precedence over homogeneous processes for the process of fine chemicals, in many of such processes, the undesirable side reactions give rise to unwanted products and impurities in varying proportions, and process development chemists instruct the graduate chemists to purify the products through various separation processes for instance distillation and recrystallization. Ultimately, an idea emerged in the minds of academicians to invent processes by combining the advantages of homogeneous and heterogeneous catalysis, which is the 'immobilization or heterogenization,' employing tailor-made transition metal catalytic materials. The concept of heterogenization technology has been attempted since the late 1960s [6], and it is continuously growing towards the success; now, there are many commercial processes that are operative with this technology for the production of fine chemicals to pharmaceutically important compounds. Among all the possible ways available, here in this chapter, the discussion will focus its attention on the exploitation of ionic liquids [7] as support for transition metal catalysts.

Ionic liquids are the modern technological materials made up of molten organic and inorganic salts with a melting point below 100°C, and they are loosely bound; therefore, they are liquid under conventional ambient conditions, and they can be tailor-made according to the needs and demands, so they are also known as 'Designer Solvents' and/or 'Solvents of the future'. The design of room temperature ionic liquids has been achieved through appropriate selection of weakly coordinating cations and anions; theoretically, as per the principles of crystallography, the lattice energy has to be brought down as low as possible, and hence, the design lies on the *antithesis* of assembling pure crystal/s, and/or also symmetry in the ions should be destroyed by all means. Chemists around the globe realised the importance of ionic liquids in technology; in synthetic organic chemistry, they designed innumerable classes of ionic liquids, and mathematically through various permutations and combinations, millions of ionic liquids can be synthesized. Unlike other field of research, the field and phenomenon of ionic liquids progressed in stages, and during its progress, more and more technological applications were realized. In the year 1914, the salts of alkylammonium nitrates were identified as liquid in nature , and then chloroaluminates were identified as ionic liquids. In the 1990s, aluminates were replaced with other anions such as tetrafluoroborate $[BF_4]^-$ or hexafluorophosphate $[PF_6]^-$, and then the field of ionic liquids started gaining familiarity and popularity because of convenient recyclability. After 1990, numerous ionic liquids with variety of unique anions and cations were reported in the literature, and this field received recognition in developing neoteric technologies, very importantly, synthetic organic chemists found that the ionic liquids were useful as recyclable solvent; with this discovery, this field became appealing and convincing to synthetic organic chemists. Almost all name reactions were carried out in the ionic liquids, and the bench chemist has meticulously screened and succeeded through *trial and error* or *hit and miss* methodologies. In relevance to the title of the chapter, the discussion will revolve around the use of transition metal catalysis in ionic liquids towards organic functional group transformations. As the catalytic materials derived from d-block and rare earths remain expensive and yield unwarranted side effects when consumed by human, the chemists and technologists invented a variety of techniques to immobilize the metallic catalytic systems in various ways through high levels of creativity. Among them, ionic liquids proved to be an excellent matrix to retain the charged, neutral or polar transition metal complexes; consequently, the catalyst was recycled and reused without the loss of catalytic activity and selectivity; moreover, the separation of the product seems to be easy from the reaction mixture, the product reaches the extracting solvent, and the catalyst was held through electrostatic forces in the ionic liquid matrix. Consequently, from the last decade, ionic liquids (ILs) have been employed and found to be effective for the immobilization of homogeneous catalytic materials, and new technology has emerged, which is known as supported ionic liquids phase (SILP). In SILP, mostly an IL film is immobilized on a high-surface area porous solid through either physisorption or chemisorption, and the homogeneous catalyst is immersed in an IL matrix. This unique concept has been employed for the immobilization of homogeneous catalytic materials, absorbents, and other functional materials; consequently, this concept is expected to benefit the scientific community to simplify the lives of people at large. With the added advantages, SILP in this era is much exploited in the immobilization of homogenous catalytic systems; ultimately, it will be surely impossible to picturize the complete overview of this area. In this chapter, we decided to limit our discussion to the most important contributions for brevity and conclude with recommendations, future perspectives, and outlook.

2. Supported ionic liquid phase (SILP)

In view of recycling and reusing the transition metal catalytic materials, the technology of immobilization or heterogenization was exploited, and the notable methodologies include polymer supports, macrocellular silica-based foams, magnetic Fe_3O_4, carbon nanotubes (CNT), multiwalled carbon nanotubes (MWCNT), mesoporous solids, functionalized magnetic Fe_3O_4, metal organic framework (MOF), monolithic ionogels, and SILP. Among them, the SILP technology is the most recent one; there are debates that the use of ILs is not cost-effective; however, the whole SILP system is reusable and recyclable and hence the cost can be abruptly reduced; the technology ranks to be green in nature since there is no escape of volatile organic compounds (VOCs). Metal in the form of oxides, ions, metal complexes coordinated to the cation or anion, or as nanoparticles was supported in an IL matrix through a thin film. In the SILP system, IL is a part of the support material; accordingly, the bulk properties of IL such as solvation strength, conductivity, viscosity and density are altered. ILs with hydrophobic in nature can offer excellent surface support. ILs can also act as nanoparticles stabilizers. Even though the stabilization mechanism has not been proved, ILs seem to create an electrostatic and steric barrier between nanoparticles; a stabilization type Derjaguin-Landau-Verwey-Overbeek (DLVO) has been proposed due to their polymeric structure.

2.1 SILP in macrocellular silica

The preparation of macrocellular silica is quite simple; the transition metal compounds of interest is suspended in a solution employing an ionic liquid in an volatile organic solvent. Upon stripping the volatile organic compound, the free-flowing immobilized catalytic material can be obtained where the transition metals are trapped within the SILP system. The materials thus trapped within the SILPs were found to be useful in promoting the organic functional group transformations catalyzed by transition and/or noble metals. Mainly the aryl-X activation (Mizoroki-Heck reaction) was carried out using palladium containing SILPs (in imidazolium ILs). With low palladium loading, the Heck-type coupling reaction was performed, and the reaction of iodobenzene with cylohexyl acrylate yielded the corresponding cinnamates with high turnover numbers (TON) and turnover frequency (TOF) [8]. Palladium acetate $Pd(OAc)_2$ supported on amorphous silica immobilized in [bmim] PF_6 (1-butyl-3-methylimidazolium hexafluorophosphate) was reported to be highly efficient towards the promotion of the Heck-type coupling reaction without any additional ligands, gave the trans-cinnamates in preparative yields [9], and the TON and TOF reached 68,400 and 8000 h^{-1}, respectively.

2.2 Heterogeneous ionic liquid (HIL)-encapsulated IL on silica

A variation for the synthesis of heterogeneous IL-based palladium-containing catalytic system was developed by Shi et al. [10], through a one pot synthesis of heterogeneous ionic liquid encapsulated on silica involving sol-gel method. These catalysts provided excellent activity for the carbonylation of aniline and nitrobenzene to synthesize diphenyl urea in (Pd or Rh)-EMImBF$_4$/silica gel (1-ethyl-3-methylimidazolium tetrafluoroborate) or (Pd or Rh)-DMImBF$_4$/silica gel (0.1 g, 50–60 mesh), (1-decyl-3-methylimidazolium tetrafluoroborate), or DMImBF$_4$ ionic liquid (2.5 g) containing Rh complex (1 mg) or 0.11 wt%Rh 35 wt% DMImBF$_4$/SiO$_2$ (0.1 g, for the purpose of comparison). The selectivity, yield, and turnover number (TON) were excellent. In the same publication, the identical catalytic system was proved to be excellent for the carbonylation of aniline to synthesize carbamates and also oxime transfer between cyclohexanone oxime and acetone.

Yokoyama et al. [11] reported that the Mizoroki-Heck reaction of iodobenzene and 4-methyl-iodobenzene with olefins using silica-supported palladium complex catalysts was fruitful in 1-butyl-3-methylimidazolium hexafluorophosphate ([bmim]PF_6), exhibiting higher activities than in DMF (N,N-dimethylformamide) in addition to easy product isolation and catalyst recycling. Silica-supported Pd catalysts were synthesized in 1-butylimidazolium hexafluorophosphate with low palladium loadings (0.35 and 0.08%); fascinatingly, both samples thus synthesized proved to be effective catalytic system for Heck-type coupling reactions between bromo- and chloroarenes with ethylacrylate. They also reported that the presence of electron withdrawing groups attached to the aromatic moiety increases the conversion and selectivity towards trans-selectivity. They could recycle the catalyst but pointed out that there is some loss of activity of the catalytic material upon recycling; the hot filtration measurements proved the leaching of Palladium [12]. A multiple-layered SILP system was developed for a hydroformylation process and found to be effective in the hydroformylation reactions [13].

2.3 Palladacycles on macroporous aluminosilicate

In a variation, oxime Pd cycle was derivatized to increase its ionophilicity by a tethered imidazolium group; the resulting IL forms an imidazolium complex which is supported on high surface area Al/MCM-41, aluminosilicate (1-butyl-3-methyl-imidazolium bromide is tethered onto oxime Pd cycle which is then heterogenized on Al/MCM-41). This is a solid active catalyst for the Mizoroki-Heck and Suzuki coupling. Results are moderate to good yields. Ionophilicity of the oxime Pd cycle can be significantly increased, limited by the need for basic conditions. Catalytic reuse showed increase in homocoupling [14].

2.3.1 Palladium on clay composite carrying phosphinite-functionalized ionic liquid moieties (CCPIL)

A clay composite comprising of phosphinite-functionalized ionic liquid moieties was synthesized, in which palladium nanoparticles were uniformly ligated, and this system is called as clay carrying phosphinite-functionalized ionic liquid (CCPIL) moieties or systems. The preparation involves the synthesis of an epoxy functionalized clay, which is reacted with 1-methyl imidazole and chlorodiphenyl phosphine yielding the required CCPIL. The CCPIL material thus obtained was separated by simple filtration to which later palladium was complexed. This novel catalytic material combines the advantages of homogeneous IL phase, phosphorylated ligands, and a new heterogeneous clay support for Pd catalyzed C-C bond formation in green media. The catalytic material is pictorially represented as **Figure 1**.

The heterogeneous catalyst was successfully exploited for the Suzuki coupling of aryl iodides in aqueous media to give biaryls in high yields. Bromide and chloride substrates too gave the biaryls but required longer reaction times. Recycling was repeated for five runs with little leaching observed where each run required the same reaction times. These catalysts were also used for solvent-free MH reaction. Further, copper-free Sonogashira-Hagihara reaction of aryl iodides, bromides, and chlorides with phenylacetylene was also catalyzed by these Pd on CCPIL. Recycling was also possible with little depletion, and the added advantage is that the catalyst was air and moisture stable [15].

2.4 SILP on magnetic nanoparticles

Creativity is often exploited by scientists; interestingly, iron-containing catalytic materials were developed in order to separate the catalytic material through magnetic separation techniques from the reaction mixture. In this context, few magnetically separable catalytic materials were reported in SILP, which enabled the

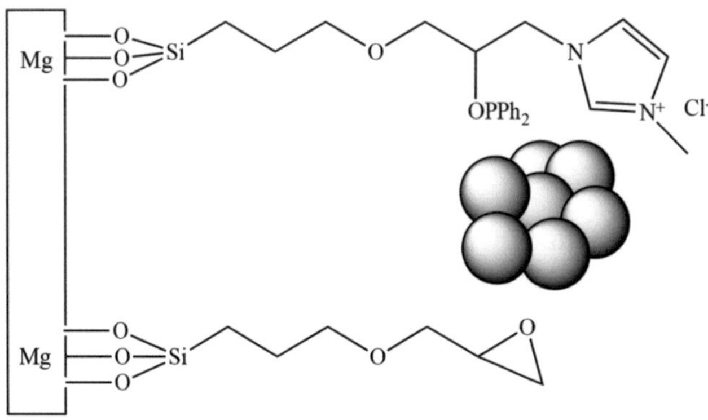

Figure 1.
Clay carrying phosphinite-functionalized ionic liquid [15].

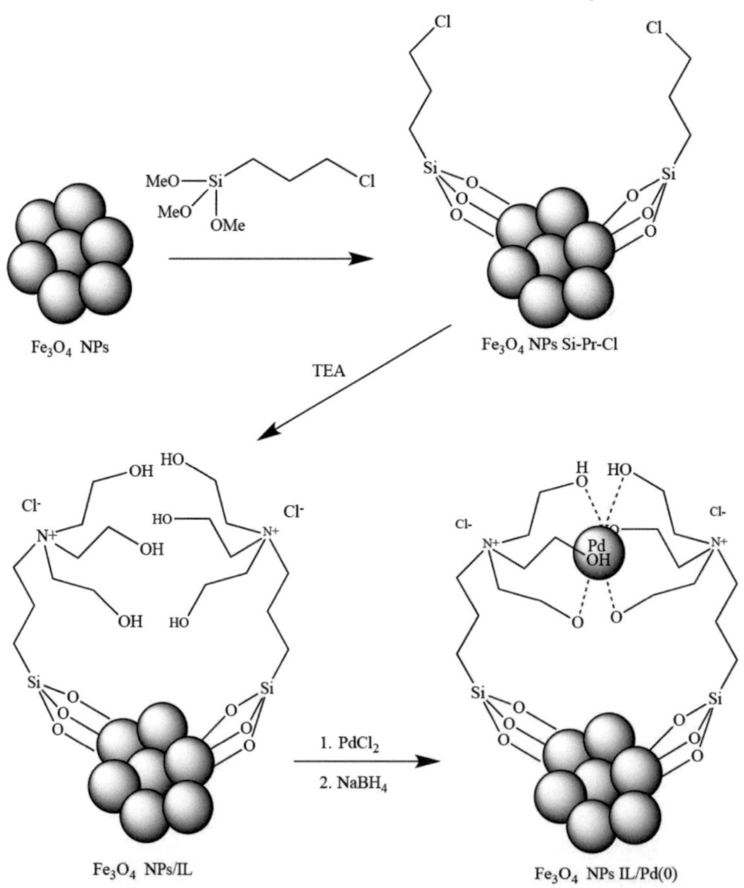

Figure 2.
SILP on magnetic nanoparticles [16].

recovery of the catalytic materials. A hybrid magnetic nanocomposite synthesis was reported with ionic nature, and the methodology was depicted as in **Figure 2**.

2.4.1 Palladium supported on Fe_3O_4

Magnetic Fe_3O_4 was functionalized with 3-chloropropyltriethoxysilane to obtain chloropropyl functionalized magnetic nanoparticles {MNPs (Fe_3O_4/Si-Pr-Cl)}. The reaction of the MNPs with triethanolamine, triethylammonium chloride made their surfaces ionic in nature. The Fe_3O_4 nanoparticles (NPs) bonded on triethylammonium chloride (IL) was then impregnated with $PdCl_2$ followed by reduction with $NaBH_4$ to give reduced Pd NP supported on the magnetic NPs (Fe_3O_4/IL/Pd). Thus, it is a multistep synthetic procedure where palladium containing NPs were immobilized on the functionalized magnetic Fe_3O_4 compound. Catalytic activity was tested for the Suzuki coupling, and high yields were obtained in the reaction between bromobenzene and phenyl boronic acid to yield the corresponding biphenyl derivatives. Reactions were carried out in ethanol water solvent system. The catalyst was separated, dried, and could be reused for eight cycles with no loss of activity [16]; the methodology is depicted in **Figure 3**.

2.4.2 Functionalized magnetic nanoparticles

Gamma (γ) Fe_2O_3 MNPs immobilized with Pd-DABCO (*1,4-diazabicyclo[2.2.2] octane*) complex were prepared where the nitrogen atoms coordinates with

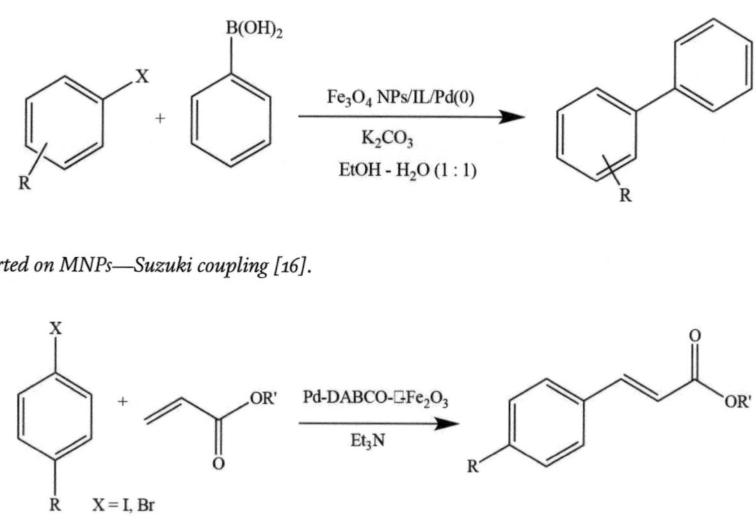

Figure 3.
Pd supported on MNPs—Suzuki coupling [16].

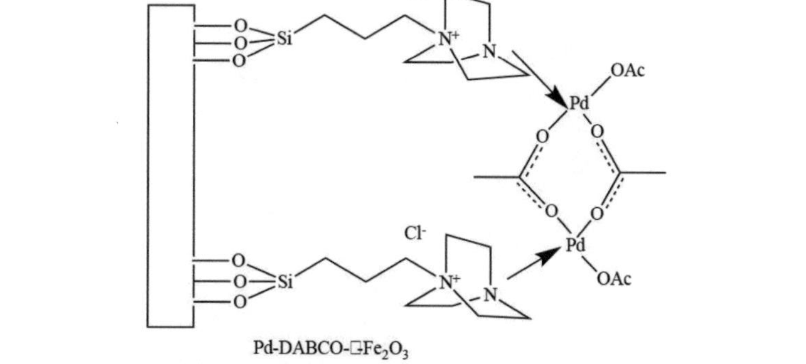

Figure 4.
Pd-DABCO supported on magnetic nanoparticles, Mizoroki-Heck reaction [17].

Palladium, forming complex (**Figure 4**). Chloro-functionalized γ-Fe_2O_3 was prepared by reacting Fe_2O_3 with chloropropyltrimethoxysilane. Treatment with DABCO gave the DABCO-γ-Fe_2O_3. Further reaction with $Pd(OAc)_2$ gave the desired Pd-DABCO-γ-Fe_2O_3. This catalyst was proven to be a recyclable catalyst for Heck-type coupling reactions.

Heck reactions were carried out with aryl iodobenzene to give high yields of the products in 30 min. Aryl bromide and chlorides also underwent the Heck-type coupling reactions but requiring longer reaction times of 12–24 h. The average isolated yield of 90% was retained for five consecutive runs [17]. An external magnet was used to separate the catalyst from the reaction mixture and reused at least five times without significant degradation in its catalytic activity. The pictorial representation of the catalytic preparation and the reaction is depicted in **Figure 4.**

2.4.3 Palladium supported on amine-functionalized magnetic nanoparticles

The issue of leaching could not be fully resolved in the SILP systems too; as a result, a convenient protocol was developed to recover the palladium catalyst magnetically. Magnetic nanoparticles (NPs) were functionalized to impart stronger bonding to the metal. Towards this end amine functionalized ILs were synthesized. Fe_3O_4 was coated with silica. This was then functionalized with (3-chloropropyl)triethoxysilane. The ionic liquid moiety was then easily anchored onto the surface of the SiO_2/Fe_3O_4 to obtain amine-functionalized ionic liquid-modified magnetic NPs (IL-$NH_2/SiO_2/Fe_3O_4$). Excellent results were achieved with this catalyst for the Suzuki coupling of various aryl iodides and bromides (**Figure 5**). The authors claimed that the catalyst was well dispersed in the reaction medium, magnetically recovered from reaction mixture, and reused for several times without significant loss of activity. All these advantages make the protocol to be a green and convenient process for other metal catalyzed important reactions [18].

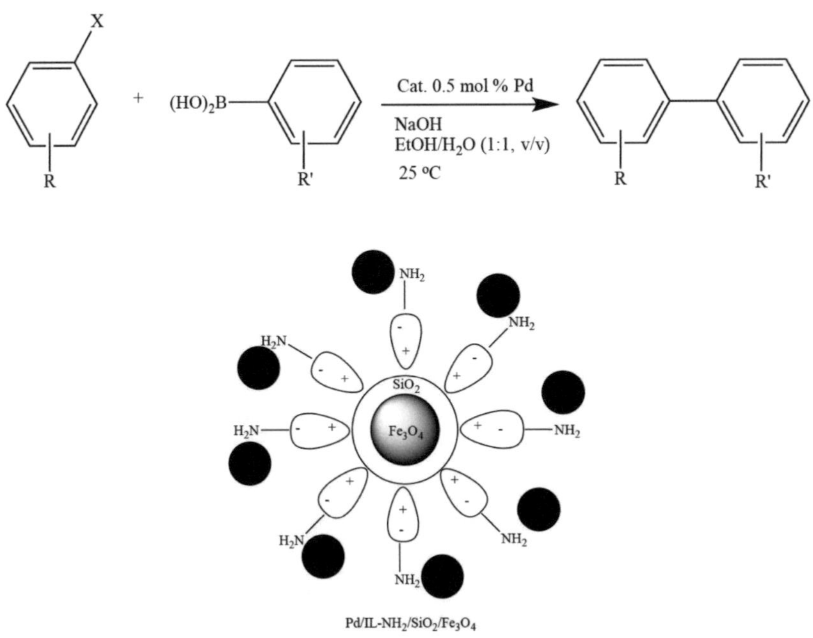

Figure 5.
Pd supported on amine-functionalized magnetic nanoparticles—Suzuki coupling [18].

2.4.4 Phosphinite-functionalized magnetic nanoparticles (MNPs)

In continuation of research on magnetite-supported palladium catalysts, the synthesis and characterization of phosphinite-functionalized magnetic nanoparticles containing imidazolium ionic liquid moiety for stabilization of palladium nanoparticles and its application as a catalyst in Suzuki-Miyaura coupling reactions were examined.

Fe_3O_4 NPs were coated with a thin layer of silica using tetraethyl orthosilicate to provide core shell Fe_3O_4 NP (SiO_2@Fe_3O_4 NP). These NPs were then treated with glycidoxypropyltrimethoxysilane to afford epoxy functionalized SiO_2@ Fe_3O_4. Further reaction with 1-methyl imidazole and chlorodiphenyl phosphine gave the Im-Phos-SiO_2-@Fe_3O_4. The catalyst was then prepared by treating with $PdCl_2$ (**Figure 6**). The catalyst was characterized by various methods. The obtained compound was characterized by SEM, TEM, EDX, solid UV, VSM, XRD, XPS, FT-IR and N_2 adsorption-desorption analyses. Aryl bromides gave high yields in EtOH:water solvent. For aryl chlorides, the solvent was changed to DMF and, at 120°C, it gave high yields. Hot filtration test was carried out and showed that the catalyst was mostly heterogeneous. The catalyst was recyclable for at least eight times with little depletion of activity [19]. The scheme route of the catalyst is presented in **Figure 6**.

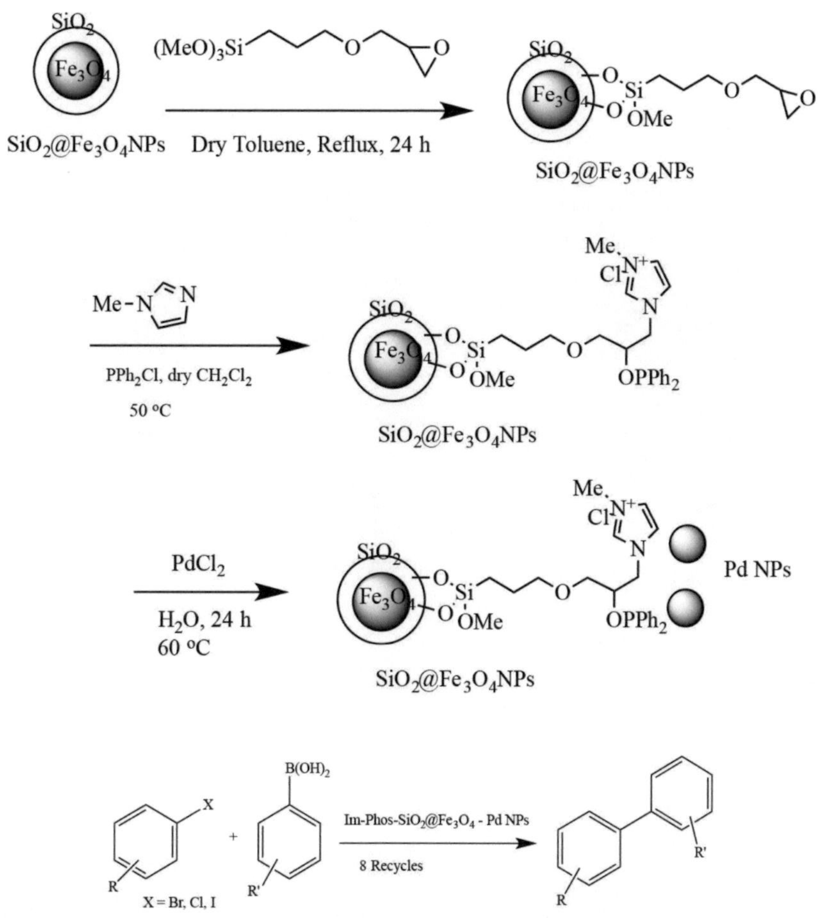

Figure 6.
Phosphinite-functionalized magnetic nanoparticles-Suzuki coupling [19].

2.4.5 Ni supported on magnetic nanoparticles (MNPs)

Ni^{2+} ion containing 1-methyl-3 (trimethyoxsilylpropyl)imidazolium chloride IL was impregnated on magnetic F_3O_4 NP. Heck reaction was conducted at 100°C (**Figure 7**). IL catalyst separated by magnet can be reused several times after washing without loss of activity. IL immobilized on magnetic particles is an excellent technology for catalytic reactions and separation technologies with substantive progress. The combination of MNPs and ILS gives a magnetic supported IL, which exhibits the dual properties of IL as well as facile separation by magnetic external field [20].

2.5 Polymer-supported HIL

A polymer-supported catalyst was prepared, which exhibited high catalytic activity. Copolymerization of 1-vinyl-3-butylimidazolium with styrene gave polymer supports to immobilize $Pd(OAc)_2$ using a method of alcohol reduction. The Pd existed as NP of less than 6 nm in these copolymers. This copolymer-supported Pd was efficient and a reusable catalyst for the Heck reaction in aqueous media in the absence of any phase transfer catalysts (PTCs) or ligands (**Figure 8**). The catalyst could be reused for three cycles without depletion of yield [21].

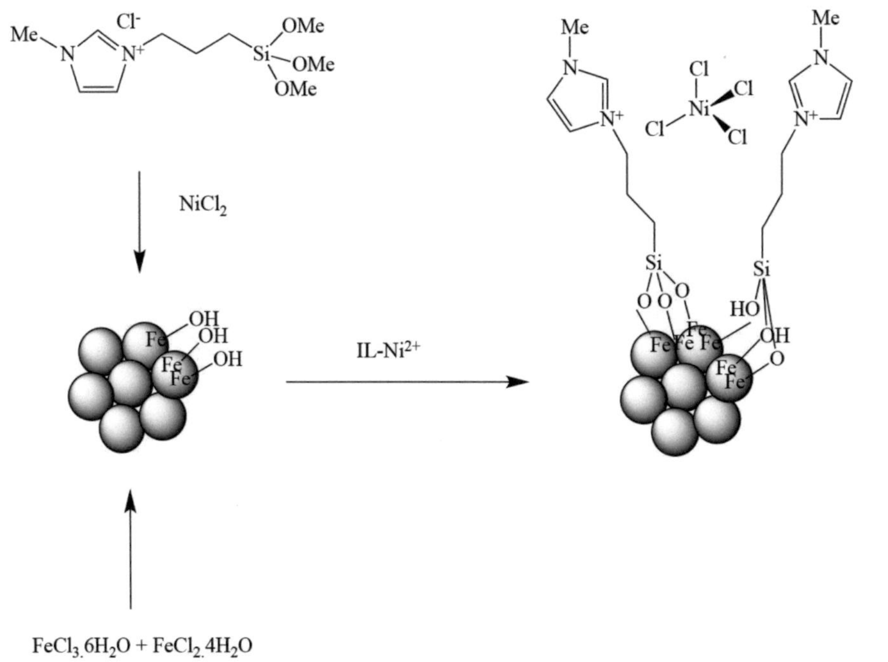

Figure 7.
Ni supported on magnetic nanoparticles [20].

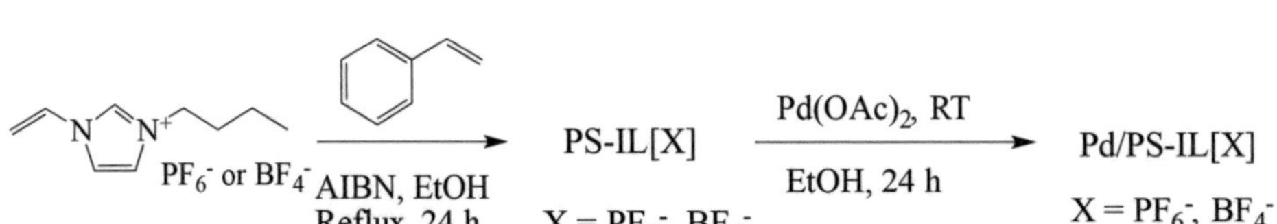

Figure 8.
Polymer-supported HIL [21].

2.6 Heterogenous ionic liquids (HILs) on carbon nanotubes (CNT) and multiwalled carbon nanotubes (MWCNT)

Ionic liquid hybrid materials comprising of either CNT or MWCNT were developed, where the CNT or MWCNT were covalently anchored with the imidazolium-based ILs. The material thus obtained as per this methodology allows specific interactions between the IL thin film and the chemical composition of the MWCNT or CNT surfaces. Typically, the catalytic material was immobilized on the IL thin film. Lee et al. [22] have functionalized MWCNTs with ionic liquid moieties. Interestingly, the change of the ionic liquid anion modulates the solubility of the nanotubes in different solvents. This property makes the modified MWCNTs soluble in ionic liquid and created an IL-based catalytic system when Pd nanoparticles are immobilized on them.

2.6.1 HIL on mesoporous solids

Palladium-containing nanoparticles were immobilized to attain heterogeneity on the mesoporous nanoparticles, namely SBA-15 using an ionic liquid, namely 1.1.3.3-tetramethylguanidinium lactate. Very interestingly, this immobilized Pd catalyst was exploited for solvent-free Heck-type coupling reactions and showed excellent activity and reusability. No deactivation was observed even after six recycles. High yields were achieved even with very low catalyst loading 0.001% of Pd, which is remarkable. Leaching occurs during the reaction, but the Pd gets redeposited on the surface at the end of the reaction with the help of TMG (1,1,3,3-tetramethylguanidine) moiety, excellent stabilizer for metallic NPs [23]. Pd acetate was immobilized on amorphous silica with the aid of an IL, namely [bmim]PF_6. This immobilized catalyst was highly efficient for the Mizoroki-Heck reaction of various aryl halides with cyclohexyl acrylate in dodecane as solvent. A TON of 68,400 and TOF of 8000 h^{-1} were achieved in the reaction of iodobenzene with cyclohexyl acrylate. The recyclability of the catalyst was lost after three cycles. Leaching studies showed loss of less than 0.24% Pd [9]. This methodology is presented in **Figure 9**.

2.6.2 Encapsulated heterogenous ionic liquids

$Pd(OAc)_2$ was immobilized on amorphous silica or alumina with the aid of an ionic liquid (Pd-SILC—Pd supported ionic liquid catalyst). The catalytic materials immobilized on N,N-diethylamino propylated (NDEAP) alumina or silica with or without bmimPF_6 gave the best results for the Suzuki-Miyaura coupling of aryl halides with arylboronic to yield the respective biphenyls. The synthesis of the encapsulated heterogeneous ionic liquid and the reaction were represented below (**Figure 10**). It was found out that the immobilization with the aid of an IL, bmimPF_6 was essential for inhibiting leaching of $Pd(OAc)_2$. The catalyst gave 95%

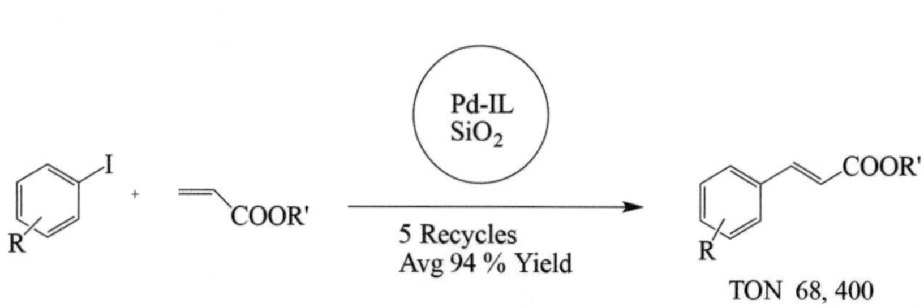

Figure 9.
HIL on mesoporous solids—Mizoroki-Heck reaction.

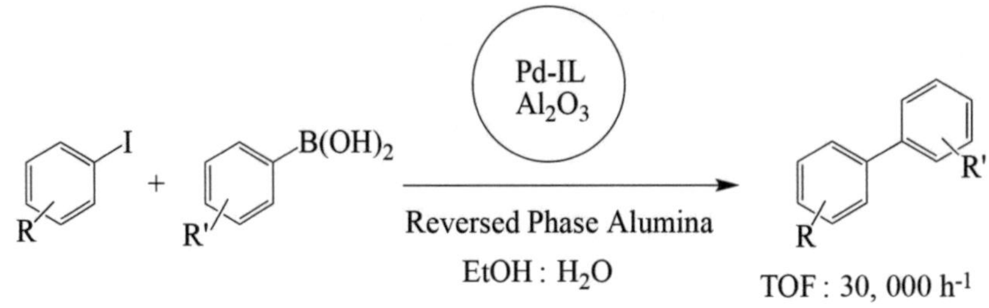

Figure 10.
Suzuki-Miyaura reaction catalyzed by immobilized Pd catalyst on reversed phase alumina [24].

average yield in reuse studies up to five times though catalyst activity was gradually lost. High efficiency of the catalyst was exhibited by a TON of 2 million and TOF of 30,000 h^{-1} in the reaction of 4-bromo acetophenone and phenyl boronic acid [24].

2.6.3 Graphene oxide-based heterogeneous ionic liquid

From the structural point of view, graphene is a single layer of sp^2 C atoms bonded in a hexagonal lattice. They have extremely large surface area, fast charge mobility, remarkably high mechanical strength, Young's modulus and chemical stability; they are inexpensive and thus are excellent candidate to disperse or immobilize catalytically active species. Silylation modification technique on graphene oxide provides graphene nanocomposites with catalytic activities. To prepare the catalyst, first, a sub-stoichiometric amount of $Pd(OAc)_2$ was reacted with an excess of 1-methyl-3-(3-(trimethoxysilyl)propyl)-1H-imidazol-3-ium chloride to afford the (NHC) N-heterocyclic carbene Pd IL. This NHC-Pd was grafted on the surface of the graphene oxide (GO) in refluxing ethanol (**Figure 11**). The catalyst was characterized by FT-IR, SEM, TEM, Raman, XPS, TGA, and EDS measurements. The catalyst provided high yields with aryl iodide and bromides but required the addition of terabutylammonium bromide (TBABr) for chloride substrates. The

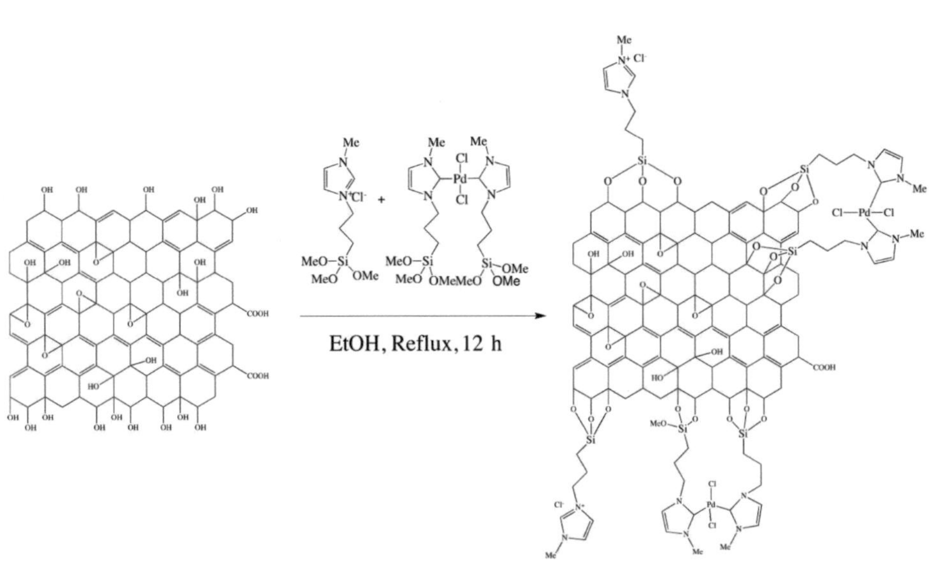

Figure 11.
HIL on graphene oxide [25].

catalyst could be recycled five times. The hot filtration test was conducted showing very little loss of Pd [25].

2.7 Silica-supported N-heterocyclic carbene-Pd

N-heterocyclic carbenes (NHCs) have been found to be excellent ligands for several aryl-X activation reactions due to the high dissociation energies of NHC-metal complexes and making them good ligands for heterogeneous systems. Polymers, silica, or NPs have been used as supports. Immobilization of Pd complexed with NHCs is a much wanted technology for the recycling of the catalyst. The immobilization was carried out by the reaction of appropriately functionalized imidazolium chlorides, triethoxysilylpropylimidazolium chloride with silica gel, or variations. The catalyst $Pd(OAc)_2$ was then anchored on this silica-supported ionic liquid. Other catalysts were supported on polymers like polystyrene, surface-grafted polystyrene resins (**Figure 12**). These could be recycled for over 10 successive runs with very high TONs exceeding 50,000 and TOFs of 5200 (**Figure 13**). Heterogeneous silica NHC-Au (I) gave good results for the Suzuki coupling of aryl iodides [26]. Pd-NHCs were immobilized on the surface of polymer or silica-coated

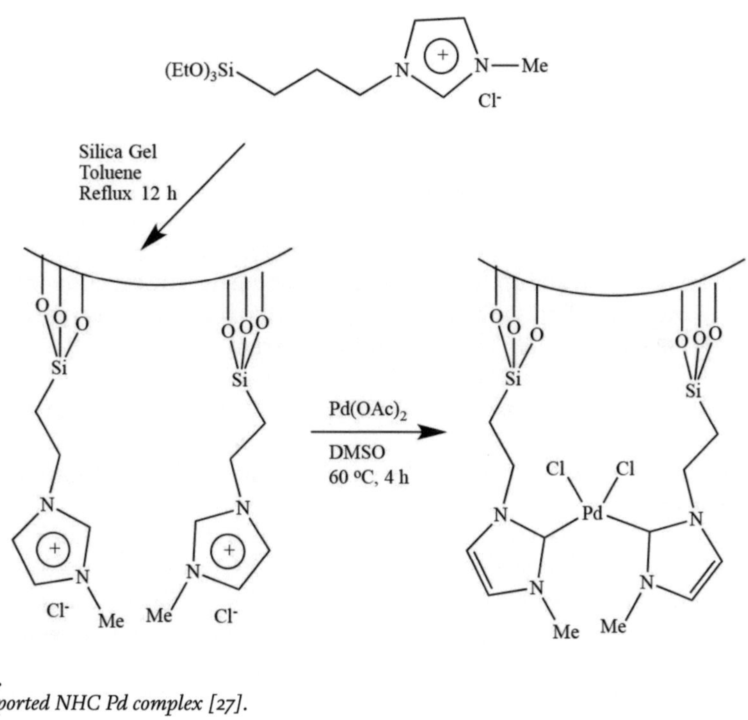

Figure 12.
Silica-supported NHC Pd complex [27].

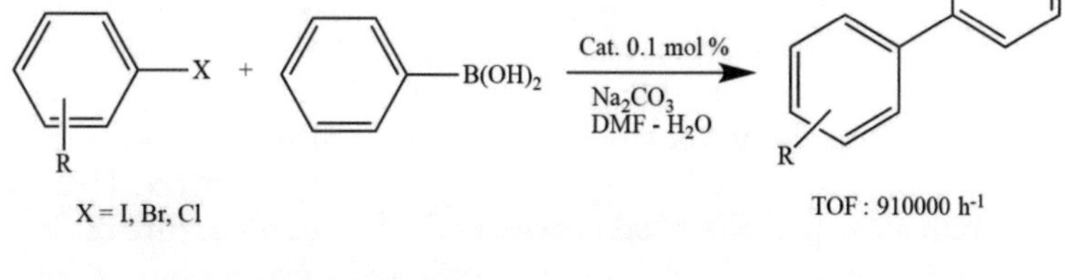

Figure 13.
Silica-supported N-heterocyclic carbene-Pd-Suzuki coupling [27].

MNPs to generate MNP-Pd-NHC complexes. Catalysts could be easily separated by external magnets [27].

The Suzuki coupling of aryl iodides was catalyzed by this heterogenized ionic liquid transition metal catalysts (HIL-TMCs), and reactions were complete in 0.5 h. Bromides were equally active while substituted aryl chlorides gave lower yields and required longer reaction times mixed with some homocoupling of the aryl boronic acids as byproduct. The catalyst could be reused for six times without significant loss of activity. Atomic absorption spectroscopy (AAS) showed no leaching of Pd [28].

3. Conclusion and recommendations

The bird's eye view on the immobilization technologies in IL matrix shows an amazing picture; using SILP technologies, a rich chemistry of organic functional transformations has been tremendously developed over the recent decades. From the point of view of recyclability and reusability, these novel technologies represent that the metallic species can be as much as possibly retained in the IL matrix. Though the success was achieved in many of the reported research articles, theoretically, academically and technically, many queries need to be addressed, and they are recommended as below:

1. The cohesive force that is operative in retaining the metallic species in IL matrix needs to be understood in detail using the principle of physical organic chemistry.

2. The nature of the catalytic material/s that is undergoing reaction in the IL matrix is not yet understood, and studies need to be carried out in this regard.

3. Not much is known about the role of IL matrix during the reaction, more answer is needed whether the IL is acting in only dissolving the reactants or alters the oxidation states of the catalytic material.

4. The ILs can be broadly classified as hydrophobic (phosphonium ILS—ionic liquid supported) and hydrophilic (nitrogen-based ILs), and hence, we need to identify whether the hydrophilic ILs can deactivate or slower the kinetics in the consecutive runs.

5. The main advantage of IL matrix is that we can carry out the reaction at elevated temperatures depending on the thermal stability of ILs but not much literature reports are known on kinetic measurements at elevated temperatures.

6. More exploration is needed to estimate the nature of side products and impurities in the recycle experiments, if any.

This multidisciplinary research has united many professionals and entrepreneurs to work together; though this chapter has highlighted on the transition metal catalysis, this surface coating of solid materials in thin film and using them in IL matrix does not limit to organic synthesis alone. The summary of the recent research article findings show tremendous advantages of the individual homogeneous and heterogeneous catalytic systems reported in IL matrix. This area of research will soon end up in chemical processes for the manufacture of fine chemicals and so on with the fundamental understanding of various physicochemical properties of SILP systems with improved activity, selectivity and recyclability.

DOI: http://dx.doi.org/10.5772/intechopen.86846

Author details

Alwar Ramani[1]*, Suresh Iyer[2]* and Murugesan Muthu[3]

1 Heriot Watt University, Edinburgh, United Kingdom

2 National Chemical laboratory, Pune, India

3 Imperial College London, United Kingdom

*Address all correspondence to: alwar.ramani@gmail.com and drsureshiyer@yahoo.com

IntechOpen

© 2019 The Author(s). Licensee IntechOpen. This chapter is distributed under the terms of the Creative Commons Attribution License (http://creativecommons.org/licenses/by/3.0), which permits unrestricted use, distribution, and reproduction in any medium, provided the original work is properly cited.

References

[1] Biffis A, Centomo P, Del Zotto A, Zecca M. Pd metal catalysts for cross-couplings and related reactions in the 21st century: A critical review. Chemical Reviews. 2018;**118**(4):2249-2295. DOI: 10.1021/acs.chemrev.7b00443

[2] Tsuji J. Transition Metal Reagents and Catalysts: Innovations in Organic Synthesis. Wiley; 2002. DOI: 10.1002/0470854766

[3] Crabtree RH. The Organometallic Chemistry of the Transition Metals. 6th ed. New York: Wiley; 2014. DOI: 10.1002/aoc.3241 ISBN: 978-1-118-13807

[4] Hosseini MJ, Jafarian I, Farahani S, Khodadadi R, Tagavi SH, Naserzadeh P, et al. New mechanistic approach of inorganic palladium toxicity: Impairment in mitochondrial electron transfer. Metallomics. 2016;**8**:252-259. DOI: 10.1039/C5MT00249D

[5] Liu L, Corma A. Metal catalysts for heterogeneous catalysis: From single atoms to nanoclusters and nanoparticles. Chemical Reviews. 2018;**118**(10):4981-5079. DOI: 10.1021/acs.chemrev.7b00776

[6] Kemball C, Dowden DA, Scurrell MS. Heterogenized homogeneous catalysts. Catalysis. 1978;**2**:215-242. DOI: 10.1039/9781847553157

[7] Welton T. Room-temperature ionic liquids. Chemical Reviews. 1999;**99**(8):2071-2084. DOI: 10.1021/cr980032t. PMID 11849019

[8] Brun N, Hesemann P, Laurent G, Sanchez C, Birot M, Deleuze H, et al. Macrocellular Pd@ionic liquid@organo-Si(HIPE) heterogeneous catalysts and their use for heck coupling reactions. New Journal of Chemistry. 2013;**37**:157-168. DOI: 10.1039/C2NJ40527J

[9] Hagiwara H, Sugawara Y, Isobe K, Hoshi T, Suzuki T. Immobilization of $Pd(OAc)_2$ in ionic liquid on silica: Application to sustainable Mizoroki-heck reaction. Organic Letters. 2004;**6**(14):2325-2328. DOI: 10.1021/ol049343i

[10] Shi F, Zhang Q, Li D, Deng Y. Silica-gel-confined ionic liquids: A new attempt for the development of supported nanoliquid catalysis. Chemistry: A European Journal. 2005;**11**(18):5279-5288. DOI: 10.1002/chem.200500107

[11] Okubo K, Shirabi M, Yokoyama C. Heck reactions in a non-aqueous ionic liquid using silica supported palladium complex catalysts. Tetrahedron Letters. 2002;**43**(39):7115-7118. DOI: 10.1016/S0040-4039(02)01320-5

[12] Bucsi I, Mastalir A, Molnar A, Levent K, Juhasz L, Kunfi A. Heck coupling reactions catalysed by Pd particles generated in silica in the presence of an ionic liquid. Structural Chemistry. 2016;***28***(2):501-509. DOI: 10.1007/s11224-016-0892-9

[13] Mehnert P, Cook RA, Dispenziere NC, Afeworki MJ. Supported ionic liquid catalysis—A new concept for homogeneous hydroformylation catalysis. Journal of the American Chemical Society. 2002;**124**(44):12932-12933. DOI: 10.1021/ja0279242

[14] Corma A, García H, Leyva A. An imidazolium ionic liquid having covalently attached an oxime carbapalladacycle complex as ionophilic heterogeneous catalysts for the heck and Suzuki–Miyaura cross-coupling. Tetrahedron. 2004;**60**(38):8553-8560. DOI: 10.1016/j.tet.2004.06.121

[15] Firouzabadi H, Iranpoor N, Ghaderi A, Gholinejad M, Rahimi S,

Jokar S. Design and synthesis of a new phosphinite-functionalized clay composite for the stabilization of palladium nanoparticles. Application as a recoverable catalyst for C■C bond formation reactions. RSC Advances. 2014;**4**(53):27674-27682. DOI: 10.1039/C4RA03645J

[16] Veisi H, Pirhayati M, Kakanejadifar A. Immobilization of palladium nanoparticles on ionic liquid-triethylammonium chloride functionalized magnetic nanoparticles: As a magnetically separable, stable and recyclable catalyst for Suzuki-Miyaura cross-coupling reactions. Tetrahedron Letters. 2017;**58**(45):4269-4276. DOI: 10.1016/j.tetlet.2017.09.078

[17] Sobhani S, Pakdin-Parizi Z. Palladium-DABCO complex supported on γ-Fe_2O_3 magnetic nanoparticles: A new catalyst for CC bond formation via Mizoroki-heck cross-coupling reaction. Applied Catalysis A: General. 2014;**479**:112-120. DOI: 10.1016/j.apcata.2014.04.028

[18] Wang J, Xu B, Sun H, Song G. Palladium nanoparticles supported on functional ionic liquid modified magnetic nanoparticles as recyclable catalyst for room temperature Suzuki reaction. Tetrahedron Letters. 2013;**54**(3):238-241. DOI: 10.1016/j.tetlet.2012.11.009

[19] Gholinejad M, Razeghi M, Ghaderi A, Biji P. Palladium supported on phosphinite functionalized Fe_3O_4 nanoparticles as a new magnetically separable catalyst for Suzuki–Miyaura coupling reactions in aqueous media. Catalysis Science & Technology. 2016;**6**(9):3117-3127. DOI: 10.1039/C5CY00821B

[20] Safari J, Zarnegar Z. Ni ion-containing immobilized ionic liquid on magnetic Fe_3O_4 nanoparticles: An effective catalyst for the heck reaction. Comptes Rendus Chimie. 2013;**16**(9):821-828. DOI: 10.1016/j.crci.2013.03.018

[21] Qiao K, Sugimura R, Bao Q, Tomida D, Yokoyama C. An efficient heck reaction in water catalyzed by palladium nanoparticles immobilized on imidazolium–styrene copolymers. Catalysis Communications. 2008;**9**(15):2470-2474. DOI: 10.1016/j.catcom.2008.06.016

[22] Rodriguez PL. PhD Thesis. University of Toulouse; 2009. ethesis.inp-14 Dec 2009 toulouse.fr/archive/00001063/01/rodriguez_perez.pdf

[23] Ma X, Zhou Y, Zhang J, Zhu A, Jiang T, Han B. Solvent-free heck reaction catalyzed by a recyclable Pd catalyst supported on SBA-15 via an ionic liquid. Green Chemistry. 2008;**10**(1):59-66. DOI: 10.1039/B712627A

[24] Hagiwara H, Ko KH, Hoshi T, Suzuki T. Supported ionic liquid catalyst (Pd-SILC) for highly efficient and recyclable Suzuki–Miyaura reaction. Chemical Communications. 2007;**27**:2838-2840. DOI: 10.1039/B704098A

[25] Movahed SK, Esmatpoursalmani R, Bazgir A. N-heterocyclic carbene palladium complex supported on ionic liquid-modified graphene oxide as an efficient and recyclable catalyst for Suzuki reaction. RSC Advances. 2014;**4**(28):14586. DOI: 10.1039/c3ra46056h

[26] Gholinejad M, Razeghi M, Ghaderi A, Biji P. Palladium supported on phosphinite functionalized Fe_3O_4 nanoparticles as a new magnetically separable catalyst for Suzuki-Miyaura coupling reaction in aqueous media. Catalysis Science & Technology. 2016;**6**(9):3117-3127. DOI: 10.1039/c5cy00821b

[27] Qiu H, Sarkar SM, Lee D-H, Jin M-J. Highly effective silica

gel-supported N-heterocyclic carbene–Pd catalyst for Suzuki–Miyaura coupling reaction. Green Chemistry. 2008;**10**(1):37-40. DOI: 10.1039/b712624g

[28] Ranganath KVS, Onitsuka S, Kumar AK, Inanaga J. Recent progress of N-heterocyclic carbenes in heterogeneous catalysis. Catalysis Science & Technology. 2013;**3**(9):216. DOI: 10.1039/c3cy00118k

Chapter 8

Pre-Screening of Ionic Liquids as Gas Hydrate Inhibitor via Application of COSMO-RS for Methane Hydrate

Muhammad Saad Khan and Bhajan Lal

Abstract

Ionic liquids (ILs) due to their potential dual functionality to shift hydrate equilibrium curve and retard hydrate nucleation are considered as a very promising gas hydrate inhibitor. However, experimental testing alone is insufficient to examine all potential ILs combinations due to a high number of cation and anion to form ILs. In this context, four fundamental properties of IL-hydrate system, namely, sigma profile, hydrogen bonding energies, activity coefficient, and solubility, were stimulated through conductor-like screening model for real solvent (COSMO-RS). ILs were then analyzed to determine if they can be correlated with IL inhibition ability. Among them, sigma profile and hydrogen bonding energies, which later upgraded to total interaction energies, exhibit a significant relationship with IL inhibition ability. Total interaction energies of ions, on the other hand, have successfully been applied to develop a model. The model can predict the thermodynamic inhibition ability in terms of average temperature depression. The correlation was further validated with experimental values from literature with an average error of 20.49%. Finally, using sigma profile graph and developed correlation, the inhibition ability of 20 ammonium-based ILs (AILs) have been predicted. Tetramethylammonium hydroxide (TMA-OH), due to its short alkyl chain length cation and highly electronegative anion, has shown the most promising inhibition ability among the considered system.

Keywords: ionic liquids, gas hydrate, thermodynamic inhibitors, COSMO-RS, methane hydrate

1. Introduction

Gas hydrates are icelike crystalline solid compounds that could form in the presence of water and gas under favorable thermodynamic temperature-pressure condition [1]. At low-temperature and high-pressure conditions, water molecules (host) will surround the gas molecules (guest) and encapsulate the gas in a hydrogen-bonded solid lattice [2]. Depending on the gases trapped, different structures of gas hydrates can be formed. The structure I hydrate trapped methane (CH_4), ethane (C_2H_6), and carbon dioxide (CO_2) gases. Structure II usually forms for propane (C_3H_8) gas, while a mixture of CH_4 and butane (C_4H_{10}) and other hydrocarbons can be captured by structure H hydrates [3].

IntechOpen

In recent decades, hydrates have received plenty of attention, because of its potential to capture and store gas [4–21]. Also, it is discovered that gas hydrates located in subsea as well as permafrost region are a potential source of energy too [15]. However, the formation of natural gas hydrates in oil and gas pipeline is never applauded [4, 15, 18, 22]. This is because hydrate formation in pipelines has resulted in blockage and affected flow assurance of natural gas [23]. In spite of the economic losses caused by the blockage, ecological disasters could occur in severe cases too [24]. To prevent hydrate formation, several methods including isobaric thermal heating, water removal, depressurization, and chemical inhibitor injection [25] have been implemented. The three former methods, however, are not feasible and costly. As a result, chemical inhibitors have been researched and developed a lot in recent years to control the growth of hydrates.

There are generally three types of inhibitors, which are thermodynamic hydrate inhibitor (THI), kinetic hydrate inhibitor (KHI), and anti-agglomerates (AA). THI prevents the formation of the hydrate by shifting the thermodynamic equilibrium curve of gas hydrate to a lower temperature and higher pressure [25]. KHI, on the other hand, does not inhibit hydration formation, but it slows down their nucleation and growth of hydrate. It works on the principle of lengthening the formation time of hydrate to be longer than the residence time of the gas in pipelines [26]. Finally, AA, also a low-dosage inhibitor, allows the formation of hydrate but, through perturbation of water molecules, prevents the hydrate molecules from accumulating and growing larger [27].

Some common THI inhibitors include methanol and sodium chloride. To be effective, THI normally needs to be injected in a high concentration of around 10–50 wt% [28], which leads to high operational cost. Furthermore, sodium chloride corrodes oil and gas pipelines [29]. While KHI inhibitors were able to work effectively at a lower dosage (<1 wt%), Kelland reported that as exploration operation goes into the deeper sea, KHI still has to work together with THI to effectively inhibit hydrate formation [27]. These limitations signify that existing chemical inhibitors are still not performing well, and there is a strong need to develop more effective inhibitor [29, 30].

This leads the oil and gas industry toward ILs which was initially introduced as inhibitors by Chen et al. [31] in 2008, as the team discussed the effect of 1-butyl-3-methylimidazolium tetrafluoroborate in inhibiting CO_2 hydrate formation. A year later, Xiao and Adidharma [29] suggested the dual function of ILs inhibitors. The results showed that IL is not only able to shift the hydrate thermodynamic equilibrium curve, but it also retards the formation of the hydrate. Since then, numerous experimental works have been carried out to study the effect of ILs in inhibiting gas hydrates formation, mainly using imidazolium- and pyridinium-based ILs [2, 3, 25, 32]. The targeted ILs of this context are ammonium-based ILs (AILs), which are cheaper and easier to synthesis, but not being studied intensively. Therefore, due to cost economics and more environmentally friendly, AILs are chosen to be studied in this work.

To date, all the testing work of ILs effectiveness is done using an experimental method, which is by measuring the average depression temperature for thermodynamic hydrate inhibitors and by measuring induction time for kinetic hydrate inhibitors. There are generally no other methods available to validate the experimental work or to pre-screen ILs in a shorter time. Due to this reason, it is very desirable if a theoretical method to predict ILs effectiveness as hydrate inhibitors could be established just by analyzing their fundamental properties. And to obtain these fundamental properties, COSMO-RS, a thermodynamic properties predictive tool, is the best option available in the market.

For this purpose, COSMO-RS, which can estimate the fundamental properties of ILs system, has been selected. COSMO-RS is a novel method to predict the thermodynamic properties of ILs based on quantum chemistry model [33]. COSMO-RS first calculates the charge density of individual molecules based on the structure of each molecule [34]. The charge density will then be distributed onto the entire molecule surface. This distribution will then be described by a one-dimensional probability density [35], or more famously known as sigma profile, $P(\sigma)$. Lastly, from the charge density, chemical potential, μ, will be calculated, and it will act as the basis for all other calculations to predict thermodynamic properties such as Henry's law constant and activity coefficient [36]. The calculated properties will then try to be correlated to IL inhibition ability to develop a prediction model that could predict the inhibition ability of ILs.

Throughout the years, COSMO-RS model has been successfully applied in numerous works to predict the thermodynamic properties of systems containing ILs, such as liquid-liquid equilibrium [37, 38] and activity coefficient [34, 39]. Therefore, this has prompted a lot of screening efforts of ILs through COSMO-RS for different purposes such as determining extraction solvent and improving separating process [37, 40–42]. Grabda et al. [43], for example, has used COSMO-RS to carry out a screening process for ILs that is used as an extraction solvent for neodymium chloride and dysprosium chloride. Kurnia and Mutalib [44], on the other hand, had screened imidazolium-based ILs for the separation process of benzene from n-hexane through COSMO-RS. Other than screening work, comparison and validation work have been conducted too. Calvar et al. [37], for instance, have compared COSMO-RS prediction of LLE values of ILs with their experimental data and found out that the result is satisfactory. In 2007, Palomar et al. [45] reinforced the applicability of COSMO-RS in predicting density and molar volume of imidazolium-based IL when their predicted values laid close to the experimental data.

To support the application of COSMO-RS in this work, it is found out that many other applications involving ammonium-based and bionic ILs have already been conducted through COSMO-RS [43]. In 2010, Sumon and Henni [46] performed a COSMO-RS study on the properties of ILs for CO_2 capture. In this study, 12 ammonium-based cations such as tetramethylammonium (TMA), tetraethylammonium, and tetrabutylammonium (TBS) cations are used to derive ammonium-based ILs to be studied. In 2014, Grabda et al. [43] studied the effectiveness of 4400 ILs for $NdCl_3$ and $DyCl_3$ extraction. Among the many cations used are tetra-n-butylammonium, tetraethylammonium, tetramethylammonium, etc. Dodecyl-dimethyl-3-sulfopropylammonium cation, which is a type of ammonium-based cation, was concluded as the best performing cation in decreasing the chemical potential of $NdCl_3$ and $DyCl_3$, thus increasing their solubility and easing the extraction process. In the same year, Pilli et al. [47] screened out the best ILs to extract phthalic acid from aqueous solution using COSMO-RS. Although ammonium-based cation ILs in this simulation do not give the highest selectivity, they, however, have the highest activity coefficient. Next, through COSMO-RS, Machanová et al. [48] also obtained well-predicted values of excess molar volumes and excess enthalpy for N-alkyl-triethylammonium-based ILs.

As it observed from literature, screening of ILs for gas hydrate inhibition through COSMO-RS is a relatively new and fresh concept, yet, based on the successfulness of previous works [6, 18, 21, 22, 37] in predicting thermodynamic properties which provide the way for this current work.

2. Methodology

The research methodology comprises several activities described in even detailed and specific manner.

2.1 Extracting experimental IL inhibition ability

As a relatively new study, it is very important to gain acknowledgment and recognition from peers. Hence, as mentioned earlier, the experimental value of IL inhibition ability will be obtained from several past studies that are highly recognized. For instance, paper from Xiao et al. is chosen as it is the pioneer of IL inhibitor research. The full list of papers that were chosen for development or correlation and later for validation work is shown in **Table 1**.

As observed from table, experimental values from four papers will be collected. All of them studied hydrate formation in the presence of methane gas for the thermodynamic hydrate inhibitor. In all these papers, the effectiveness of an IL as THI was reported in the form of IL-hydrate equilibrium curve. Generally, a larger temperature depression signifies that the IL is good in inhibiting and shifting the equilibrium curve. However, since IL-hydrate equilibrium curve is not quantifiable and thus is not possible to develop correlation, average temperature depression will be used to represent IL inhibition ability in our work. This average temperature depression value can be calculated through the following equation [9, 13, 30]:

$$\overline{T} = \frac{\Sigma \Delta T}{n} = \frac{\sum_{i=1}^{n}(T_{0,\text{pi}} - T_{1,\text{pi}})}{n} \quad (1)$$

where $T_{0,\text{pi}}$ is the dissociation temperature of methane in a blank sample without IL and $T_{1,\text{pi}}$ is the dissociation temperature of methane in a sample with IL inhibitor. The values of both dissociation temperatures should be obtained from

No.	Authors	Gas	Tested for
1.	Xiao et al. [30]	CH_4	THI
2.	Sabil et al. [25]	CH_4	THI
3.	Keshavarz et al. [50]	CH_4	THI
4.	Zare et al. [51]	CH_4	THI

Table 1.
Chosen papers for experimental values for this work.

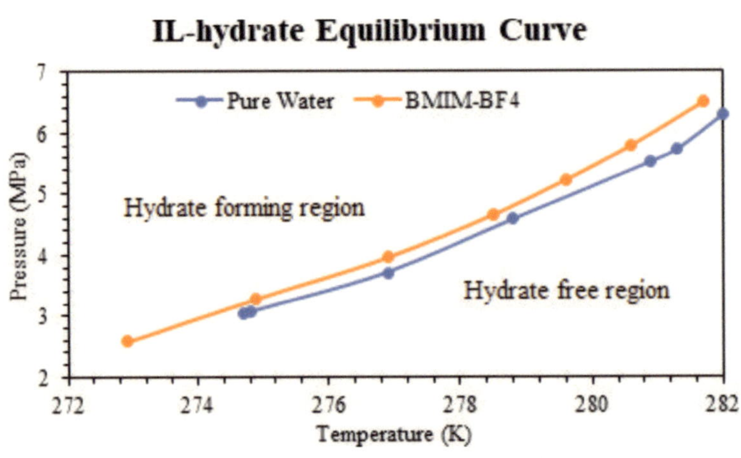

Figure 1.
Hydrate-IL equilibrium curve from the work of Keshavarz et al.

the same p_i, and n refers to the number of pressure point considered. For example, **Figure 1** shows the IL-hydrate equilibrium curve from Keshavarz et al. [49] for blank hydrate system (without IL) and hydrate system with 1-butyl-3-methyl-imidazolium tetrafluoroborate (BMIM-BF_4). Now, it is seen that with IL that acts as an inhibitor, the region of hydrate formation has reduced. It is also clear that the favorable pressure for hydrate to form has increased and the favorable temperature has reduced. This, in turn, made it hard for hydrate to form. Now to calculate average temperature depression, for instance, at 4 MPa, $T_{0,\,pi}$ is equal to the temperature of blank hydrate without IL; the temperature would be around 277.5 K. On the other hand, $T_{1,\,pi}$ that refers to the temperature of IL-hydrate system will be around 277 K. The difference between these two values is then the temperature depression. Several temperature depression values will be collected at different pressure points along the curve. Lastly, the average of these values will become the average temperature depression value.

2.2 Simulation of fundamental properties value in COSMO-RS

After obtaining the data of IL inhibition ability, now it is the time to collect another set of data, which is the fundamental property value of IL-hydrate system. Here, COSMO-RS software will be used to carry out the simulation. In COSMO-RS, all calculation works are performed based on density functional theory (DFT), utilizing the triple-zeta valence polarized (TZVP) basis set [50]. **Figure 2** shows the entire computational method of COSMO-RS.

As regards **Figure 2**, COSMO-RS first requires the input of molecular structure [51]. After this, the charge density of a segment on each molecule surface will be calculated in a virtual conductor. The distribution of this charge density on the entire surface of the molecule will then generate a sigma profile (σ-profile) through the use of COSMOtherm software [52]. Then, the σ-profile will now be used as the basis by COSMO-RS to predict the desired thermodynamic properties. Nevertheless, it is to be noted that among the computational process being shown in **Figure 2**, a user is only required to insert the input, while all the computational process will be carried out by the software itself. Therefore, it is utmost important to input the right information to extract the desired output.

The input or simulation method of COSMO-RS in this work has been conducted by referring to the work of Kurnia et al. [39, 53]. **Figure 3** shows the required input for calculating hydrogen bonding value before a proper simulation could be run.

As observed from **Figure 3**, the required inputs are temperature and the mole fraction of IL-hydrate system. For this work, the temperature is fixed at

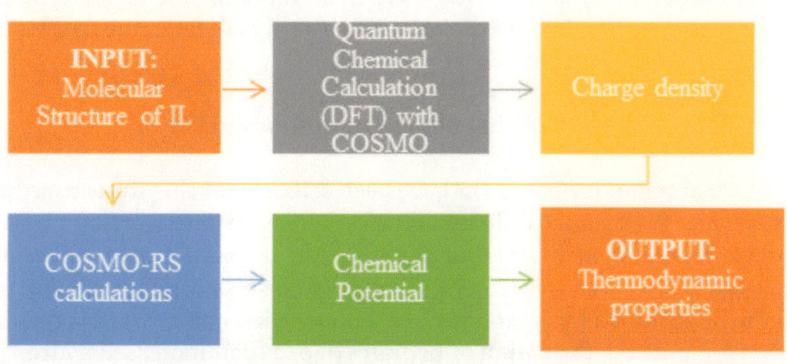

Figure 2.
Flowchart of predicting thermodynamic properties through COSMO-RS.

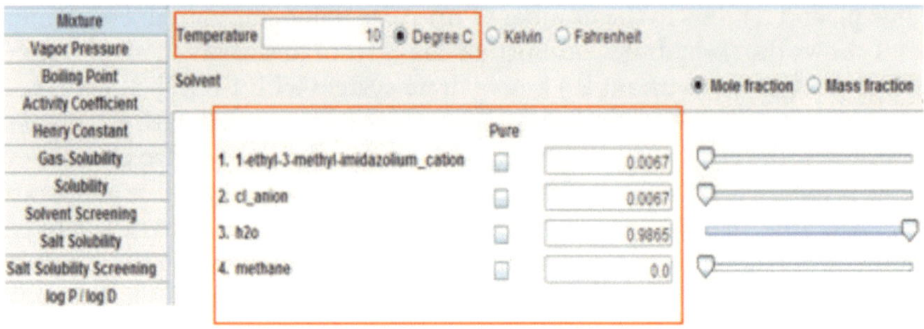

Figure 3.
Inputs required to run an IL-hydrate system simulation in COSMO-RS.

The molar mass of BMIM-BF_4	226.03g/mol
Molar mass of water	18g/mol
Mole of water	90g / (18g/mol*) = 5mol
Mole of BMIM-BF_4	10g / (226.03g/mol) = 0.044mol
Mol fraction of water	5/(5+0.044) = 0.9912
Mol fraction of BMIM-BF_4	0.044/(5+0.044)=0.0088
Mol fraction of anion / cation	0.0088/2=0.0044

**Assuming 100 g of mixture and IL is inserted at a mass fraction of 10 wt%, then 90 g will be water and 10 g will be IL.*

Table 2.
Example of calculation of mole fraction.

10°C, which is the normal temperature where hydrate will start to form. The effect of temperature is also proven not to be significant in this work, which will be explained later in the section of result and discussion. Next, the right value of mole fraction has to be entered for all four components including cation, anion, water, and involved gas. These mole fraction values need to be calculated beforehand as shown in **Table 2**. Similar to an experimental method that has been carried out by the chosen papers [25, 30, 49, 54], this simulation also considers that IL is inserted into the water at a mass fraction 10 wt%. Besides, since COSMO-RS considers IL is made up of equimolar cation and anion, a mole of IL will be divided equally into half a mole of cation and half a mole of the anion in the calculation [36, 55, 56].

When all inputs are inserted, the simulation can now be run. Similar simulation method is applied for all other desired properties including sigma profile, activity coefficient, and solubility of IL in water. When all fundamental property value is collected, the next step is the identification of pattern and, later, the development of correlation using multiple regression analysis.

2.3 Prediction of inhibition ability of ammonium-based ILs

In total, 20 ammonium-based ILs have been selected for this study based on literature review. For cations, only shorter alkyl chains cations starting from tetramethylammonium up to tetrabutylammonium cations are chosen because longer cations are not effective [29, 30]. This might be because shorter alkyl chains are easier to be adsorbed by crystal surface. Longer alkyl chain, on the other hand, might even promote the formation of hydrates due to their increased hydrophobicity to react with water [57]. On the other hand, anions are made up of halide group

No.	Name of IL	Molecular Formula
Ammonium based ionic liquids (AILs)		
1	Tetramethylammonium hydroxide (TMA-OH)	$C_4H_{13}NO$
2	Tetraethylammonium hydroxide (TEA-OH)	$C_8H_{21}NO$
3	Tetrapropylammonium hydroxide (TPA-OH)	$C_{12}H_{29}NO$
4	Tetrabutylammonium hydroxide (TBA-OH)	$C_{16}H_{37}NO$
5	Tetramethylammonium tetrafluoroborate (TMA-BF$_4$)	$C_4H_{12}BF_4N$
6	Tetraethylammonium tetrafluoroborate (TEA-BF$_4$)	$C_8H_{20}BF_4N$
7	Tetrapropylammonium tetrafluoroborate (TPA-BF$_4$)	$C_{12}H_{28}BF_4N$
8	Tetrabutylammonium tetrafluoroborate (TBA BF$_4$)	$C_{16}H_{36}BF_4N$
9	Tetramethylammonium chloride (TMA-Cl)	$C_4H_{12}ClN$
10	Tetraethylammonium chloride (TEA-Cl)	$C_8H_{20}ClN$
11	Tetrapropylammonium chloride (TPA-Cl)	$C_{12}H_{28}ClN$
12	Tetrabutylammonium chloride (TBA-Cl)	$C_{16}H_{36}ClN$
13	Tetramethylammonium bromide (TMA-Br)	$C_4H_{12}BrN$
14	Tetraethylammonium bromide (TEA-Br)	$C_8H_{20}BrN$
15	Tetrapropylammonium bromide (TPA-Br)	$C_{12}H_{28}BrN$
16	Tetrabutylammonium bromide (TBA-Br)	$C_{16}H_{36}BrN$
17	Tetramethylammonium iodide (TMA-I)	$C_4H_{12}IN$
18	Tetraethylammonium iodide (TEA-I)	$C_8H_{20}IN$
19	Tetrapropylammonium iodide (TPA I)	$C_{12}H_{28}IN$
20	Tetrabutylammonium iodide (TBA-I)	$C_{16}H_{36}IN$

Table 3.
List of ammonium-based ILs being predicted.

and tetrafluoroborate $[BF4]^-$ and hydroxide $[OH]^-$ ions due to their strong electrostatic charges and tendency to form hydrogen bonding with water [30] (**Table 3**).

All of the above chemicals will be simulated and calculated in COSMO-RS, which the calculations were carried out using TURBOMOLE6.1. The quantum chemical calculation follows the DFT, using the BP functional B88-86 with a TZVP basis set and the resolution of identity standard (RI) approximation.

3. Progress and discussion

3.1 Correlation development and validation

Using the four fundamental properties that have been identified earlier, an effort to relate them with the effectiveness of IL as a hydrate inhibitor has been carried out. These four properties are sigma profile, hydrogen bonding energy, activity coefficient, and solubility of IL in water. The following sections now thoroughly report and discuss if these four fundamental properties have successfully been related to IL inhibition ability.

3.1.1 Interpretation of sigma profile graphs

A sigma profile graph in COSMO-RS allows us to understand certain aspects of an IL-water system. The main information we can obtain from the graph is to learn about the hydrophobicity of IL and the tendency of IL to act as a hydrogen bond donor or hydrogen bond acceptor. According to Klamt [5, 58], the sigma profile graph can be divided into three regions. The first region is the hydrogen bond donor region (at the left of −1.0 e/nm^2), the second region is nonpolar region (between −1.0 and 1.0 e/nm^2), and the thirdly region is the acceptor region (at the right of 1.0 e/nm^2). By judging at which region the peak of an IL locates, the tendency of IL to act as hydrogen bond donor or acceptor would be identified. Generally, a peak that locates at the right side of the sigma profile graph indicates the more electronegative area and acts as an H-bond acceptor.

Now, **Figure 4** shows the sigma profile graph of EMIM-Cl, BMIM-Br, and water molecules. Looking at the sigma profile of water molecules as shown in **Figure 4**, it is observed that water has two high peaks, one in the hydrogen bond donor region and another in the acceptor region [39]. This indicates that water has a high affinity toward both acceptor and donor. Furthermore, **Figure 4** shows the sigma profile of two ILs, which are EMIM-Cl and BMIM-Br. From the figure, it is observed that cations EMIM and BMIM both have their peak in the nonpolar region. However, water molecules which have peaks in the polar region tend to have higher affinity only with strong hydrogen bond donor or acceptor, but not cation that lays its peak in the nonpolar region [59]. As a result, cations do not interact much with water molecules. Meanwhile, anions that have their peaks in hydrogen bonding acceptor region are more attractive to water molecules. Hence, this inferred that anion is the main ion that interacts with water molecules to prevent hydrate formation, whereas cation merely contributes very slightly in the process [57].

Moreover, we can see that EMIM, which has a shorter alkyl chain length, has its peak nearer to the polar region than BMIM. As consequences, EMIM is also more polarized and hydrophilic than BMIM, which is a desired characteristic of a good hydrate inhibitor. This also proves that a cation with shorter alkyl chain length is preferable during the tuning of IL inhibitor, as a shorter cation is less bulky and hence can more effectively interact with water molecules [49, 60]. For anion, Cl^- proves itself to be a better H-bond acceptor as it has a peak at the right side of the graph, which is the indication of its further electronegative. This at the same time means that Cl^- will be more effective in accepting H-bond from water molecules than Br^-. Therefore, this makes Cl^- more hydrophilic and serves as a better anion for hydrate inhibitor.

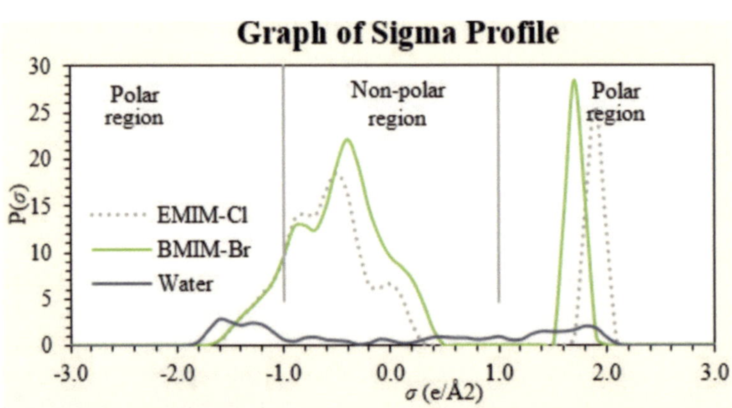

Figure 4.
Sigma profile graph of EMIM-Cl and BMIM-Br.

In short, referring to **Figure 4**, we can see that EMIM-Cl is the best combination of ions among the two types of ILs. Due to its lower alkyl chain length cation and a more electronegative anion, it should perform the best among the four ILs. This deduction is supported by the work of Xiao et al. [30], which reported the order of IL effectiveness as EMIM-Cl > EMIM-Br > BMIM-Cl > BMIM-Br.

3.1.2 Hydrogen bonding

Although hydrogen bonding strength has been widely quoted to have a relationship with the effectiveness of IL as hydrate inhibitor [29, 30], so far, no work has been conducted to prove this relationship. In this work, validation is done and has successfully proven that a linear relationship exists between hydrogen bonding strength and the effectiveness of IL as hydrate. This linearity is validated through four different sets of data that comes from three papers [25, 30, 61]. All four sets of data show good linearity relationship, with the highest regression value as $R^2 = 1$ and the lowest as $R^2 = 0.8926$. As a result, this implies that the prediction of IL effectiveness could be made through the comparison of hydrogen bonding strength.

Besides proving this relationship, several interesting findings have also been observed throughout the process. Firstly, computation of COSMO-RS, in total, will calculate three kinds of energy value for an IL, namely, misfit energy (E_{MF}), hydrogen bonding energy (E_{HB}), and van der Waals energy (E_{vdW}). The summation of these three energies leads to the value of total interaction energy (E_{int}). Although hydrogen bonding strength is known to affect the effectiveness of IL, the significance of other energies could not be neglected yet. Hence, in **Figure 8**, all types of predicted energies including E_{MF}, E_{vdW}, E_{HB}, and E_{INT} are plotted against average depression temperature to determine if these energies could also affect the effectiveness of ILs as hydrate inhibitor.

Figure 5 demonstrates that for ILs with BMIM cation, it is evidently shown the anion contributes more to the total interaction energy than the cation. The reason behind this is virtually consistent; van der Waals energies are nearly constant for all of the tested ILs and have thus no effect on the temperature depression. The contribution of misfit energy, having only a regression value of 0.2247, is also negligible. This leaves the hydrogen bonding energy to be the only energy that plays an essential role in affecting the effectiveness of BMIM-ILs. Furthermore, the relationship between total interaction energy (E_{INT}) and temperature depression is also not convincing. This graph hence supports the earlier statement that hydrogen

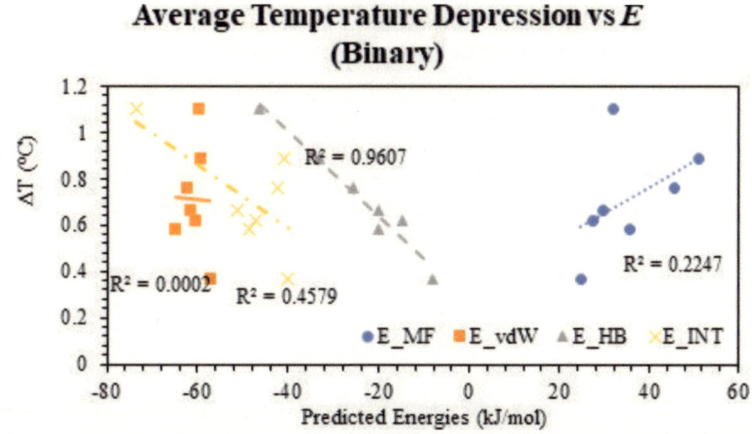

Figure 5.
Average temperature depression from Sabil et al. [25] work vs. types of predicted energy (binary components).

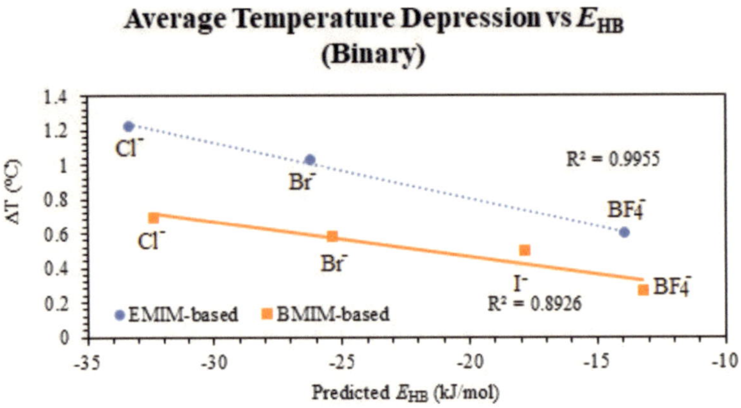

Figure 6.
Average temperature depression against predicted hydrogen bonding energy for both EMIM- and BMIM-based ILs (binary components).

bonding strength between cation and anion is the most important type of energy that regulates IL interaction with water molecules [29, 30, 62]. The same pattern of relationship is then also observed in another two data sets from the work of Xiao et al. [30]. **Figure 6** now shows the relationship between average temperature depression of ILs and the predicted hydrogen bonding energy from COSMO-RS.

Clearly, the graph shows that the temperature depression value of IL-hydrate system is directly proportional to the hydrogen bonding energy (E_{HB}) for both EMIM-based and BMIM-based ILs. The larger the absolute value of E_{HB}, the higher the temperature depression of a hydrate system. For instance, for BMIM-based ILs in this graph, the rank of E_{HB} from highest to lowest is as BMIM-Cl > BMIM-Br > B MIM-I > BMIM-BF_4. The same ranking occurred to the average temperature depression as well, where BMIM-Cl has the highest temperature depression and BMIM-BF_4 has the lowest depression. This ranking could be explained by the fact that among four anions, Cl^- anion has the highest polarized charge and thus acts as the best hydrogen bond acceptor. BF_4^- anion, on the other hand, has the lowest polarized charge after Br^- and I^- anion and thus shows the lowest hydrogen bond strength because it is the weakest hydrogen bond acceptor among all. This graph, however, also displays an interesting finding, which is the separation of EMIM- and BMIM-based ILs into two different data sets, instead of one. This step is necessary as the combination of all ILs into one data set may lower the linearity of relationship. This statement is supported by **Figure 10**, which shows a graph of average temperature depression against predicted hydrogen bonding energy.

Figure 7 inferred that linear relationship only exists when ILs with the same cation are compared. An early deduction is that to ensure a linear relationship for a set of data, only one single ion, which is either cation or anion, can vary, while another one must be fixed. The relationship could not be applied to predict ILs with different cations and anions. This deduction is supported by **Figure 8**, which shows the regression value between average depression temperature and hydrogen bonding strength for a set of ILs with different cations but same Cl^- anion.

With the regression value as high as 0.8976 from **Figure 8**, this supports our deduction earlier, where one ion must be fixed and another one could be varied to see the relationship. Furthermore, it is noticeable that when ILs with fixed anion but different cations are measured, the relationship between hydrogen bonding strength and average depression temperature is inversely proportional as before. The higher the absolute value of hydrogen bonding energy, the lower the average temperature depression. This could be explained by the sigma potential graph that has been

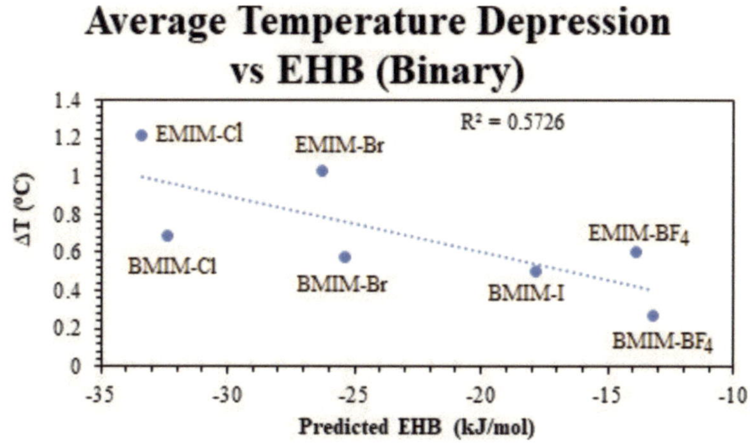

Figure 7.
Average temperature depression against predicted hydrogen bonding energy for a single data set consisting of both EMIM- and BMIM-based ILs (binary components).

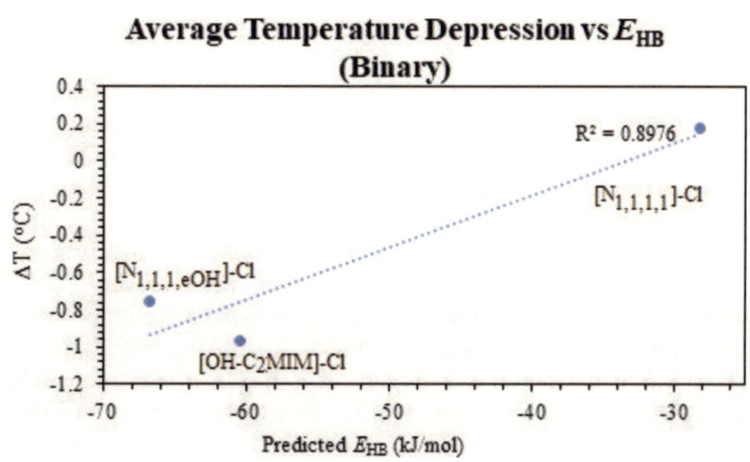

Figure 8.
Average temperature depression against predicted hydrogen bonding energy for ILs with Cl- as anion but different cations (binary components).

discussed earlier. Previously, it is explained that cations generally have their peaks located in the nonpolar region of sigma profile, which is from −1 to 1 e/nm^2.

On the other hand, water molecules show two high peaks, one at the region of hydrogen bond donor and another at hydrogen bond acceptor. As a result, water molecules tend to have higher affinity only with strong hydrogen bond donor or acceptor, but not cation that lays its peak in the nonpolar region. Hence, this inferred that anion is the main ion that interacts with water molecules to prevent hydrate formation, whereas cation merely contributes very slightly in the process [57]. Since cations have a low affinity with water molecules, this also indicates that most of the cations in water will continue to bond with anions. In that case, the excess hydrogen bonding energy provided by stronger cation (that has higher E_{HB}) is unnecessary. This stronger hydrogen bonding energy will be used by cation to bond with anion, thus reduces the number of anions that are free to interact with water molecules. As a consequence, it will bring about an inverse effect on average temperature depression and reduce the effectiveness of ILs as a hydrate inhibitor.

In short, linear relationship does exist between hydrogen bonding strength and the thermodynamic hydrate inhibition ability of an IL. For a set of ILs with fixed cation and

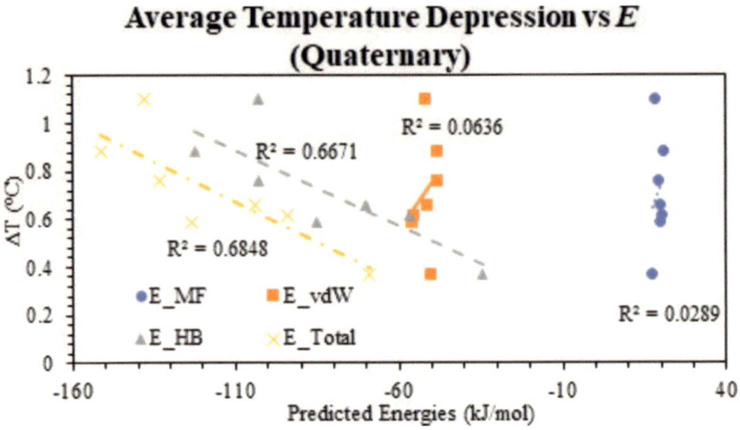

Figure 9.
Average temperature depression from Sabil et al. [25] work vs. types of predicted energy (quaternary components).

different anions, stronger hydrogen bonding between ILs lead to higher average depression temperature of an IL-hydrate system. Vice versa, for a set of ILs with fixed anion but different cations, stronger E_{HB} produces lower depression temperature. A lower depression of temperature subsequently signifies that the IL is less capable of shifting the equilibrium curve and is thus a weaker THI inhibitor. Predicted hydrogen bonding energy computed by COSMO-RS through a binary system consisting of only cation and anion has thus proven to be useful in predicting the effectiveness of ILs as inhibitors.

However, the above method of computation in COSMO-RS involves only the interaction between cation and anion, and it does not represent the hydrate system fully. Thus, the second computation of the quaternary system containing cation, anion, water, and methane gas has been conducted. Similar graphs have been plotted to find out how consistent E_{HB} is in predicting the effectiveness of IL. **Figure 9** shows the graph of average depression temperature plotted against a different type of predicted energies.

As observed from **Figure 12**, when quaternary components are involved, which include cations, anions, water, and methane, it is still obvious that hydrogen bonding energy (E_{HB}) is the main energy that influences the hydrate inhibition effect. Meanwhile, misfit energy and van der Waals energy have only a low regression value that is below 0.10. However, it is noticed that total interaction energy (E_{INT}) provides a slightly higher regression value than E_{HB} which is 0.6848 than 0.6671, which does not occur in binary component simulation. This could be because while involving more components such as methane and water, the van der Waals energy and misfit energy between different components are now more significant and influential. As compared to binary component regression value, the highest regression value that is obtained here is only 0.6848, which is extracted from the E_{INT}. Nevertheless, this low regression value could be improved to 0.8276 by removing the outlier which is BMIM-HSO_4 (1-butyl-3-methylimidazolium hydrogen sulfate) as shown in **Figure 10**. This is because of the nature of HSO_4^- anion, which has an extra hydrogen bonding functional group, OH^- (hydroxide), and thus resulting in stronger inhibition effect [62].

Figure 11 then shows the regression value of two more data sets from the work of Xiao et al. [30]. For both sets of data, total interaction energy (E_{INT}) gives the highest regression value too.

Similarly, from **Figure 12**, when the anions are fixed and cations are varied to study, the temperature depression value also decreases as the hydrogen bonding energy becomes more negative (stronger) which leads the increase in total

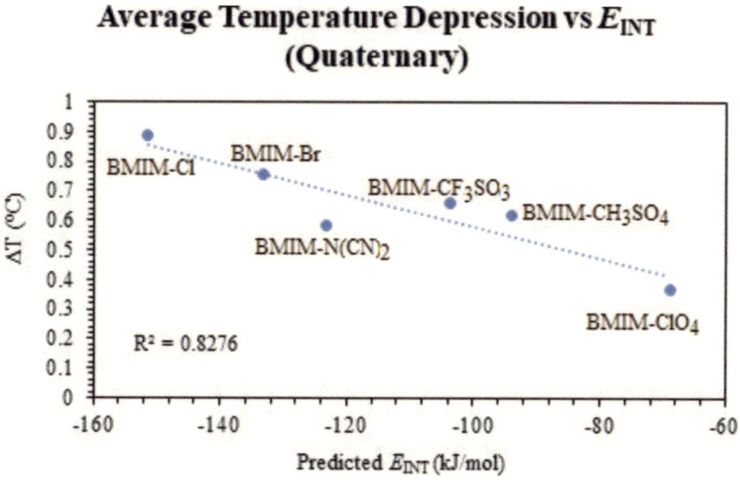

Figure 10.
Average temperature depression from Sabil et al. [25] work vs. predicted total interaction energy (quaternary components, without BMIM-HSO$_4$).

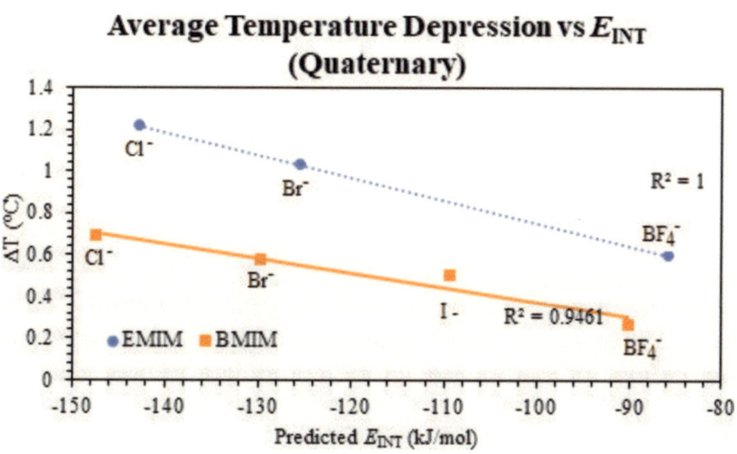

Figure 11.
Average temperature depression from Xiao et al. [30] work vs. predicted total interaction energy (quaternary components).

interaction energy. Therefore, generally, COSMO-RS simulation of binary components and quaternary components both work well as a quick prediction for the effectiveness of IL as a thermodynamic hydrate inhibitor. **Table 4** shows the regression value of both binary and quaternary components simulation.

Here, it is shown that the simulation of hydrogen bonding energy (E_{HB}) of binary components simulation provides a more consistent regression value. On the other hand, total interaction energy (E_{INT}) of quaternary components simulation more accurately reflects out the hydrated state which involves not only the IL itself but also water molecules and methane gas. To determine whether binary or quaternary components simulation is more effective in predicting ILs effectiveness, more sets of experimental data should be validated using the above approach. However, experimental work that tested ILs set with fixed anion or cation is very limited. Therefore, it is hard to conclude here whether binary or quaternary components simulation is more superior. Nevertheless, since real hydrate system consists of the interaction between water, methane, and IL, quaternary components simulation will be further studied, and correlation will be developed in this work.

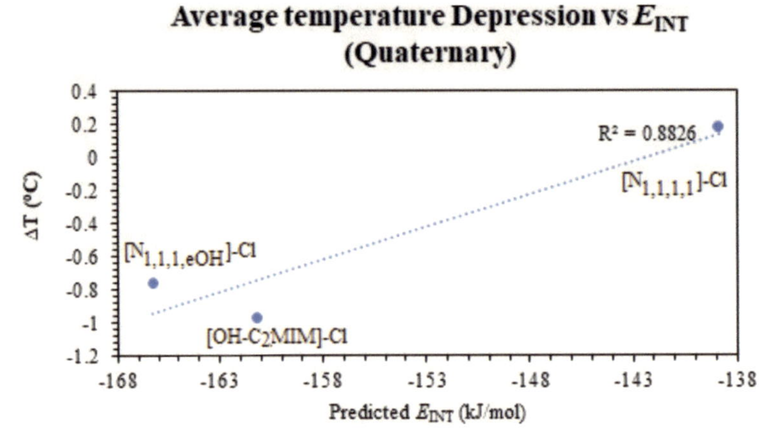

Figure 12.
Average temperature depression against predicted total interaction energy for ILs with Cl- as anion but different cations (quaternary components).

Literature	R^2 (Binary)	R^2 (Quaternary)
BMIM based ILs (Sabil et al. [25])	0.9607	0.8276
BMIM based ILs (Xiao et al.[30])	0.8926	0.9461
EMIM based ILs (Xiao et al. [30])	0.9955	1
Chloride based ILs (Keshavarz et al.) [50]	0.8976	0.8826

Table 4.
Comparison of regression values produced by binary and quaternary components simulation.

From the previous analysis for quaternary simulation, it is observed that the total interaction energy of anion and cation has a different effect on average temperature depression. Anion with higher interaction energy shows a higher average temperature depression, while the stronger interaction energy of cation reduces the average temperature depression. Due to the opposite effect of these two types of interaction energies (cation and anion), it is thus a must to consider them separately during the development of correlation. This results in the splitting of total interaction energy (E_{INT}) into two variables, which are E_{INT} contributed by anion ($E_{INT,\,A}$) and E_{INT} contributed by cation ($E_{INT,\,C}$). Both of them are available and obtainable from COSMO-RS simulation. **Table 5** shows an example of $E_{INT,\,C}$, $E_{INT,\,A}$, and E_{INT} calculated by COSMO-RS for the ILs from the work of Sabil et al. [25].

From **Table 5**, it is clear that the summation of $E_{INT,\,A}$ and $E_{INT,\,C}$ would result in the value of E_{INT}. In comparison, it is also evidently shown that anion contributes more to the total interaction energy than the cation. Now after obtaining the two variables, Minitab is used to assist in developing a suitable correlation for the prediction of average temperature reduced by each IL. Generally, the model could be described as.

$$Y = \beta_1 + \beta_2 X_2 + \beta_3 X_3 + \ldots \quad (2)$$

$$\Delta T = \beta_1 + \beta_2 E_{INT,C} + \beta_3 E_{INT,A} \quad (3)$$

Among the many equations that have been tested, the best equation is listed below. It involves both $E_{INT,\,A}$ and $E_{INT,\,C}$ as independent variables.

ILs	$E_{INT, C}$	$E_{INT, A}$	E_{INT}
BMIM-Cl	-25.61552	-125.76895	-151.384
BMIM-Br	-25.86398	-107.45348	-133.317
BMIM-DCA	-26.37615	-77.31303	-103.689
BMIM-CF3SO3	-26.24535	-67.59179	-93.8371
BMIM-CH3SO4	-26.26126	-96.99914	-123.26
BMIM-ClO4	-26.20864	-42.78859	-68.9972
BMIM-HSO4	-26.19087	-111.6787	-137.87

Table 5.
Type of interaction energies predicted by COSMO-RS for the work of Sabil et al. [25].

Model:

$$\Delta T = 1.758 + 0.0643 E_{INT,C} - 0.00559 E_{INT,\ A} \quad (4)$$

Table 6 then shows the experimental value obtained from the literature review, as well as the predicted temperature using the above equation. It listed 25 ILs with their experimental average temperature depression value from 4 literature review [25, 30, 49, 54]. Using the values of ions' interaction energy ($E_{INT,A}$ and $E_{INT,C}$) obtained from COSMO-RS, average temperature depression has been predicted for each IL. Absolute error between the experimental and predicted value is then calculated and shown at the last row of the table, without considering the three extreme outliers that are highlighted in red.

Table 6 shows several interesting findings and limitations of the model. First, regarding the three extreme outliers, all three of them are substituted cations that have a hydroxyl (OH^-) group. This type of substituted cation, as calculated by COSMO-RS, has an overly high $E_{INT,\ C}$ (42.17 kJ/mol for [OH-C2MIM]-Cl as compared to 20.60 kJ/mol for EMIM-Cl), which is supposed to reduce their inhibition ability. But, in truth, hydroxyl group-substituted cation has constantly performed better than common cation because the OH^- serves as a strong hydrogen bond donor that will react with water [61, 63]. The increased interaction with water molecules will thus improve the average temperature depression [64]. Due to this reason, a large discrepancy is observed between experimental and predicted temperature depression for OH^--substituted cation-based ILs. This also signifies that the model developed earlier does not apply to hydroxyl group-substituted cations or possibly any other substituted cations ILs.

Next, a pure error which is caused by inconsistency between experiments has also limited the accuracy of this model. **Table 7** shows the simplified list of ILs which have different experimental average temperature depression value obtained from the literature review.

As observed from **Table 8**, the experimental value obtained from literature review does not agree with each other. They are inconsistent, and this has thus hindered the development of a fully accurate model that could predict the inhibition ability of ILs as THI inhibitors. For instance, the inhibition ability of BMIM-BF4 was reported in three different papers, and the difference of experimental value from each paper is fairly large, ranging from 0.270 to 0.858°C. Nevertheless, Zare et al. [54] reported an experimental value of 0.460°C, which only presents a 2.92% error when compared to the predicted value.

Next, looking at BMIM-HSO_4 and EMIM-HSO_4, it is experimentally proven that EMIM-HSO_4, which has a smaller alkyl chain length for cation, would serve as a better inhibitor [49, 60, 62]. However, because experimental values are obtained from two different papers [25, 54], BMIM-HSO_4 recorded a higher average

Paper	ILs	Experimental ΔT (°C)	Predicted ΔT (°C)	Absolute Error (%)
Xiao et al. [30]	EMIM-Cl	1.220	1.125	7.75
	EMIM-Br	1.030	1.001	2.77
	EMIM-BF4	0.600	0.742	23.68
	BMIM-Cl	0.690	0.843	22.24
	BMIM-Br	0.580	0.727	25.32
	BMIM-I	0.500	0.593	18.64
	BMIM-BF4	0.270	0.473	75.35
	PMIM-I	0.800	0.731	8.66
Sabil et al. [25]	BMIM-Cl	0.887	0.843	4.91
	BMIM-Br	0.758	0.727	4.11
	BMIM-DCA	0.663	0.527	20.49
	BMIM-CF3SO3	0.617	0.482	21.93
	BMIM-MeSO4	0.585	0.644	10.03
	BMIM-ClO4	0.370	0.348	5.88
	BMIM-HSO4	1.103	0.729	33.88
	[OH-C2MIM]-Cl	1.329	-0.260	119.58
	[OH-C2MIM]-Br	0.960	-0.373	138.88
Keshavarz et al. [50]	BMIM-BF4	0.858	0.473	44.79
	BMIM-DCA	0.720	0.527	26.82
	N2,2,2,2-Cl	1.080	1.218	12.80
Zare et al. [51]	BMIM-BF4	0.460	0.473	2.92
	EMIM-EtSO4	0.670	0.938	40.07
	EMIM-HSO4	0.990	0.999	0.95
	BMIM-MeSO4	1.020	0.644	36.90
	OH-EMIM-BF4	1.100	-0.625	156.78
		Average:		**20.49 %**

Table 6.
Experimental and predicted average temperature depression of ILs for selected literature review.

temperature depression. This contradiction due to inconsistency again hardened the process of model development. Two factors could probably explain this inconsistency between experimental values: (i) purity of ILs being used in an experiment and (ii) experimental procedure and atmospheric condition.

In short, hydrogen bonding energy is the main type of energy that affects the interaction of ions with water and subsequently the inhibition ability of ILs. For a quaternary component simulation, however, total interaction energy shows a better linear relationship with average temperature depression. The model developed which considers cation interaction energy and anion interaction energy sufficiently predicts average temperature depression with an average error of 20.49%. It is to be noted that to a certain degree, the inconsistency between experimental values also contributed to the average error. **Table 9** shows the regression statistics and *P*-value from the ANOVA test for the equation developed. The confidence level for the model is set at 95%, and thus a *P*-value of 0.000 (<0.05) signifies a reliable model.

ILs	Literature Review	Experimental ΔT (°C)	Predicted ΔT (°C)	Absolute Error (%)
BMIM-Cl	Xiao and Adidharma [30]	0.690	0.843	22.24
	Sabil et al. [25]	0.887		4.91
BMIM-Br	Xiao and Adidharma [30]	0.580	0.727	25.32
	Sabil et al. [25]	0.758		4.11
$BMIM-BF_4$	Xiao and Adidharma [30]	0.270	0.473	75.35
	Keshavarz et al. [50]	0.858		44.79
	Zare et al. [51]	0.460		2.92
$BMIM-MeSO_4$	Sabil et al. [25]	0.585	0.644	10.03
	Zare et al. [51]	1.020		36.90
$BMIM-HSO_4$	Sabil et al. [25]	1.103	0.729	33.88
$EMIM-HSO_4$	Zare et al. [51]	0.990	0.999	0.95

Table 7.
ILs with inconsistent experimental average temperature depression.

Model Summary					
R^2	: 78.23%				
R^2 (Adjusted)	: 73.88%				
Standard Error	: 0.131049				
ANOVA					
Source	DF	Adjusted SS	Adjusted MS	F-value	P-value
Regression	2	0.6173	0.30865	17.97	0.000
Residual	10	0.1717	0.01717		
Total	12	0.7890			

Table 8.
Model summary and ANOVA for the model developed.

3.1.3 Effect of temperature on predicted inhibition ability

From the earlier section, it is mentioned that the simulation work in this study is fixed at a temperature of 10°C, which is a common temperature where hydrates start to form. In this section, the effect of temperature is further examined to investigate if the predicted inhibition ability of ILs changes dramatically with temperature. **Figure 13** shows the graph of predicted average temperature depression against simulation temperature.

Nevertheless, if the percentage of difference is calculated out, it will be noticed that the effect of temperature is very insignificant. For example, for EMIM-Cl, using simulation of 10°C as reference state, the percentage difference for each temperature is shown at the table.

As shown in the table, the range of predicted average temperature depression is between 1.079 and 1.151°C, where the difference is really small. Furthermore, it is found out that most experimental studies involve only temperature range of −3.15

Temperature (°C)	-10	-5	0	5	10	15	20
Predicted ΔT (°C)	1.08	1.09	1.10	1.11	1.12	1.13	1.15
Percentage difference (%)	4.10	3.12	2.11	1.07	0.00	1.10	2.24

Table 9.
Percentage difference of predicted ΔT for EMIM-Cl due to temperature difference.

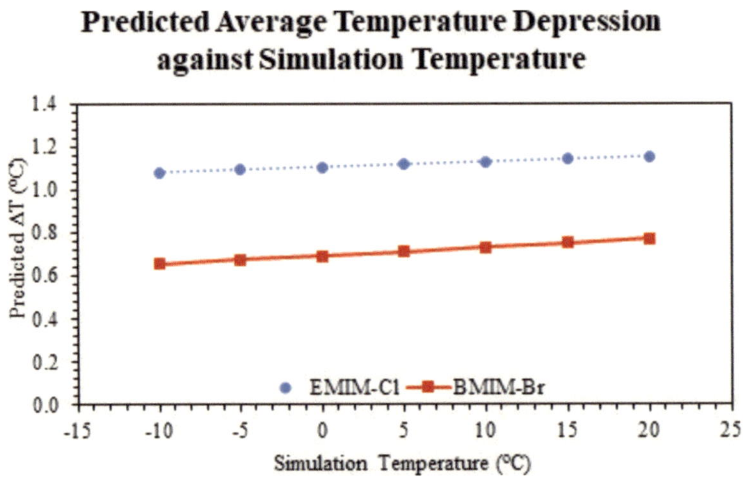

Figure 13.
Graph of predicted average temperature depression against simulation temperature.

to 16.85°C (270–290 K) [3, 29, 30, 49, 63]. This means that the highest percentage difference is just around 3.12% (for −5°C). Hence, it can be concluded that the effect of temperature is insignificant and would not affect the screening process of ILs using the correlation.

3.1.4 Activity coefficient

As discussed by Kurnia et al. [39], the lower the activity coefficient of a water-IL mixture, the higher the interaction between components in the mixture. Khan et al. also explain that for a water-IL mixture, activity coefficient below 1 signifies favorable interaction between water and ILs in the mixture [34]. When ILs interact well with water, supposedly, less water will be free to bond with each other to form hydrate. Theoretically, the activity coefficient could then reflect out the inhibition ability of IL. Therefore, validation effort was made through four sets of data [25, 30, 61] to find out if the relationship between activity coefficient and average temperature depression exists. **Figure 14** shows the graph of average temperature depression against the natural logarithm of activity coefficient.

As shown in **Figure 14**, the highest regression value is observed for BMIM-based ILs from the work of Xiao et al., which is a mere 0.6658. Meanwhile, another two sets of data record unacceptably low regression value of only 0.0045 and 0.2989. Hence, regrettably, these three sets of data could not exhibit any significant relationship between these two variables. Nevertheless, a general pattern of decreasing average temperature depression is observed when the natural logarithm of activity coefficient increases (activity coefficient increases). The incapability of the activity coefficient in reflecting the inhibition ability of IL since the calculations of activity coefficient

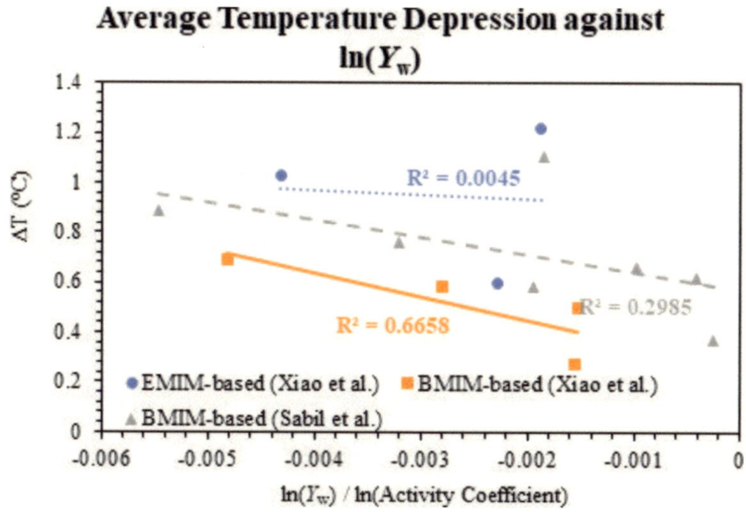

Figure 14.
Graph of average temperature depression against ln(Y$_W$) for three data sets.

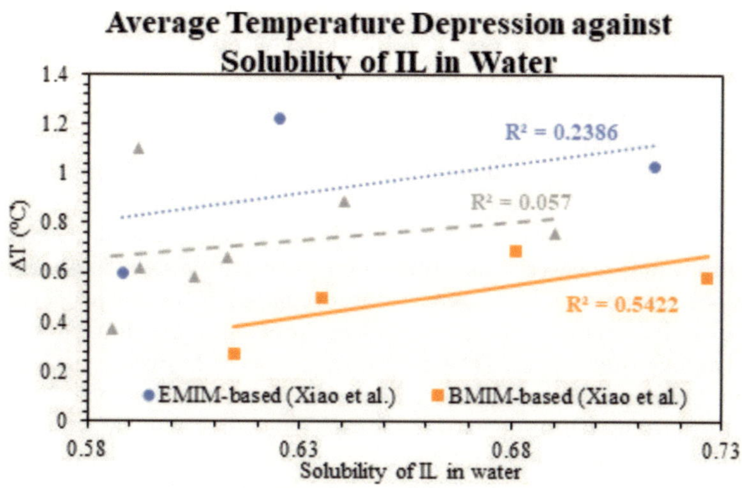

Figure 15.
Graph of average temperature depression against solubility of IL in water.

in COSMO-RS considers only the input of temperature, but no input of pressure is allowed. Meanwhile, in reality, hydrate occurs at low temperature (around 10°C) but high pressure. This kind of special nature of hydrate formation has thus made it hard for COSMO-RS to accurately predict out the activity coefficient water for a system of low temperature yet high pressure.

3.1.5 Solubility

A more soluble IL in water signifies that the IL can easily dissolve itself and interact with water molecules. Supposedly, a good IL should have high solubility in water, to bond with other water molecules and reduce the possibility of free water molecules from forming hydrate. To test the validity of the statement, four sets of data [25, 30, 61] were studied to find out if the relationship between the solubility of IL in water and average temperature depression exists. **Figure 15** shows the graph of average temperature depression against the solubility of IL in water.

Here, the regression values for all three data sets are very low as well, with the lowest regression value of 0.057. Hence, similarly to the activity coefficient, no significant relationship could be deduced from this variable.

3.2 Prediction of inhibition ability of ammonium-based ILs

From the validation part, it has been confirmed that sigma profile and total interaction energy of ILs can be correlated to the effectiveness of an IL as THI inhibitor. Hence, in this section, prediction work will be conducted on 20 ammonium-based ILs (refer to **Table 3**) to determine their ability as hydrate inhibitor, through the study of their sigma profile and total interaction energies.

3.2.1 Sigma profile

Although sigma profile could not directly compute a value to represent the effectiveness of an IL as a hydrate inhibitor, it does show the affinity of an IL toward the water. The higher the affinity of IL toward the water, the more hydrophilic it is, and the easier it could interact with water. This will then result in a more effective hydrate inhibitor. Hence, in this section, three sigma profile graphs will be used to determine the affinity of each ammonium-based ILs toward the water. The first figure, **Figure 16**, displays the sigma profile of the four types of cations involved here, which range from TMA to TBA.

From Section 3.1.1, it is discussed that the sigma profile graphs could be divided into three regions: hydrogen bond donor region (at the left of −1.0 e/nm^2), nonpolar region (between −1.0 and 1.0 e/nm^2), and hydrogen bond acceptor region (at the right of 1.0 e/nm^2). Here, all tetraalkylammonium-based cations have their peaks within the nonpolar region and are thus deduced to have a low affinity with water. This is because water molecules have only peaks within the hydrogen bond donor and hydrogen acceptor region. Due to this property, they do not interact well with ions that have a peak in the nonpolar region. However, when compared among themselves, TMA cation which has its highest peak at around −0.9 e/nm^2 performs the best because its peak is nearest to the polar region and thus has the highest affinity toward the water. This is because TMA has the lowest alkyl chain length, thus is less bulky and can easily interact with water molecules [34]. This makes TMA the most suitable cation among the four to be tuned as a hydrate inhibitor.

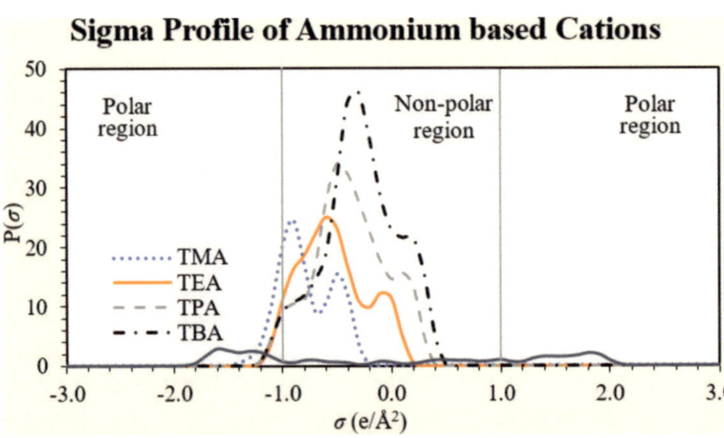

Figure 16.
Sigma profile of ammonium-based cations.

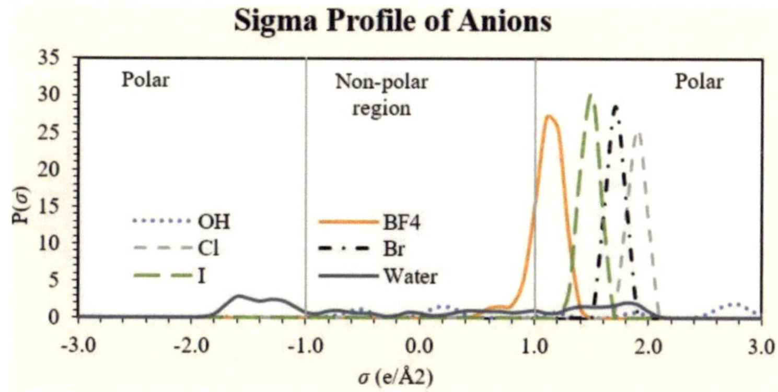

Figure 17.
Sigma profile of anions.

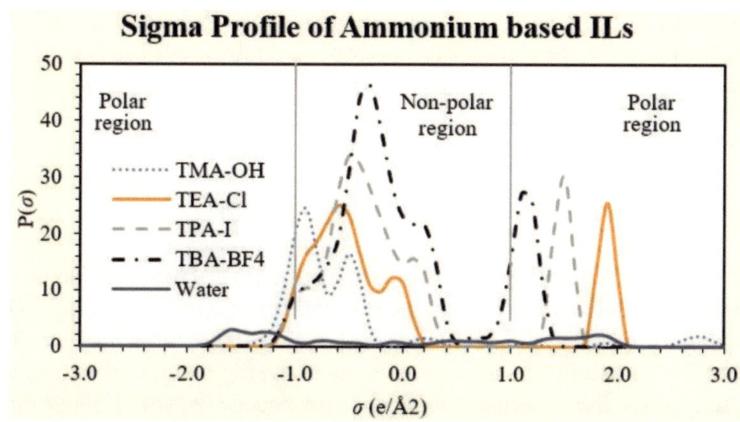

Figure 18.
Sigma profile of several ammonium-based ILs.

Figure 17 shows the sigma profile of five types of different anions. From this graph, it is observed that all anions have their peaks located in the polar region at the right side, which is the hydrogen bond acceptor region [33]. This indicates that all of them are electronegative and has a lone pair ready to share with another hydrogen bond donor. Due to their readiness to interact with hydrogen bond donor, they have high affinity with water molecules and tend to bond well with water molecules. The highest tendency of interaction goes to OH^- ion, which has its peak at 3.6 e/nm^2. In general, an anion that lays its peak further at the right side of the sigma profile graph is effective in inhibiting as it has high affinity with water molecules. This is because the further the peak to the right, the larger the sigma value and, thus, the more electronegative an anion is. The high electronegativity then results in higher interaction energy and thus interacts better with the water molecules. Meanwhile, BF_4^- ion that has its peak close to the nonpolar region is not an effective inhibitor anion because of its low polarized charge.

Lastly, the third figure, **Figure 18**, has selectively displayed the sigma profile graph for four ILs, including TMA-OH, TEA-Cl, TPA-I, and TBA-BF_4. The idea of this graph is to showcase several possible combinations of ILs by tuning the cation and anion. Here, it is easily observed that all cations show their peak in the nonpolar region. TMA cation shows its peak closest to the polar region and is thus the most suitable cation, due to its higher affinity with water. This could be explained by its short alkyl chain length as compared to others, which makes it more hydrophilic. Meanwhile, all anions lay in the polar region on the right side. The most electronegative anion is OH^- ion that has its peak furthest at the right. Due to its highest electronegativity and hence high interaction with water, it serves as the best anion to be used for an inhibitor. Therefore, from the graph, it is identifiable that TMA-OH is the best combination of all. This is followed by TEA-Cl, TPA-Br, and, finally, TBA-BF_4. From this graph, it is inferred that to choose the right anion for the hydrate inhibitor, its peak should be located as far as possible at the right side of the graph. This indicates a highly electronegative anion that can bond well with water molecules. Meanwhile, it is reported that most of the ILs cations have their peaks located in the nonpolar region. This characteristic causes cations to behave as nonpolar molecules that are hydrophobic and does not interact well with water molecules [58]. Therefore, a cation with the lowest hydrophobicity should be chosen to be tuned as a hydrate inhibitor, so that it will not hinder interactions between IL and water molecules. This, in turn, signifies that the most recommendable cation should have its peak closest to the left polar region.

3.2.2 Total interaction energies

In Section 3.1.2, a correlation has been developed to describe the relationship between the average depression temperature of IL-hydrate system and the total interaction energies. It is found out that both cation and anion interaction energy have a different effect on IL inhibition ability. High anion interaction energy is preferable, while high cation interaction energy will reduce an IL inhibition ability.

$$\Delta T = 1.758 + 0.0643 E_{\mathrm{INT,C}} - 0.00559 E_{\mathrm{INT,A}} \qquad (5)$$

Using the above correlation, the ability of ammonium-based ILs has been predicted through the calculation of average temperature depression. **Table 10** shows the list of ammonium-based ILs together with their total interaction energies and predicted inhibition ability measured in terms of average temperature depression.

Table 10 shows a list of tetraalkylammonium-based ILs, which range from cation tetramethylammonium to tetrabutylammonium paired with five types of different anions that are hydroxide ion, a tetrafluoroborate ion, chloride ion, bromide ion, and iodide ion. From this table, it is observed that when the anion is fixed, an increase in cation interaction energy, which is caused by the increase in alkyl chain length, will reduce average temperature depression. This again agrees to the earlier statement which explained that the longer alkyl chain length of cation, the bulkier it is and thus harder for it to interact with water molecules [49, 60]. This, as a result, increases its hydrophobicity, reduces its ability to bond with water, and is thus a less effective thermodynamic hydrate inhibitor [62]. In fact, among the five TBA ionic liquids (ILs), three of them show negative temperature depression. This is because of the poor combination of the bulky cation (TBA) and weak electronegativity anion (Br^-, I^-, BF_4^-), resulting in a super ineffective inhibitor. A negative temperature depression signifies that instead of serving as hydrate inhibitor, they have now become hydrate promoter that favors the formation of the hydrate phase.

In terms of the effect of anion, we can see that the higher the interaction energy of anion, the higher the average temperature depression is. Here, the rank of E_{INT} is as $OH^- > Cl^- > Br^- > I^- > BF_4^-$. This resulted in the average temperature depression to follow the same pattern. For instance, looking at tetramethylammonium ILs, inhibition ability rank is as TMA-OH > TMA-Cl > TMA-Br > TMA-I > TMA-BF_4. Hence, this again proves that the interaction energy provided by anion plays a crucial role in determining its inhibition ability. Also, this prediction agrees well with work reported by Tariq et al. [62]. In his review work, he reported that for a methylimidazolium-based IL, the order of efficiency follows as such C_2C_1im-Cl > C_2C_1im-Br > C_2C_1im-I > C_2C_1im-BF_4. Regrettably, OH^- ILs are not studied in Tariq's work; yet, the whole ranking ranging from Cl^- to BF_4^- is similar to the predicted ranking. This proves that the

ILs	$H_{INT,cation}$	$H_{INT,anion}$	$T_{predicted}$ (°C)
TMA-OH	-14.52	-204.20	1.97
TEA-OH	-18.60	-205.11	1.71
TPA-OH	-27.76	-205.51	1.12
TBA-OH	-37.78	-205.74	0.48
TMA-BF_4	-16.96	-64.84	1.03
TEA-BF_4	-20.28	-65.40	0.82
TPA-BF_4	-28.77	-65.80	0.28
TBA-BF_4	-38.32	-66.01	-0.34
TMA-Cl	-15.54	-123.28	1.45
TEA-Cl	-19.19	-124.10	1.22
TPA-Cl	-28.12	-124.45	0.65
TBA-Cl	-38.00	-124.64	0.01
TMA-Br	-16.12	-105.40	1.31
TEA-Br	-19.66	-106.10	1.09
TPA-Br	-28.33	-106.43	0.53
TBA-Br	-38.01	-106.60	-0.09
TMA-I	-16.60	-84.49	1.16
TEA-I	-20.06	-85.01	0.94
TPA-I	-28.55	-85.32	0.40
TBA-I	-38.07	-85.50	-0.21

Table 10.
Predicted average temperature depression of AILs.

developed correlation is performing outstandingly in predicting the inhibition ability of ILs. Lastly, from this model, TMA-OH is identified to show the strongest ability as THI, with the highest depression temperature of 1.97°C. This is due to the highly electronegative OH^- anion that bonds well with water molecules and a short alkyl chain length TMA cation that does not hinder the IL interaction with water molecules.

4. Conclusions

In conclusion, among the four identified fundamental properties, sigma profile and hydrogen bonding energy have been successfully correlated to the inhibition ability of IL. Sigma profile provides a qualitative understanding of each IL in the sense of their affinity toward water molecules. Meanwhile, hydrogen bonding energy, or later upgraded to total interaction energy, has been able to satisfactorily predict out a quantitative value of average temperature depression provided by each IL. This value will then tell us the effectiveness of each IL as a thermodynamic hydrate inhibitor. The correlation developed is validated with open literature and is found out to have an average error of 20.49%. From the predicted data, it is observed that TMA-OH depresses the temperature of IL-hydrate system by 1.97°C, whereas the widely studied EMIM-Cl can only experimentally depress the system by 1.22°C. TMA-OH has shown the highest inhibition ability due to the combination of its short alkyl chain length cation and a highly electronegative OH^- anion. Findings, however, show that this correlation is not suitable to be used for substituted cations, as the introduced functional group such as hydroxyl group will provide extra H bonding with water molecules. COSMO-RS simulation, on the other hand, has been proven to be applicable in computing fundamental properties of IL-hydrate system. Simulation of COSMO-RS in calculating fundamental properties paired with the correlation developed in this work could now serve as a pre-screening tool of ILs inhibition ability. This helps to narrow down the scope of ILs to be focused during experimental work and thus speeds up the rate of potential ILs being tested and applied to industrial processes.

Acknowledgements

YUTP financially supports this work under the Grant no. 015LCO-154.

Acronyms and abbreviations

ILs	ionic liquids
LDHI	low-dosage hydrate inhibitor
TMA-OH	tetramethylammonium hydroxide
TMA-Cl	tetramethylammonium chloride
TMA-Br	tetramethylammonium bromide
CDA	carbamyldicyanomethanide
$N_{2,2,2,2}$-Cl	tetraethylammonium chloride
BMIM-ClO_4	1-butyl-3-methylimidazolium perchlorate
BMIM-Cl	1-butyl-3-methylimidazolium chloride
BMIM-$N(CN)_2$	1-butyl-3-methylimidazolium dicyanamide
KHI	kinetic hydrate inhibitors
PVP	polyvinyl pyrrolidinium
ΔT (°C)	average suppression temperature
COSMO-RS	conductor-like screening model for real solvent
CH_4	methane
BMIM-BF_4	1-butyl-3-methylimidazolium tetrafluoroborate
$N_{1,1,1,1}$-Cl	tetramethylammonium chloride
BMIM-CF_3SO_3	1-butyl-3-methylimidazolium trifluoromethanesulfonate
BMIM-CH_3SO_4	1-butyl-3-methylimidazolium methylsulfate

Author details

Muhammad Saad Khan[1,2] and Bhajan Lal[1,2*]

1 Chemical Engineering Department, Universiti Teknologi PETRONAS, Bandar Seri Iskandar, Perak, Malaysia

2 CO_2 Research Centre (CO_2RES), Universiti Teknologi PETRONAS, Bandar Seri Iskandar, Perak, Malaysia

*Address all correspondence to: bhajan.lal@utp.edu.my

IntechOpen

© 2019 The Author(s). Licensee IntechOpen. This chapter is distributed under the terms of the Creative Commons Attribution License (http://creativecommons.org/licenses/by/3.0), which permits unrestricted use, distribution, and reproduction in any medium, provided the original work is properly cited.

References

[1] Li B, Li XS, Li G, Wang Y, Feng JC. Kinetic behaviors of methane hydrate formation in porous media in different hydrate deposits. Industrial and Engineering Chemistry Research. 21 Mar 2014;**53**(13):5464-5474

[2] Nazari K, Moradi MR, Ahmadi AN. Kinetic modeling of methane hydrate formation in the presence of low-dosage water-soluble ionic liquids. Chemical Engineering and Technology. 2013;**36**(11):1915-1923

[3] Avula VR, Gardas RL, Sangwai JS. An efficient model for the prediction of CO_2 hydrate phase stability conditions in the presence of inhibitors and their mixtures. The Journal of Chemical Thermodynamics. 2015;**85**:163-170

[4] Khan MS, Bavoh CB, Partoon B, Nashed O, Lal B, Mellon NB. Impacts of ammonium based ionic liquids alkyl chain on thermodynamic hydrate inhibition for carbon dioxide rich binary gas. Journal of Molecular Liquids. 2018;**261**:283-290

[5] Bavoh CB, Khan MS, Lal B, Bt Abdul Ghaniri NI, Sabil KM. New methane hydrate phase boundary data in the presence of aqueous amino acids. Fluid Phase Equilibria. 2018;**478**(December):129-133

[6] Bavoh CB, Lal B, Nashed O, Khan MS, Lau KK, Bustam MA. COSMO-RS: An ionic liquid prescreening tool for gas hydrate mitigation. Chinese Journal of Chemical Engineering. 2016;**24**(11):1619-1624

[7] Khan MS, Bavoh CB, Partoon B, Lal B, Bustam MA, Shariff AM. Thermodynamic effect of ammonium based ionic liquids on CO_2 hydrates phase boundary. Journal of Molecular Liquids. 2017;**238**(July):533-539

[8] Khan MS, Cornelius BB, Lal B, Bustam MA. Kinetic Assessment of tetramethyl ammonium hydroxide (ionic liquid) for carbon dioxide, methane and binary mix gas hydrates. In: Rahman MM, editor. Recent Advances in Ionic Liquids. London, UK: IntechOpen; 2018. pp. 159-179

[9] Qasim A, Khan MS, Lal B, Shariff AM. Phase equilibrium measurement and modeling approach to quaternary ammonium salts with and without monoethylene glycol for carbon dioxide hydrates. Journal of Molecular Liquids. 2019;**282**:106-114

[10] Bavoh CB, Lal B, Khan MS, Osei H, Ayuob M. Combined inhibition effect of 1-ethyl-3-methy-limidazolium chloride + glycine on methane hydrate. Journal of Physics Conference Series. 2018;**1123**:012060

[11] Khan MS et al. Experimental equipment validation for methane (CH_4) and carbon dioxide (CO_2) hydrates. IOP Conference Series: Materials Science and Engineering. 2018;**344**:1-10

[12] Khan MS, Lal B, Sabil KM, Ahmed I. Desalination of seawater through gas hydrate process: An overview. Journal of Advanced Research in Fluid Mechanics and Thermal Sciences. 2019;**55**(1):65-73

[13] Khan MS, Lal B, Shariff AM, Mukhtar H. Ammonium hydroxide ILs as dual-functional gas hydrate inhibitors for binary mixed gas (carbon dioxide and methane) hydrates. Journal of Molecular Liquids. 2019;**274**(January):33-44

[14] Kassim Z, Khan MS, Lal B, Partoon B, Shariff AM. Evaluation of tetraethylammonium chloride on methane gas hydrate phase conditions. IOP Conference Series: Materials Science and Engineering. 2018;**458**:012071

[15] Khan MS, Lal B, Bavoh CB, Keong LK, Bustam A. Influence of ammonium

based compounds for gas hydrate mitigation: A short review. Indian Journal of Science and Technology. 2017;**10**(5):1-6

[16] Foo KS, Khan MS, Lal B, Sufian S. Semi-clathratic impact of tetrabutylammonium hydroxide on the carbon dioxide hydrates. IOP Conference Series: Materials Science and Engineering. 2018;**458**:012060

[17] Bavoh CB et al. The effect of acidic gases and thermodynamic inhibitors on the hydrates phase boundary of synthetic malaysia natural gas. IOP Conference Series: Materials Science and Engineering. 2018;**458**:012016

[18] Khan MS, Partoon B, Bavoh CB, Lal B, Mellon NB. Influence of tetramethylammonium hydroxide on methane and carbon dioxide gas hydrate phase equilibrium conditions. Fluid Phase Equilibria. 2017, 2017;**440**(May):1-8

[19] Nashed O, Dadebayev D, Khan MS, Bavoh CB, Lal B, Shariff AM. Experimental and modelling studies on thermodynamic methane hydrate inhibition in the presence of ionic liquids. Journal of Molecular Liquids. 2018;**249**:886-891

[20] Khan MS, Lal B, Partoon B, Keong LK, Bustam MA, Mellon NB. Experimental evaluation of a novel thermodynamic inhibitor for CH_4 and CO_2 hydrates. Procedia Engineering. 2016;**148**:932-940

[21] Khan MS, Liew CS, Kurnia KA, Cornelius B, Lal B. Application of COSMO-RS in investigating ionic liquid as thermodynamic hydrate inhibitor for methane hydrate. Procedia Engineering. 2016;**148**:862-869

[22] Khan MS, Lal B, Keong LK, Sabil KM. Experimental evaluation and thermodynamic modelling of AILs alkyl chain elongation on methane riched gas hydrate system. Fluid Phase Equilibria. 2018;**473**(October):300-309

[23] Kim K-S, Kang JW, Kang S-P. Tuning ionic liquids for hydrate inhibition. Chemical Communications. 2011;**47**(22):6341-6343

[24] Sloan ED et al. Clathrate Hydrates of Natural Gases. 3rd ed. Vol. 87(13-14). Boca Raton; London; New york: CRC Press Taylor & Francis; 2008

[25] Sabil KM, Nashed O, Lal B, Ismail L, Japper-Jaafar A. Experimental investigation on the dissociation conditions of methane hydrate in the presence of imidazolium-based ionic liquids. The Journal of Chemical Thermodynamics. 2015;**84**:7-13

[26] Zeng H, Wilson LD, Walker VK, Ripmeester JA. Effect of antifreeze proteins on the nucleation, growth, and the memory effect during tetrahydrofuran clathrate hydrate formation. Journal of the American Chemical Society. 2006;**128**(9):2844-2850

[27] Kelland MA. History of the development of low dosage hydrate inhibitors. Energy and Fuels. 2006;**20**(3):825-847

[28] Hong SY, Il Lim J, Kim JH, Lee JD. Kinetic studies on methane hydrate formation in the presence of kinetic inhibitor via in situ Raman spectroscopy. Energy & Fuels. 2012;**26**:7045-7050

[29] Xiao C, Adidharma H. Dual function inhibitors for methane hydrate. Chemical Engineering Science. 2009;**64**(7):1522-1527

[30] Xiao C, Wibisono N, Adidharma H.Dialkylimidazolium halide ionic liquids as dual function inhibitors for methane hydrate. Chemical Engineering Science. 2010;**65**(10):3080-3087

[31] Chen Q et al. Effect of 1-butyl-3-methylimidazolium tetrafluoroborate on the formation rate of CO_2 hydrate. Journal of Natural Gas Chemistry. 2008;**17**(3):264-267

[32] Richard AR, Adidharma H. The performance of ionic liquids and their mixtures in inhibiting methane hydrate formation. Chemical Engineering Science. 2013;**87**:270-276

[33] Klamt A. COSMO-RS: From Quantum Chemistry to Fluid Phase Thermodynamics and Drug Design. 1st ed. Amsterdam: Elsevier; 2005

[34] Khan I, Kurnia KA, Sintra TE, Saraiva JA, Pinho SP, Coutinho JAP. Assessing the activity coefficients of water in cholinium-based ionic liquids: Experimental measurements and COSMO-RS modeling. Fluid Phase Equilibria. 2014;**361**(January):16-22

[35] Franke R, Hannebauer B, Jung S. Accurate pre-calculation of limiting activity coefficients by COSMO-RS with molecular-class based parameterization. Fluid Phase Equilibria. 2013;**340**:11-14

[36] Diedenhofen M, Klamt A. COSMO-RS as a tool for property prediction of IL mixtures—A review. Fluid Phase Equilibria. 2010;**294**(1-2):31-38

[37] Calvar N, Domínguez I, Gómez E, Palomar J, Domínguez Á. Evaluation of ionic liquids as solvent for aromatic extraction: Experimental, correlation and COSMO-RS predictions. The Journal of Chemical Thermodynamics. 2013;**67**:5-12

[38] Freire MG, Ventura SPM, Santos LMNBF, Marrucho IM, Coutinho JAP. Evaluation of COSMO-RS for the prediction of LLE and VLE of water and ionic liquids binary systems. Fluid Phase Equilibria. 2008;**268**(1-2):74-84

[39] Kurnia KA, Pinho SP, Coutinho JAP. Evaluation of the conductor-like screening model for real solvents for the prediction of the water activity coefficient at infinite dilution in ionic liquids. Industrial and Engineering Chemistry Research. 2014;**53**(31):12466-12475

[40] Domínguez I, González EJ, Palomar J, Domínguez Á. Phase behavior of ternary mixtures {aliphatic hydrocarbon + aromatic hydrocarbon + ionic liquid}: Experimental LLE data and their modeling by COSMO-RS. Journal of Chemical Thermodynamics. 2014;**77**:222-229

[41] Anantharaj R, Banerjee T. COSMO-RS-based screening of ionic liquids as green solvents in denitrification studies. Industrial and Engineering Chemistry Research. 2010;**49**(18):8705-8725

[42] Grabda M et al. Theoretical selection of most effective ionic liquids for liquid-liquid extraction of NdF_3. Computational & Theoretical Chemistry. 2015;**1061**:72-79

[43] Grabda M et al. COSMO-RS screening for efficient ionic liquid extraction solvents for $NdCl_3$ and $DyCl_3$. Fluid Phase Equilibria. 2014;**383**:134-143

[44] Kurnia KA, Mutalib A, Ibrahim M, Man Z, Bustam MA. Selection of ILs for separation of benzene from n-hexane using COSMO-RS. A quantum chemical approach. InKey Engineering Materials. Trans Tech Publications; 2013;**553**:35-40

[45] Palomar J, Ferro VR, Torrecilla JS, Rodríguez F. Density and molar volume predictions using COSMO-RS for ionic liquids. An approach to solvent design. Industrial and Engineering Chemistry Research. 2007;**46**(18):6041-6048

[46] Sumon KZ, Henni A. Ionic liquids for CO_2 capture using COSMO-RS: Effect of structure, properties and molecular interactions on solubility and selectivity. Fluid Phase Equilibria. 2011;**310**(1-2):39-55

[47] Pilli S, Mohanty K, Banerjee T. Extraction of phthalic acid from aqueous solution by using ionic liquids: A quantum chemical approach. International Journal of Thermodynamics. 2014;**17**(1):42-51

[48] Machanová K, Troncoso J, Jacquemin J, Bendová M. Excess molar volumes and excess molar enthalpies in binary systems N-alkyl-triethylammonium bis(trifluoromethylsulfonyl)imide + methanol. Fluid Phase Equilibria. 2014;**363**:156-166

[49] Keshavarz L, Javanmardi J, Eslamimanesh A, Mohammadi AH. Experimental measurement and thermodynamic modeling of methane hydrate dissociation conditions in the presence of aqueous solution of ionic liquid. Fluid Phase Equilibria. 2013;**354**(02):312-318

[50] Schäfer A, Huber C, Ahlrichs R. Fully optimized contracted Gaussian basis sets of triple zeta valence quality for atoms Li to Kr. The Journal of Chemical Physics. 1994;**100**(8):5829

[51] Jaapar SZS, Iwai Y, Morad NA. Effect of co-solvent on the solubility of ginger bioactive compounds in water using COSMO-RS calculations. Applied Mechanics and Materials. 2014;**624**(June):174-178

[52] Ingram T, Gerlach T, Mehling T, Smirnova I. Extension of COSMO-RS for monoatomic electrolytes: Modeling of liquid-liquid equilibria in presence of salts. Fluid Phase Equilibria. 2012;**314**:29-37

[53] Kurnia KA, Coutinho JAP. Overview of the excess enthalpies of the binary mixtures composed of molecular solvents and ionic liquids and their modeling using COSMO-RS. Industrial and Engineering Chemistry Research. 2013;**52**(38):13862-13874

[54] Zare M, Haghtalab A, Ahmadi AN, Nazari K. Experiment and thermodynamic modeling of methane hydrate equilibria in the presence of aqueous imidazolium-based ionic liquid solutions using electrolyte cubic square well equation of state. Fluid Phase Equilibria. 2013;**341**:61-69

[55] Diedenhofen M et al. Compounds in ionic liquids using COSMO-RS. Engineering. 2003;**48**(1):475-479

[56] Diedenhofen M, Eckert F, Klamt A. Theoretical prediction for the infinite dilution activity coefficients of organic compounds in ionic liquids. Journal of Chemical & Engineering Data. 2003;**48**(3):475-479

[57] Peng X, Hu Y, Liu Y, Jin C, Lin H. Separation of ionic liquids from dilute aqueous solutions using the method based on CO_2 hydrates. Journal of Natural Gas Chemistry. 2010;**19**(1):81-85

[58] Klamt A. COSMO-RS for aqueous solvation and interfaces. Fluid Phase Equilibria. 2016;**407**:152-158

[59] Gonfa G, Bustam MA, Sharif AM, Mohamad N, Ullah S. Tuning ionic liquids for natural gas dehydration using COSMO-RS methodology. Journal of Natural Gas Science and Engineering. 2015;**27**:1141-1148

[60] Avula VR, Gardas RL, Sangwai JS. An improved model for the phase equilibrium of methane hydrate inhibition in the presence of ionic liquids. Fluid Phase Equilibria. 2014;**382**:187-196

[61] Sen Li X, Liu YJ, Zeng ZY, Chen ZY, Li G, Wu HJ. Equilibrium hydrate formation conditions for the mixtures of methane + ionic liquids + water. Journal of Chemical & Engineering Data. 2011;**56**(1):119-123

[62] Tariq M, Rooney D, Othman E, Aparicio S, Atilhan M, Khraisheh M.

Gas hydrate inhibition: A review of the role of ionic liquids. Industrial and Engineering Chemistry Research. 2014;**53**(46):17855-17868

[63] Partoon B, Wong NMS, Sabil KM, Nasrifar K, Ahmad MR. A study on thermodynamics effect of [EMIM]-Cl and [OH-C2MIM]-Cl on methane hydrate equilibrium line. Fluid Phase Equilibria. 2013;**337**:26-31

[64] Kim K, Kang S-P. Investigation of pyrrolidinium- and morpholinium-based ionic liquids into kinetic hydrate inhibitors on structure I methane hydrate. In: 7th International Conference on Gas Hydrates (ICGH). 2011. pp. 17-21

Chapter 9

Ionic Liquids as Environmental Benign Solvents for Cellulose Chemistry: A Review

Indra Bahadur and Ronewa Phadagi

Abstract

The application of cellulose and its derivatives is restricted because of their limited solubility in water and many organic solvents. Recently, several attempts are being made to dissolve them in inorganic and organic solvents. The solubility of these polymeric materials mainly depends upon their molecular weight, pH, and source of origin. Nowadays, there has been a new breakthrough of applying ionic liquids (ILs; designer solvents) in the field of cellulose solvent chemistry. Association of ionic liquids with several salient features such as high thermal, chemical and low vapor pressure, and so on makes them ideal environmentally green solvents to be used for cellulose. The present chapter deals with a collection of some major works in which ionic liquids have been used as solvents for cellulose dissolution. The articles also describe the works illustrating the use of ionic liquids as cosolvents (organic aprotic solvents) for the better increase of the solvent activity (solubility).

Keywords: ionic liquids, cellulose chemistry, biocompatibility, green chemistry, degree of polymerization, polysaccharides

1. Introduction

Cellulose is defined as a natural homopolymer which is obtained mainly from living organisms including fungi, bacteria, algae, and animals [1, 2]. The cellulose is constructed by repetition of D-glucose units in which two glucose units are linked together by β(C1 → C4)-glycosidic bond [1] as shown in **Figure 1**. Additionally, degree of polymerization of cellulose highly depends upon the raw material; for native cellulose it can be up to 15,000 [3]. Cellulose has become one of the most widely used natural polymers because of its various fascinating physical and structural properties along with its biocompatibility and wide range of availability [4]. The several magnificent properties of cellulose arise due to the presence of multiple inter- and intrahydrogen bonding interactions as illustrated in **Figure 2**. The cellulose is an unbranched homopolymer because it exists in semicrystalline form containing both crystalline and amorphous phases [5].

It is clear from **Figure 1** that cellulose consisted of two types of hydroxyl (—OH) substituents, one is primary hydroxyl substituent and the other is secondary hydroxyl substituent [5].

The discovery of cellulose was made in late 1830s by Anselme Paven and Hyatt. In the year 1870, manufacturing company took a positive outlook in cellulose

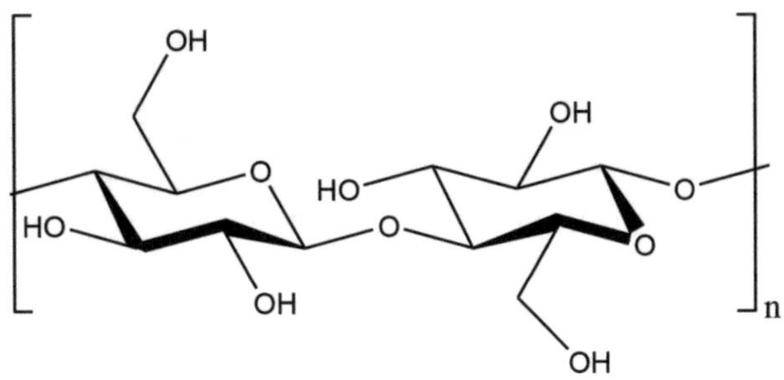

Figure 1.
Chemical structure of cellulose.

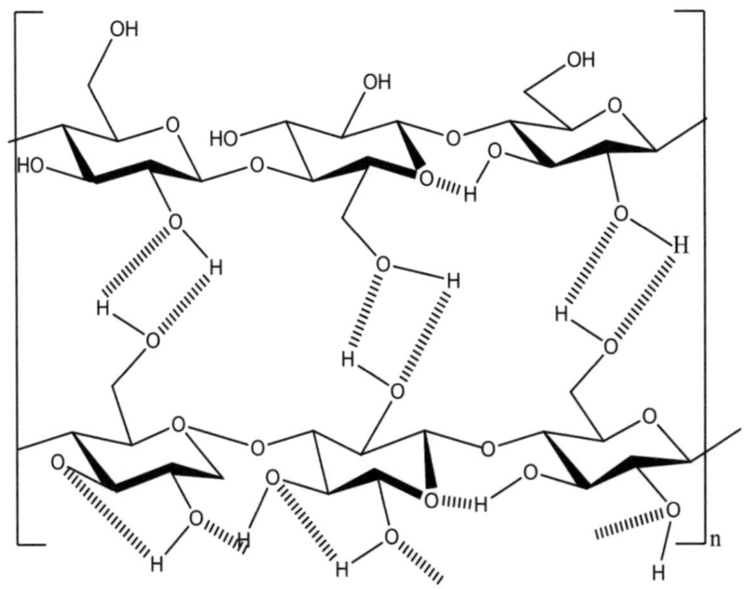

Figure 2.
Pictorial presentation of inter- and intramolecular H bonding in cellulose.

research and produced first cellulosic thermoplastics [4, 6, 7]. The research of cellulose modification has been developing, from one process to another, trying to find as the most eco-friendly and eco-efficient method as possible from the time when it was discovered as cellulose is the most abundant natural material on earth.

2. Cellulose dissolution history

Ever since cellulose has been discovered, many researches are being carried out on the most efficient ways to dissolve this natural polymer in various common solvents [8]. The first attempt toward cellulose dissolution was made by Hyatt Manufacturing Co. in which cellulose was converted into nitrocellulose with the aid of nitric acid. The nitric acid treatment of the cellulose gives cellulosic-based thermoplastic polymer that has several industrial and biological applications [4]. As time progresses a new and a better method, namely, "viscose," was developed which became effective from the 1890s and is the most widely used method in processing cellulose to date. Approximately 3.2 million tons of cellulose are processed per year using this method [9–11]. This method makes the use of alkali (NaOH) and carbon disulfide (CS_2) and produces cellulose xanthate. The detail methodology about the viscose process has been described elsewhere [12–14]. During the late 1960s, lyocell (amine oxides) technique was invented mainly to overcome the challenges of viscose method which is found to be more effective in processing of cellulose. Lyocell technique employs the use of N-methylmorpholine N-oxide (NMMO). NMMO dissolves cellulose directly due to its strong N-O dipole [15]. However lyocell technique was not as effective as "viscose" technique as it had significant engineering complications; hence viscose remained to be the most widely used technique [16]. Both of those cellulose dissolution techniques retain major environmental complications (some of which are listed below).

Viscose	Lyocell
• For every kilogram of cellulose produced, there is two kilograms of waste [11, 17] • Makes use of carbon disulfide and dihydrogen sulfide which are both environmental rivals and also not sustainable [10]	• Poor thermal stability of NMMO [17, 18] which leads to major investment in safety measures of the industry and difficulties in recycling the solvent [16] • NMMO requires high temperatures which leads to degradation of cellulose [19, 20]

Those challenges lead to further research in cellulose dissolution process in search of solvents which can be as environmental friendly as possible. Therefore the new solvents should be associated with low volatile, easily recyclable, high thermal stability, nontoxic, non-derivative, etc. [21]. The solvents which were found to be most effective were NaOH/thiourea and urea aqueous solvents as well as molten salts. The NaOH/thiourea was found to dissolve cellulose at low temperatures ranging from −8 to −5°C as described elsewhere in detail [22, 23]. However this method was found to produce cellulose fibers with high degree of crystallinity but lower degree of crystal orientation when compared to fibers obtained from viscose process [23, 24]. In order to advance this method, thiourea was replaced by urea forming a new solvent system NaOH/urea [12], and this solvent was found to have enhanced cellulose solubility by precooling the solution to −12°C. The advantage of the method is based on the fact that the by-products so formed were nontoxic. The methodologies related to the preparation of NaOH/Urea system can be elsewhere [11, 12, 24]. Another eco-friendly solvent that is still in the academia is ionic liquids (designated by ILs). Recently, the ILs have been identified as the most universal solvents for the future due to their several fascinating properties such as high polarity, negligible vapor pressure, nonflammability, low melting point, good thermal stability, tunable viscosity, broad liquid range, high thermal conductivity, good dissolution properties, etc. [25–27]. Because of the association of ILs with these fascinating properties, they (ILs) can be considered as most important, cost-effective, and environmental benign solvents for cellulose.

3. Ionic liquids

3.1 Brief summary of ionic liquids

Ionic liquids are commonly defined as molten salts; these salts are composed of inorganic or organic anions and organic cations. In ILs, the oppositely charged ions are held together by columbic forces [28–30]. Most widely studied ILs are room temperature ionic liquids (RTILs) having melting point of below 100°C [31–34]. The ILs are liquids at room temperature because they are composed of polyatomic, bulk, asymmetric organic cations and charge-diffuse ions [17, 35], and their ions are not packed well; hence, they remain liquid [36] at room temperature. The

S. no.	Cations		Anions
	Name	Structure	
1	Imidazolium	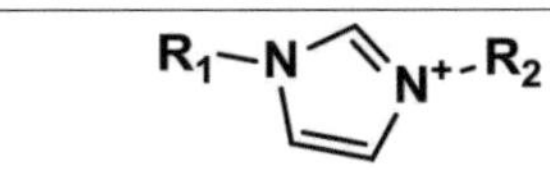 $[BMIM]^+$, $[AMIM]^+$, $[MMIM]^+$, $[EMIM]^+$, $[AMIM]^+$, $[HEMIM]^+$, $[HMIM]^+$, $[OMIM]^+$, $[C_2MIM]^+$, $[EMMIM]^+$	Cl^-, Br^-, SCN^-, $[PF_6]^-$, $[BF_4]^-$ $[CF_3SO_3]^-$, $[BF_4]^-$, $[PF_6]^-$, $[I]^-$, $[CH_3SO_3]^-$, $[Ac]^-$, $[HSCH_2COO]^-$, $[HCOO]^-$, $[(C_6H_5]COO]^-$, $[H2NCH_2COO]^-$, $[HOCH_2COO]^-$, $[CH_3CHOHCOO]^-$, $[Fmt]^-$, $[OAc]^-$, $[SCN]^-$, $[Tos]^-$, $[N(CN)_2]^-$, $[CH_3CO_2]^-$
2	Pyridinium	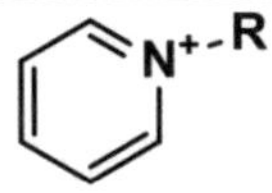 $[C_4MPY]^+$, $[BMPY]^+$, ($[C_nMPy]$ where n = 2–10), $[AMPy]^+$	
3	Ammonium	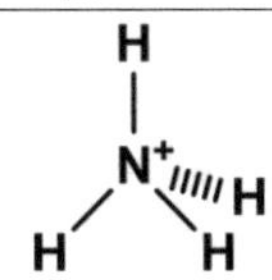 $[BDTAC]^+$	
4	Phosphonium	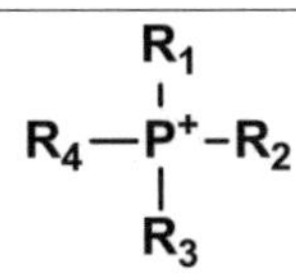	

Table 1.
Structures and abbreviations of some common cations and anions of ionic liquids employed as solvents.

RTILs are considered to be designer solvents since one can alter the physicochemical properties of an IL by simply varying anions' type or alky chain length of the cations [37–40]. The RTILs are composed of nitrogen or phosphorus containing dissymmetrical organic cations such as imidazolium, pyridinium, or ammonium or phosphonium cations and wide variety of simple anions such as chloride, nitrate, bromide, tetrafluoroborate, acetate, triflate, etc. [29, 41, 42] as shown in **Table 1**. So far, the ILs have been found to have numerous applications in various areas including solvent science for manufacturing of different materials including dissolution of biomass, electrochemistry for electrolytes in batteries, polymer chemistry for plasticizers, and separation technology for extractions and separations [26, 36, 37, 43]. Since ILs have been discovered, they are widely becoming advanced; in addition, the efficient methods of synthesis ILs are being developed. The application of ILs is vastly increasing yearly; recently, ILs have been started to be researched toward biomass processing, particularly cellulose, since cellulose is the most abundant biomaterial on earth [1, 17]; therefore, the use of ILs in cellulose chemistry can bring about economic sufficient developments archiving of the United Nation (UN) mission of moving away from depending on fossils fuels as well as using of harmful convectional solvents.

3.2 Cellulose dissolution using ionic liquids

The very first cellulose dissolution research was carried out by Richard P. Swatloski in the year 2002 [44]. Swatloski along with his coworkers reported that 1-butyl-3-methylimidazolium chloride ($[C_4MIM]^+$) with Cl^-, Br^-, and SCN^- could dissolve cellulose whereas with $[BF_4]^-$ and $[PF_6]^-$ could not. They were further observed that as increase in the alkyl chain length of the ILs decreases the solubility of the investigated ionic liquids as [C_4MIM][Cl], [C_6MIM][Cl], and [C_8MIM][Cl] since [C_4MIM][Cl] dissolved 10 wt%, [C_6MIM][Cl] dissolved 5 wt% and [C_8MIM][Cl] were slightly soluble. In addition the dissolution could be significantly enhanced by heating the solution using oil bath, microwave, or ultrasonic technique. This research indicated that "ionic liquids" can be used as solvents for cellulose and opened a new horizon in green chemistry [45]. Ever since Swatloski et al. [44] breakthrough in cellulose chemistry, many researchers proceed with his outcomes to further understand the dynamics and factors associated with the dissolution of cellulose using ionic liquids. To determine dynamics and factors that play a role in cellulose dissolution using ILs, properties such as different structures of the ionic liquids, degree of polymerization of cellulose, dissolution time, temperature, water content, and cosolvent were investigated starting from Zhang et al. [46] to Meenatchi et al. [47] including Kosan et al. [48], Heinze et al. [49], Lee et al. [50], Kilpeläinen et al. [51], Erdmenger et al. [52], Zavrel et al. [53], Sun et al. [54], Vits et al. [55], Fukaya et al. [21, 56], Sashina et al. [43], Xu et al. [57], Sescousse et al. [58], Zhen et al. [59], and Freire et al. [60], and the detailed summary of the factors affecting cellulose solubility is given in **Table 2**. It can be concluded that most of the studied ionic liquids are imidazolium-based ionic liquids; this is due to a fact that imidazolium ILs have the finest properties among other ILs; they have the lowest melting points; many of them are liquid at room temperature; they have high conductivity and a wide electrical window stability that makes them suitable for a variety of applications including solvents for the dissolution of cellulose [61]. Among the most widely studied imidazolium-based ILs [BMIM][Cl], [BMIM][Ac], [BMIM][Fmt], and [BMIM][OAc] together with [AMIM][Cl] and [AMIM][Fmt]

Factor	Explanation	Reference
Presence of water	Water content should be very low in both IL and cellulose, typically less than 1%	Swatloski et al. [44], Muhammad et al. [62]
Anions type	ILs containing anions that have strong electronegativity such as halides, e.g., Cl^-, have better dissolution properties unlike noncoordinating anions such as [BF_4] and [PF_6]	Holm et al. [63], Dadi et al. [64], Swatloski et al. [44]
Alkyl chain length	Methylimidazolium cations with even number of carbons were found to have high solubilizing power than the one with odd number and pyridinium-based ionic liquids showed that as alkyl chain length of the cation increases, the solubility decreases	Erdmenger et al. [52], Sashina et al. [43], Olivier-Bourdigou et al. [65]
Degree of polymerization	Solubility rate of cellulose decreases as degree of polymerization increases	Kosan et al. [48], Zhang et al. [46]
Dissolution time	Dissolution time should be short typically around 12 h at low heating temperature	Kilpeläinen et al. [51]
IL viscosity	Low viscosity promotes higher dissolution since it promotes greater ions mobility	Tywabi [24], Fort et al. [66], Kilpelainen et al. [51]
Cosolvent	Polar aprotic cosolvents promote higher and quicker dissolution at low temperature Polar protic is non-solvent which causes precipitation of the cellulose from the IL solution	Rinaldi et al. [67], Zhao et al. [68], Xu et al. [69, 70], Xu and Zhang [71], Bengtsson [72], Andanson et al. [73], Holding et al. [74], Swatloski et al. [44]

Table 2.
Detailed summary of factors affecting cellulose solubility.

as well as [EMIM][Cl], [EMIM][OAc], and [EMIM][MP] have been documented to be one of the most effective ILs for cellulose dissolution, with [BMIM][Cl] able to dissolve up to 25% wt of cellulose having a DP = 1000 using a microwave heating method. During dissolution process, it has been shown by aid of various analytical techniques that the strong inter- and intramolecular hydrogen bonding of cellulose is broken up by formation of intense hydrogen bonding between ionic liquid anions and cellulose hydroxyls making cellulose to be soluble; hence, ILs with noncoordinating anions are nonsolvents [44].

3.3 Cellulose dissolution using ionic liquids and cosolvents (organic solvents)

The use of organic solvents in the cellulose chemistry has been vain for many decades, which lead to the use of solvents like oxide amines (NMMO), NaOH/urea, LiCl/DMI, ILs ([BMIM][Cl], [AMIM][Cl]), etc. However every solvent has its own rewards and drawbacks; so far the greenest solvents proposed are the ILs. However, the use of ILs is still not yet practical and faces a lot of industrial challenges since ILs are available in small amount, are relatively costly, and have very high viscosity compared to other common aqueous thermochemical pretreatment reagents, and it deactivates regular cellulolytic enzymes [75–77]. To overcome these challenges faced by ILs, Renaldi [67] created a solvent system, which is a bicomponent containing both ionic liquid and cosolvent (polar aprotic solvents such as DMSO, DMF DMI, etc.) that significantly lowers the IL viscosity, which increases ionic mobility, thereby promoting higher cellulose dissolving rate than net ILs [57, 67] at ambient temperature [78]. Furthermore, Renaldi [60] in his studies conducted an experiment which revealed that the use of solvent system [BMIM][Cl]/DMI dissolves more cellulose (10 wt% in few minutes) than convectional solvent LiCl/DMI (2 wt% at 150°C for 30 min). Many researches today have further carried experiments to understand the cellulose chemistry using bicomponent IL with aprotic solvent [68–71, 73–75]. The results of studied cosolvents (DMSO, DMF, DMA, and DMI) with certain ILs showed that during the dissolution of cellulose, the aprotic cosolvent does not slightly interact with the hydroxyl of the cellulose; however, it decreases the association of the IL cation with the anion making more free ions to be available for the interaction of the cellulose hydroxyls; hence, more cellulose becomes readily soluble at ambient temperatures; furthermore, as more amount of the cosolvent is added to the solution, more cellulose continues to dissolve since more IL anions become available [67–71, 73–75, 78, 79]; in addition, Xu et al. [70] further indicated that the best aprotic cosolvent are the ones which have the highest dipole moment; hence, as the dipole moment decreases, the effectiveness of the aprotic cosolvent weakens.

4. Conclusion

In this review we summarized the major solvents that are applied in the cellulose chemistry. The ILs are implemented as new solvents for the dissolution of cellulose and its derivatives having several biological and industrial applications. The cellulose can be modified such as ether, ester, sulfate, amine, hemiacetal, carbanilate, etc., and the resulting material can be used for other applications. The cellulose derivatives can be used for various purposes such as textile, medicine, agriculture, biofuels, etc. In order for ionic liquids to be applied industrially, there are major problems still to be addressed such as:

- Development of the efficient ways to synthesis ionic liquids so that ILs can be available in high quantity.

- The need to do further research in solvent recovery for both ILs and for ILs with a cosolvent.
- To do research on IL chemical and toxicological data with cellulose.
- The dissolution mechanism of cellulose with ILs in details.

Author details

Indra Bahadur* and Ronewa Phadagi
Department of Chemistry and Materials Science Innovation, Modelling Research Focus Area, School of Mathematical and Physical Sciences, Faculty of Agriculture, Science and Technology, North-West University, Mmabatho, South Africa

*Address all correspondence to: bahadur.indra@gmail.com

IntechOpen

© 2019 The Author(s). Licensee IntechOpen. This chapter is distributed under the terms of the Creative Commons Attribution License (http://creativecommons.org/licenses/by/3.0), which permits unrestricted use, distribution, and reproduction in any medium, provided the original work is properly cited.

References

[1] O'sullivan AC. Cellulose. The structure slowly unravels. Cellulose. 1997;**4**:173

[2] Borbély É. Lyocell, the new generation of regenerated cellulose. Acta Polytechnica Hungarica. 2008;**5**:11-18

[3] Sjostrom E. Wood Chemistry: Fundamentals and Applications. 2nd ed, Elsevier Publication; 2013

[4] Isik M, Sardon H, Mecerreyes D. Ionic liquids and cellulose: Dissolution, chemical modification and preparation of new cellulosic materials. International Journal of Molecular Sciences. 2014;**15**:11922-11940

[5] Gürdağ G, Sarmad S. Cellulose Graft Copolymers: Synthesis, Properties, and Applications, Olysaccharide Based Graft Copolymers. Berlin, Heidelberg: Springer; 2013. p. 15

[6] Schönbein C. Notiz Über Eine Veränderung Der Pflanzenfaser Und Einiger Andern Organischen Substanzen, Berichte der Naturforschenden Gesellschaft (Basel). 1847;**7**:27

[7] Hyatt. US Patent. Google Patents. 1880;**232**:037

[8] Heinze T, Koschella A. Solvents applied in the field of Cellulose Chemistry: A mini review. Polímeros. 2005;**15**:84-90

[9] Ramamoorthy SK, Skrifvars M, Persson A. A review of natural fibers used in biocomposites: Plant, animal and regenerated cellulose fibers. Polymer Reviews. 2015;**55**:107-162

[10] Heinze T, Liebert T. Unconventional methods in cellulose functionalization. Progress in Polymer Science. 2001;**26**:1689

[11] Liebert T. Cellulose Solvents: Remarkable history, bright future, cellulose solvents. For analysis, shaping and chemical modification (chapter 1st). American Chemical Society; 2010. p. 3-54

[12] Liebert T, Heinze T. Interaction of ionic liquids with polysaccharides. 5. Solvents and reaction media for the modification of cellulose. BioResources. 2008;**3**:576

[13] Brown EK. Cellulose dissolution in ionic liquids and their mixtures with solvents: linking ion/solvent structure and efficacy of biomass pretreatment. North Carolina State University; 2012

[14] Sixta H. Handbook of Pulp. Wiley Online Library; 2006

[15] Cao Y, Wu J, Zhang J, Li H, Zhang Y, He J. Room temperature Ionic Liquids (RTILs): A new and versatile platform for cellulose processing and derivatization. Chemical Engineering Journal. 2009;**147**:13-21

[16] Hermanutz F, Gähr F, Uerdingen E, Meister F, Kosan B. New developments in dissolving and processing of cellulose in ionic liquids, macromolecular symposia. Wiley Online Library; 2008. p. 23-27

[17] Clough MT, Geyer K, Hunt PA, Son S, Vagt U, Welton T. Ionic liquids: Not always innocent solvents for cellulose. Green Chemistry. 2015;**17**:231-243

[18] Dorn S, Wendler F, Meister F, Heinze T. Interactions of ionic liquids with polysaccharides–7: thermal stability of cellulose in ionic liquids and N-Methylmorpholine-N-oxide. Macromolecular Materials and Engineering. 2008;**293**:907-913

[19] Wendler F, Graneß G, Heinze T. Evidence of autocatalytic reactions in cellulose/nmmo solutions with thermal

and spectroscopic methods. Lenzinger Berichte. 2005;**84**:92-102

[20] Rosenau T, Elder T, Potthast A, Herbert S, Kosma P. The lyocell process: Cellulose solutions in N-Methylmorpholine-N-oxide (NMMO)-degradation processes and stabilizers. In: 12th International Symposium on Wood and Pulping Chemistry. June 9-12. Madison, Wisconsin; 2003. pp. 305-308

[21] Fukaya Y, Hayashi K, Wada M, Ohno H. Cellulose dissolution with polar ionic liquids under mild conditions: Required factors for anions. Green Chemistry. 2008;**10**:44-46

[22] Zhang L, Ruan D, Gao S. Dissolution and regeneration of cellulose in naoh/thiourea aqueous solution. Journal of Polymer Science Part B: Polymer Physics. 2002;**40**:1521-1529

[23] Jiang M, Zhao M, Zhou Z, Huang T, Chen X, Wang Y. Isolation of cellulose with ionic liquid from steam exploded rice straw. Industrial Crops and Products. 2011;**33**:734-738

[24] Tywabi Z. Processing of dissolving pulp in ionic liquids, Doctoral dissertation. 2015. Available from: https://ir.dut.ac.za/handle/10321/1746

[25] Ye C, Liu W, Chen Y, Yu L. Room-temperature ionic liquids: A novel versatile lubricant. Chemical Communications. 2001:2244-2245

[26] Han D, Row KH. Recent applications of ionic liquids in separation technology. Molecules. 2010;**15**:2405-2426

[27] Baranyai KJ, Deacon GB, MacFarlane DR, Pringle JM, Scott JL. Thermal degradation of ionic liquids at elevated temperatures. australian journal of chemistry. 2004;**57**:145-147

[28] Rogers RD, Seddon KR. Ionic liquids--solvents of the future? Science. 2003;**302**:792-793

[29] Berthod A, Ruiz-Angel M, Carda-Broch S. Ionic liquids in separation techniques. Journal of Chromatography A. 2008;**1184**:6-18

[30] Heintz A. Recent Developments in thermodynamics and thermophysics of non-aqueous mixtures containing ionic liquids. A review. The Journal of Chemical Thermodynamics. 2005;**37**:525-535

[31] Yang Z, Pan W. Ionic liquids: Green solvents for nonaqueous biocatalysis. Enzyme and Microbial Technology. 2005;**37**:19-28

[32] Alvarez VH, Mattedi S, Martin-Pastor M, Aznar M, Iglesias M. Thermophysical properties of binary mixtures of {Ionic Liquid 2-Hydroxy Ethylammonium Acetate+ (Water, Methanol, or Ethanol)}. The Journal of Chemical Thermodynamics. 2011;**43**:997-1010

[33] Andreatta AE, Arce A, Rodil E, Soto A. Physical properties of binary and ternary mixtures of Ethyl Acetate, Ethanol, and 1-Octyl-3-Methyl-Imidazolium Bis (Trifluoromethylsulfonyl) Imide at 298.15 K. Journal of Chemical & Engineering Data. 2009;**54**:1022-1028

[34] Bahadur I, Kgomotso M, Ebenso EE, Redhi G. Redhi, influence of temperature on molecular interactions of imidazolium-based ionic liquids with acetophenone: Thermodynamic properties and quantum chemical studies. RSC Advances. 2016;**6**:104708-104723

[35] Laus G, Bentivoglio G, Schottenberger H, Kahlenberg V, Kopacka H, Röder T, et al. Sixta, ionic liquids: Current developments, potential and drawbacks for industrial applications. Lenzinger Berichte. 2005;**84**:71-85

[36] Keskin S, Kayrak-Talay D, Akman U, Hortaçsu Ö. A review of ionic liquids

towards supercritical fluid applications. The Journal of Supercritical Fluids. 2007;**43**:150-180

[37] Plechkova NV, Seddon KR. Applications of ionic liquids in the chemical industry. Chemical Society Reviews. 2008;**37**:123-150

[38] Rogers RD, Seddon KR, Volkov S. Green industrial applications of ionic liquids. Springer Science & Business Media; 2012

[39] Wu C-T, Marsh KN, Deev AV, Boxall JA. Liquid– liquid equilibria of room-temperature ionic liquids and butan-1-ol. Journal of Chemical & Engineering Data. 2003;**48**:486-491

[40] Holbrey J, Seddon K. Ionic liquids. Clean Products and Processes. 1999;**1**:223-236

[41] Liu J-f, Jiang G-b, Jönsson JÅ. Application of ionic liquids in analytical chemistry. TrAC Trends in Analytical Chemistry. 2005;**24**:20-27

[42] Stepnowski P. Application of chromatographic and electrophoretic methods for the analysis of imidazolium and pyridinium cations as used in ionic liquids. International Journal of Molecular Sciences. 2006;**7**:497-509

[43] Sashina E, Novoselov N, Kuz'mina O, Troshenkova S. Ionic liquids as new solvents of natural polymers. Fibre Chemistry. 2008;**40**:270-277

[44] Swatloski RP, Spear SK, Holbrey JD, Rogers RD. Dissolution of cellose with ionic liquids. Journal of the American Chemical Society. 2002;**124**:4974-4975

[45] El Seoud OA, Koschella A, Fidale LC, Dorn S, Heinze T. Applications of Ionic liquids in carbohydrate chemistry: A window of opportunities. Biomacromolecules. 2007;**8**:2629-2647

[46] Zhang H, Wu J, Zhang J, He J. 1-Allyl-3-Methylimidazolium chloride room temperature ionic liquid: A new and powerful nonderivatizing solvent for cellulose. Macromolecules. 2005;**38**:8272-8277

[47] Meenatchi B, Renuga V, Manikandan A. Cellulose dissolution and regeneration using various imidazolium based protic ionic liquids. Journal of Molecular Liquids. 2017;**238**:582-588

[48] Kosan B, Michels C, Meister F. Dissolution and forming of cellulose with ionic liquids. Cellulose. 2008;**15**:59-66

[49] Heinze T, Schwikal K, Barthel S. Ionic liquids as reaction medium in cellulose functionalization. Macromolecular Bioscience. 2005;**5**:520-525

[50] Lee SH, Doherty TV, Linhardt RJ, Dordick JS. Ionic liquid-mediated selective extraction of lignin from wood leading to enhanced enzymatic cellulose hydrolysis. Biotechnology and Bioengineering. 2009;**102**:1368-1376

[51] Kilpeläinen I, Xie H, King A, Granstrom M, Heikkinen S, Argyropoulos DS. Dissolution of wood in ionic liquids. Journal of Agricultural and Food Chemistry. 2007;**55**:9142-9148

[52] Erdmenger T, Haensch C, Hoogenboom R, Schubert US. Homogeneous tritylation of cellulose in 1-Butyl-3-Methylimidazolium chloride. Macromolecular Bioscience. 2007;**7**:440-445

[53] Zavrel M, Bross D, Funke M, Büchs J, Spiess AC. High-throughput screening for ionic liquids dissolving (Ligno-) cellulose. Bioresource Technology. 2009;**100**:2580-2587

[54] Sun N, Rahman M, Qin Y, Maxim ML, Rodríguez H, Rogers RD. Complete

dissolution and partial delignification of wood in the ionic liquid 1-Ethyl-3-Methylimidazolium acetate. Green Chemistry. 2009;**11**:646-655

[55] Vitz J, Erdmenger T, Haensch C, Schubert US. Extended dissolution studies of cellulose in imidazolium based ionic liquids. Green Chemistry. 2009;**11**:417-424

[56] Fukaya Y, Sugimoto A, Ohno H. Superior solubility of polysaccharides in low viscosity, polar, and halogen-free 1, 3-Dialkylimidazolium formates. Biomacromolecules. 2006;**7**:3295-3297

[57] Xu A, Wang J, Wang H. Effects of anionic structure and lithium salts addition on the dissolution of cellulose in 1-Butyl-3-Methylimidazolium-based ionic liquid solvent systems. Green Chemistry. 2010;**12**:268-275

[58] Sescousse R, Gavillon R, Budtova T. Aerocellulose from cellulose–ionic liquid solutions: Preparation, properties and comparison with cellulose–naoh and cellulose–nmmo routes. Carbohydrate Polymers. 2011;**83**:1766-1774

[59] Ding Z-D, Chi Z, Gu W-X, Gu S-M, Liu J-H, Wang H-J. Theoretical and experimental investigation on dissolution and regeneration of cellulose in ionic liquid. Carbohydrate Polymers. 2012;**89**:7-16

[60] Freire MG, Teles ARR, Rocha MA, Schröder B, Neves CM, Carvalho PJ, Evtuguin DV, Santos LM, Coutinho JA. Thermophysical characterization of ionic liquids able to dissolve biomass. Journal of Chemical & Engineering Data. 2011;**56**:4813-4822

[61] Ngo HL, LeCompte K, Hargens L, McEwen AB. McEwen, Thermal Properties of Imidazolium Ionic Liquids. Thermochimica Acta. 2000;**357**:97-102

[62] Muhammad N, Man Z, Khalil MAB. Ionic Liquid—A future solvent for the enhanced uses of wood biomass. European Journal of Wood and Wood Products. 2012;**70**:125

[63] Holm J, Lassi U. Ionic liquids in the pretreatment of lignocellulosic biomass, ionic liquids. ionic liquids: Applications and Perspectives. InTech; 2011

[64] Dadi AP, Varanasi S, Schall CA. Schall, enhancement of cellulose saccharification kinetics using an ionic liquid pretreatment step. Biotechnology and Bioengineering. 2006;**95**:904

[65] Olivier-Bourbigou H, Magna L, Morvan D. Morvan, ionic liquids and catalysis: Recent progress from knowledge to applications. Applied Catalysis A: General. 2010;**373**:1

[66] Fort DA, Remsing RC, Swatloski RP, Moyna P, Moyna G, Rogers RD. Can ionic liquids dissolve wood? processing and analysis of lignocellulosic materials with 1-N-Butyl-3-Methylimidazolium chloride. Green Chemistry. 2007;**9**:63

[67] Rinaldi R. Instantaneous dissolution of cellulose in organic electrolyte solutions. Chemical Communications. 2011;**47**:511

[68] Zhao Y, Liu X, Wang J, Zhang S. Insight into the cosolvent effect of cellulose dissolution in imidazolium-based ionic liquid systems. The Journal of Physical Chemistry B. 2013;**117**:9042

[69] Xu A, Cao L, Wang B, Ma J. Dissolution behavior of cellulose in IL. Advances in Materials Science and Engineering. 2015

[70] Xu A, Cao L, Wang B. Acile Cellulose dissolution without heating in [C4mim][CH3COO]/DMF solvent. Carbohydrate Polymers. 2015;**125**:249

[71] Xu A, Zhang Y. Insight into dissolution mechanism of cellulose in

[C4mim][CH3COO]/DMSO solvent by 13C NMR spectra. Journal of Molecular Structure. 2015;**1088**:101

[72] Bengtsson J, De Clerck KP, Persson AC. Evaluating recyclability and suitability of tetrabutylammonium acetate: Dimethyl sulfoxide as a solvent for cellulose. 2016

[73] Andanson J-M, Bordes E, Devémy J, Leroux F, Pádua AA, Gomes MFC. Understanding the role of co-solvents in the dissolution of cellulose in ionic liquids. Green Chemistry. 2014;**16**:2528

[74] Holding AJ, Parviainen A, Kilpeläinen I, Soto A, King AW, Rodríguez H. Efficiency of hydrophobic phosphonium ionic liquids and DMSO as recyclable cellulose dissolution and regeneration media. RSC Advances. 2017;**7**:17451

[75] Wu L, Lee S-H, Endo T. Effect of dimethyl sulfoxide on ionic liquid 1-Ethyl-3-Methylimidazolium acetate pretreatment of eucalyptus wood for enzymatic hydrolysis. Bioresource Technology. 2013;**140**:90

[76] Engel P, Mladenov R, Wulfhorst H, Jäger G, Spiess AC. Point by point analysis: How ionic liquid affects the enzymatic hydrolysis of native and modified cellulose. Green Chemistry. 2010;**12**:1959

[77] Turner MB, Spear SK, Huddleston JG, Holbrey JD, Rogers RD. Ionic liquid salt-induced inactivation and unfolding of cellulase from trichoderma reesei. Green Chemistry. 2003;**5**:443

[78] Xu A, Zhang Y, Zhao Y, Wang J. Cellulose dissolution at ambient temperature: Role of preferential solvation of cations of ionic liquids by a cosolvent. Carbohydrate Polymers. 2013;**92**:540

[79] Rein DM, Khalfin R, Szekely N, Cohen Y. True molecular solutions of natural cellulose in the binary ionic liquid-containing solvent mixtures. Carbohydrate Polymers. 2014;**112**:125

Section 3

Solvent Effects

Chapter 10

Solvent Effects on Dye Sensitizers Derived from Anthocyanidins for Applications in Photocatalysis

Diana Barraza-Jiménez, Azael Martínez-De la Cruz, Leticia Saucedo-Mendiola, Sandra Iliana Torres-Herrera, Adolfo Padilla Mendiola, Elva Marcela Coria Quiñones, Raúl Armando Olvera Corral, María Estela Frías-Zepeda and Manuel Alberto Flores-Hidalgo

Abstract

Anthocyanidins under the effects of solvents water, ethanol, n-hexane, and methanol are interesting due to their suitability as natural dyes for photocatalytic applications. In this chapter, DFT and TDDFT methodologies are used to study their electronic structure. The results displayed include HOMO, LUMO, HOMO-LUMO gap, chemical properties, and reorganization energies for the ground states, and excited state data are also displayed. Malvidin in gas phase has lower gap energy. After addition of solvents, gap energy increases in all cases but malvidin with n-hexane presents narrower gap. Conceptual DFT results show that cyanidin and malvidin may have good charge transfer. Cyanidin presented lower electron reorganization energy (λ_e) using solvent water; however, ethanol and methanol had similar values. TDDFT is used to calculate excited states, and absorption data show wavelength main peak between 479.1 and 536.4 nm. UV-Vis absorption spectra were generated and solvent effects on each molecule is discussed. Anthocyanidins work well in the visible region with the stronger peak at the green region. These pigments are good options for photocatalysis application and cyanidin and malvidin, in this order, may be the best choices for dye sensitization applications.

Keywords: anthocyanidins, dyes, solvent effects, DSSC, TDDFT

1. Introduction

Organic pigments have raised great interest in late years, may be driven by their potential in renewable energy applications which has been reinvigorated with the invention of dye-sensitized solar cells (DSSCs). Dye-sensitized solar cells (DSCs) are an attractive solar energy conversion technology and present advantages that include low cost of manufacture, ease of fabrication, and modifiable features such as color and transparency [1–5]. First DSSCs employed ruthenium (II)-based dyes in conjunction with iodide-based electrolytes to achieve an 11.9% solar-to-electric power conversion efficiency (PCE) [6]. A new generation of DSSCs based on naturally obtained pigments is a great option as dye sensitizers, and this is the reason of a revitalized interest in these pigments [6, 12]. For example, porphyrin-based dyes have been tested as viable options, and they displayed great flexibility to work as panchromatic sensitizers [6, 13, 14]. It has been reported that porphyrin chromophore has strong light absorption around 400 nm in the blue region which is known as the Soret band or Soret peak and also in the Q-bands which is a region between slightly over 500 and 620 nm, but presents weak absorption in the region between these two features [6].

Then, it may be considered an interesting green option using organic pigments in DSSC technology. Analogously, these principles may be used to decompose chemical pollutants naturally without any contaminant waste. These organic pigments possess environmentally friendly properties, easy accessibility, and high absorption in the visible region which make them good candidates [7]. An alternative organic pigment to porphyrins may be anthocyanidins, a group of flavonoids contained in different parts of plants such as fruits, leaves, and flowers. Anthocyanidins may be considered water-soluble plant pigments that usually carry colors ranging from red to blue [8]. These natural pigments have shown health benefits and are commonly used colorants in food industry [9, 10].

Researchers continue to look for viable alternatives to ruthenium-based dyes and DSSC components in order to increase efficiency [11–15]. Natural pigments represent, in particular, a good option and among them anthocyanins are within our research interest [13–15]. These pigments have shown relevant advantages in DSSC technology, for example, they are metal-free, nontoxic, widely available, and inexpensive. They also have hydroxyl groups that benefit binding with TiO_2 and have been shown to be able to inject electrons into the TiO_2 conduction band at an ultrafast rate when excited with visible light [16]. There have been several studies on DSSC using anthocyanins as a photosensitizer with promising results [17–21]. The efficiency (η) from those studies, however, was generally quite low (0.5–0.6%). Recently, one of these reports [22] was carried out using sealed solar cells with enhanced electrodes (multilayer TiO_2 film plus a scattering layer), and anthocyanins contained organic acids demonstrated an efficiency of around 1.0%.

In nature, dyes can absorb visible light to enable plant photochemical processes; many of them are able to inject an electron into the conduction band of the semiconductor which is fundamental for photocatalytic processes [23]. This property is of great interest in dye-sensitized solar cells (DSSC), where dyes are used with a photocatalyst that may be a semiconductor oxide such as TiO_2 or ZnO for example [24]. An important consideration relates to prevent the degradation of the dye on a DSSC but this may not be the case for aqueous suspension of dye and photocatalyst. In such case, it may be confusing whether the dye degradation is due to dye sensitization itself or by action of the photocatalyst or under the influence of both factors [25, 26].

In regard to chemical processes, for chemical decoloration, the oxidation method is the most used because of its easy application. This method may be found in the literature as chemical oxidation and advanced oxidation process. Both these methods achieve the degradation of chemical dyes, pesticides among other pollutants, either partially or completely under ambient conditions [27]. The advanced oxidation process may be categorized in photocatalytic oxidation (use of light for activation of catalyst) and Fenton chemistry suitable for treating wastewater in particular for processes resistant to biological treatment [27].

Then, dye-sensitized process may be used in other applications and in this chapter, we will refer to its application in photocatalysis. This process uses light to activate a photocatalyst and represents a potential application to take advantage of sunlight for diverse processes such as gas purification, H_2 production, and water treatment. There are limitations for dye-sensitized semiconductors; for example, in water purification, the organic dye may be diluted by water or at least erodes or deteriorates from the photocatalyst surface due to continuous interaction with water. Of course, the specific device configuration used during water treatment defines the disposition and interaction of the photocatalyst with the liquid and ultimately defines the severity of the fluid effects on the sensitizer layer. Dye sensitization effect may be a good choice for water treatment, but its effectiveness depends on the device disposition and the sensitizer presentation mainly when it gets in touch with the polluted water.

Among the different natural pigments, anthocyanidins represent an interesting alternative as dye sensitizing naturally obtained pigment. Since dye sensitizer and DSSC advances may be used in solar technology applications such as photocatalysis, this chapter presents interesting information related to anthocyanidins focused on its potential application in renewable energy applications and in particular when used as dye sensitizing pigment in photocatalysis. In particular, the chapter presents information related to an analysis of the effects caused by commonly used solvents to obtain anthocyanidins such as gas phase (as comparative basis), water, ethanol, n-hexane, and methanol and includes discussion on how several electronic properties of interest are subject to different effects in consequence depending on the selected solvent.

2. Anthocyanidin molecular structure

In this section, anthocyanidin structural data from published references and also our own results obtained with DFT methodology are presented. For convenience, structural data are presented in this section and computational details used in DFT calculations will be included in the next section.

2.1 Anthocyanidins structure

Anthocyanidins are natural pigments commonly found in plants with a molecular structure based on the flavylium ion or 2-phenylchromenylium (chromenylium may be referred to as benzopyrylium). These natural pigments are salt derivatives of the 2-phenylchromenylium cation, commonly known as flavylium cation. The more common anthocyanidins and their substitution pattern are shown in **Table 1**.

The phenyl group at the 2-position can carry different substituents that determine a particular anthocyanidin. With a positive charge, anthocyanidins differ from other flavonoids. Pigment molecule substituents and features are summarized in **Table 1** with a general interpretation of structural differences amongst variants, and a general scheme for anthocyanidins is displayed in **Figure 1**.

Name	Chemical formula	Substitution pattern		Color
		R1	R2	
Cyanidin	$(C_{15}H_{11}O_6)^+$	OH	OH	Orange-red
Delphinidin	$(C_{16}H_{11}O_7)^+$	OH	OH	Blue-red
Malvidin	$(C_{15}H_{13}O_5)^+$	OCH_3	OCH_3	Blue-red
Pelargonidin	$(C_{15}H_{11}O_5)^+$	H	H	Orange
Peonidin	$(C_{15}H_{13}O_6)^+$	OCH_3	H	Orange-red
Petunidin	$(C_{15}H_{12}O_6)^+$	OCH_3	OH	Blue-red

Table 1.
Six more common anthocyanidins with their variants.

The core of an anthocyanidin is a 15-carbon structure forming two aromatic rings (A and B in **Figure 1**) joined by a third ring (C) that contains an oxygen atom that provides the molecule positive charge. The presence of two C=C bonds in the C ring distinguishes anthocyanidins from other flavonoids and imparts a positive charge to the molecule, which results to be a cation (known as flavylium) in its stable form at low pH [28].

The phenylbenzopyrylium core of anthocyanins is typically modified by the addition of a wide range of chemical groups through hydroxylation, acylation, and methylation. In this section, structural data obtained with DFT geometry calculations are included as displayed in the next paragraphs.

2.2 Structure parameters for selected anthocyanidins using DFT

Structure calculations are needed in DFT methodology because every analysis by this methodology needs first of all relaxed geometries able to provide fundamental data for the molecules ground states. A ball-stick model was used to represent each of the constituent atoms (**Figure 2**).

To obtain molecular initial parameters, molecular database Chemical Entities of Biological Interest (ChEBI) [29] was consulted and three selected anthocyanidin molecules were downloaded from this database. Three of the more common anthocyanidin variants were selected for DFT calculations. These anthocyanidin models were used as initial input data for our DFT calculations. Within this section, our DFT results corresponding to geometry parameters for the selected three anthocyanidins, cyanidin, malvidin, and peonidin, respectively, are included. Bond length values, angles, and dihedral angles obtained from DFT calculations are shown in **Table 2**.

In general, C-C bond length found with the theoretical methodology used within this work is near to the typical value for the case of benzene; it is known that bonds have the same length of 140 pm. Benzene C-C bond length average value is between the generally known length of single and double C-C bonds of 154.0 and 134 pm, respectively. In average, for selected molecules, C-C bond length within this work is 139.9 pm.

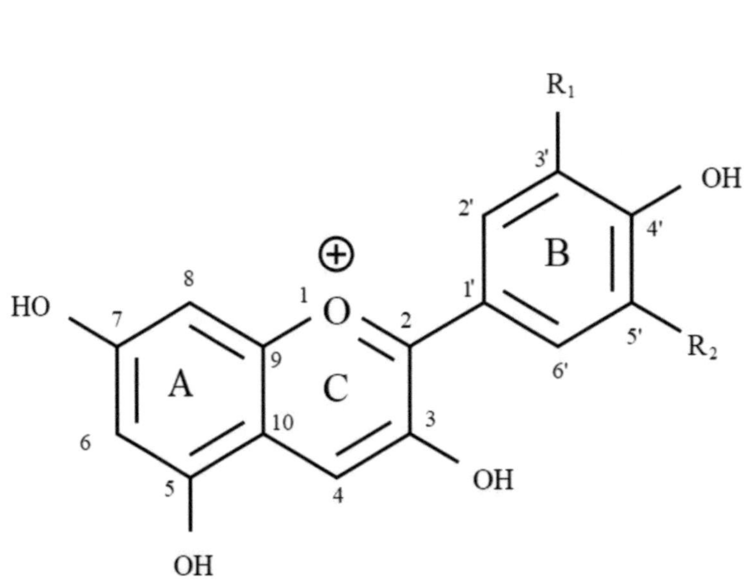

Figure 1.
Structure of anthocyanidins in their pristine form in correlation with **Table 1.**

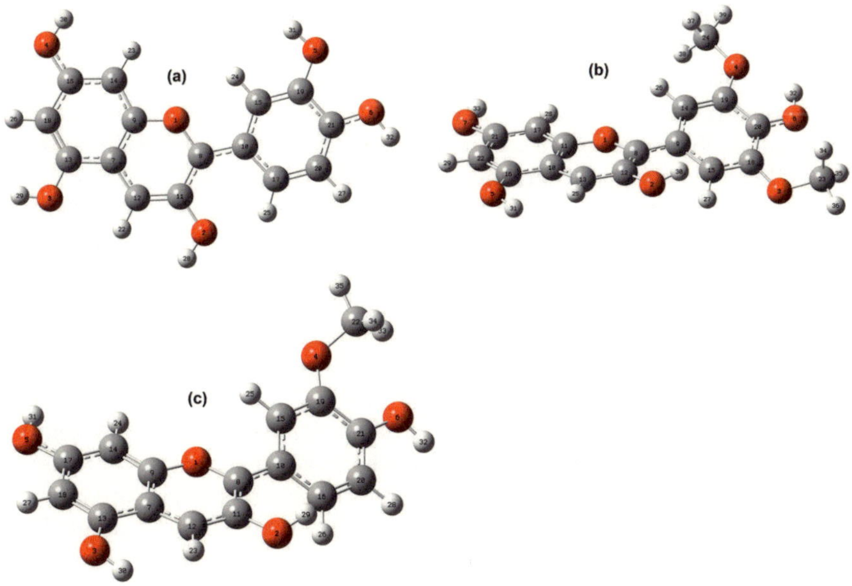

Figure 2.
Selected anthocyanidin structure in their pristine form after geometry relaxation using DFT methodology, (a) cyanidin, (b) malvidin, and (c) peonidin.

Parameter	Cyanidin	Malvidin	Peonidin	Parameter	Cyanidin	Malvidin	Peonidin
O(1)-C(2)	1.350	1.347	1.345	C(3′)-C(4′)	1.422	1.418	1.420
O(1)-C(9)	1.358	1.359	1.358	C(4′)-C(5′)	1.396	1.407	1.399
C(2)-C(3)	1.420	1.407	1.404	C(5′)-C(6′)	1.383	1.395	1.384
C(2)-C(1′)	1.436	1.444	1.447	O(1)-C(2)-C(1′)-C(6′)	180.0	151.4	150.1
C(3)-C(4)	1.382	1.388	1.390	C(3)-C(2)-C(1′)-C(2′)	180.0	149.3	149.1
C(4)-C(10)	1.403	1.399	1.397	O-C(3′)-C(4′)-C(5′)	180.0	179.3	177.4
C(5)-C(6)	1.376	1.376	1.375	H-C(5′)-C(4′)-C(3′)	180.0	175.9	178.5
C(5)-C(10)	1.427	1.435	1.436	O-C(4′)-C(3′)-C(2′)	180.0	177.0	176.7
C(6)-C(7)	1.412	1.408	1.409	O-C(4′)-C(5′)-C(6′)	180.0	178.0	177.7
C(7)-C(8)	1.395	1.398	1.398	C(8)-C(9)-C(10)-C(4)	180.0	176.3	175.9
C(8)-C(9)	1.386	1.382	1.381	O(1)-C(9)-C(10)-C(5)	180.0	178.8	178.7
C(9)-C(10)	1.409	1.421	1.423	C(8)-C(9)-O(1)-C(2)	180.0	179.9	179.4
C(1′)-C(2′)	1.422	1.414	1.409	C(5)-C(10)-C(4)-C(3)	180.0	179.3	179.5
C(1′)-C(6′)	1.414	1.407	1.411	C(9)-O(1)-C(2)-C(1′)	180.0	179.1	179.0
C(2′)-C(3′)	1.377	1.381	1.387				

Table 2.
Three selected anthocyanidins' geometric parameters, bond length, and bond angles in Å and °, respectively.

Our results for C-C bonds in average for selected anthocyanidins are within the range of 1.346–1.444 Å with <0.1 Å of difference between the larger and the shorter bonds for all cases. Literature reports for geometries include different methodologies such as B3LYP/6-31G(d) and B3LYP/6-31+G(d,p) [30–33]. All reports are in agreement that B3LYP reaches accurate results for this kind of molecules and overall C-C bond lengths are in good agreement with our results.

Dihedral angles are a good indication of the planarity in a structure; for anthocyanidins, we focused more in analyzing planarity among the three rings that form the molecule skeleton within each anthocyanidin. Also, the literature reports torsion angle as a parameter related to dihedral angles and this value may be used as a factor that helps differentiate anthocyanidins and their electronic structure behavior [30]. Dihedral values show that cyanidin is a planar molecule, selected values are 180°, and in general, all dihedrals are planar or differ with <1°. Peonidin presents more dihedrals that deviate from 180° but only a couple of dihedrals deviate by more than 5°. This last observation occurs for all the selected anthocyanidins; only a couple of dihedrals deviate in a significant amount from planarity but this small difference in the planarity determines the molecule character and its chemical properties. Then, only a few dihedrals indicate a nonplanar structure; these correspond to the relative angle variation observed in the B ring compared with the rest of the structure. These situations occur in all selected structures except in cyanidin which is a planar structure as shown by its dihedral values.

3. Solvents used to obtain anthocyanidins

In this section, literature related to the use of solvents during anthocyanidin extraction process is reviewed. Some differences in the material properties depending on the solvents used during its different chemical processes to obtain viable natural dye are expected. The same situation for anthocyanidins prevails, because different processes are used to obtain anthocyanidins in which it is needed to use different solvents resulting in behavior and property changes.

3.1 Anthocyanidin extraction

Anthocyanidin-rich extracts can be prepared from fresh, frozen, or dried plant materials. Examples of plants rich in anthocyanidins include blueberry [34], elderberry [35], and purple corn [36], among others. The particle size of source materials is an important factor during extraction; milling or grinding procedures among others are used with the goal of increasing surface area as well as the amount of compound obtained from the extraction process. Liquid nitrogen or lyophilization procedures may be complementary options during the grinding step to reduce anthocyanidin degradation. These are important recommendations given that the compounds involved may be subject to degradation caused by various factors, when carrying out the extraction procedure.

A general classification of extraction procedures is based in its phase used during the procedure such as solid or liquid extraction. Solid extraction is applied to liquid matrices, typically only during purification rather than extraction due to saturation of the absorbents. In the case of liquid extraction, a better recovery yield of anthocyanidins may be expected and for this reason, it is the more commonly used technique to extract these pigments from fruit sources. An important note captured from the literature is that there is a general practice that anthocyanidins are extracted with acidified water and polar organic solvents (methanol, ethanol, and acetonitrile) due to their hydrophilic nature [37, 38]. More recently, other solvents have been used (e.g., lactic acid-glucose and choline chloride-malic acid mixtures) in an attempt to increase green alternatives to extract anthocyanidins to avoid toxic methodologies [37].

The extraction system can also require subsequent analytical procedures, which is an important consideration. For example, it is noticed that less polar solvents (such as ethanol and acetone) used for the extraction of anthocyanidins from haskap berries compressed the Sephadex LH-20 gel used for extract purification [38]. This step has been considered responsible for favorable results such as longer retention times when the fractions were analyzed by high-performance liquid chromatography and, additionally, the co-extraction of impurities. Less favorable notes that can be mentioned from this study relate to how acetone had a low extraction efficiency and formed anthocyanidin-derived complexes (5-methyl-pyranoanthocyanin) which were not found in fresh fruits [38].

In relation to the techniques, various methods have been developed to increase the efficiency of liquid extractions, decrease the processing time, and minimize the use and exposure to organic solvents. Examples are supercritical fluid extraction [39], pressurized liquid extraction [40], microwave-assisted extraction [41], and ultrasound-assisted extraction [42].

3.2 Solvents used to obtain anthocyanidins

Undoubtedly, solvents may be considered very important in the food, agrochemical, chemical, and biotechnological, among others, process technologies. New streams of scientific research related to solvents are good alternatives for new experimentation. It may be worth mentioning the more relevant advances in this matter during the past two decades where supercritical fluids, ionic liquids, and deep eutectic solvents became the most outstanding subjects actively investigated as potential green solvents [43], in particular for the research associated with food, flavors, fragances, and medicinal plants.

Another important perspective is solvents applied in industry, where the value of the products is not only dependent on the production costs themselves but also on the way of production. A critical aspect on the solvent selection for any process including industrial processes is related to safety. An aspect is avoiding the use of chemicals that are potentially dangerous for human health but extends also to have a solvent environmentally friendly including its disposal and also of the waste products containing the solvent. Nowadays, all these aspects need to be considered when real costs of production are calculated. For these reasons and the regulations, green technology is becoming an essential part of process cost and impact estimations. Above all, green technology is a critical aspect for solvent technology and starts as the first action to improve production processes including long and short-term impact [43] and relates directly on environmental matters and thus on earth survival.

4. DFT calculations, results, and discussion

In this section, DFT calculation results from our own research are displayed and analyzed. The first set of results contains ground state data, mainly related to energy results including molecular orbitals, energy gap, and relevant chemical properties. In the second set of results are included excited state data with their corresponding molecular orbital diagrams and absorption spectra based on TDDFT calculations.

4.1 Computational methods and details

All calculations were carried out in gas phase and using four different solvents, water, ethanol, n-hexane, and methanol. These solvents were selected because they are used commonly in the process to obtain pigments in the laboratory. PCM (polarizable continuum solvation model) was employed in the present work according to its implementation in G09 program suite. Anthocyanidin geometry was relaxed with B3LYP/6-311+g(d,p), and all of them were built resembling previously reported geometric parameters but a different theoretical method was used during the set of calculations.

Geometry optimizations and vibrational frequency analyses were carried out using DFT with the well-known B3LYP approach, which includes the interchange hybrid functional from Becke in combination with the correlation functional three parameter by Lee-Yang-Parr [44] 6-311+g(d,p) basis set as implemented in the Gaussian09 program package [45]. We selected 6-311+g(d,p) because after running a set of calculations with the selected natural pigments using the reported basis set for similar organic molecules, 6-311+g(d,p) result values were comparable to the different basis sets recommended by the literature. Furthermore, several research works reported that the B3LYP/6-311+g(d,p) theoretical method provides good results with a good level of accuracy for similar organic materials [46–50]. Each geometry optimization was followed by calculations for harmonic vibrational frequencies in order to confirm that a local minimum has been reached. After vibrational frequency results are obtained, the zero-point vibrational energy (ZPVE) and the thermal correction (TC) at 298.15 K were also included to complete these calculations. Energy calculations were performed for all molecules, adiabatic energies were obtained, and with these values, global and local chemical reactivity indexes were evaluated to find the electronic properties and some of its chemical properties such as HOMO, LUMO, gap, ionization potential (IP), electronic affinity (EA), electrophilicity (ω), electronegativity (χ), and hardness (η). All calculations were carried out in gas phase and using four different solvents, water, ethanol, n-hexane, and methanol. These solvents were selected because they are used commonly in the process to obtain pigments in the laboratory. PCM (polarizable continuum solvation model) was employed in the present work according to its implementation in G09 program suite.

Our results are compared with results by other research teams that worked with the selected molecules with other methodologies or experimentally and also the generally accepted TiO_2 was used as reference in its bulk presentation [46–50] to gain insight into the pigment application as dyes. Calculations were made for several excited states, but for practical purposes, only first excited states are displayed in the result table. Excited state calculations were carried out using TDDFT with the same theoretical method, B3LYP/6-311g+(d,p). Energy graphs and excited state spectral diagrams were developed using the Chemissian code [51].

4.2 Electronic structure obtained from DFT calculations

Energy calculations for selected anthocyanidins were carried out with the B3LYP/6311+g(d,p) theoretical model for gas phase and using solvents water, ethanol, n-hexane, and methanol. To the best of our knowledge, this theoretical method has not been reported before for these specific molecules and solvents but other research groups have used other basis sets in their works. HOMO and LUMO molecular orbitals were calculated and these values are displayed in **Table 3**. The importance of molecular orbital calculation relies in the possibility that energy orbitals in these pigments may overlap with a semiconductor energy orbital such as TiO_2 or ZnO for photocatalytic applications.

HOMO and LUMO are involved in the electronic transitions because the photo-induced electron transfers from the dye excited state to the semiconductor surface. It has been reported in the literature that dye sensitizer energy levels for HOMO and LUMO are required to match the potential of the electrolyte redox and the conduction band edge level of a semiconductor such as TiO_2 [46].

Selected anthocyanidins within this work at their ground and excited states match well with the redox level of the electrolyte (−4.85 eV) and the conduction band edge for TiO_2 (−4.00 eV) respectively, according to reported literature values [46–50].

Molecular orbitals were calculated for selected anthocyanidins in gas phase and using solvents water, ethanol, n-hexane, and methanol. LUMO values for anthocyanidins are between −6.856 and −6.624 eV for gas phase LUMO molecular orbital may be the more important contribution from these pigments if used as dye sensitizers. Anthocyanidin LUMO contribution may enable molecular orbital to overlap semiconductor band gap with dye conduction band, and so, it can enable an easier charge transfer process in DSSC applications.

For molecules with solvents water, ethanol, and methanol caused similar effect in the molecular orbitals' energy and because of these solvents' value shift in around 3 eV. These three solvents had a similar effect in HOMO molecular orbital with similar shift magnitude around 3 eV. Then n-hexane causes a smaller shift in molecular orbitals with <1.5 eV in both HOMO and LUMO. HOMO and LUMO molecular orbitals and additional energy levels are displayed in **Figures 3** and **4**.

HOMO-LUMO energy difference is a good approximation to the material's band gap. For the selected anthocyanidins, energy gap was between 2.539 and 2.881 eV in gas phase with malvidin having the narrower gap.

Pigment	Solvent	H-L	HOMO	LUMO	λ_e	EEP	λ_h	HEP
$(C_{15}H_{11}O_6)^+$	Gas phase	2.664	−9.288	−6.624	0.318	5.525	0.344	10.361
	Water	2.824	−6.452	−3.628	0.262	4.064	0.284	6.038
	Ethanol	2.816	−6.528	−3.712	0.264	4.102	0.288	6.155
	n-hexane	2.712	−7.916	−5.204	0.295	4.818	0.324	8.284
	Methanol	2.818	−6.501	−3.683	0.263	4.089	0.267	6.115
$(C_{15}H_{13}O_5)^+$	Gas phase	2.539	−9.24	−6.701	0.371	5.666	0.452	10.162
	Water	2.823	−6.532	−3.709	0.294	4.172	0.46	5.946
	Ethanol	2.81	−6.61	−3.8	0.295	4.216	0.462	6.066
	n-hexane	2.657	−7.975	−5.318	0.335	4.968	0.479	8.169
	Methanol	2.815	−6.583	−3.768	0.294	4.201	0.461	6.024
$(C_{15}H_{13}O_6)^+$	Gas phase	2.691	−9.465	−6.774	0.364	5.703	0.498	10.371
	Water	2.955	−6.668	−3.713	0.293	4.173	0.527	6.019
	Ethanol	2.945	−6.748	−3.803	0.294	4.217	0.527	6.142
	n-hexane	2.815	−8.166	−5.351	0.328	4.98	0.533	8.316
	Methanol	2.948	−6.72	−3.772	0.294	4.202	0.527	6.100

H-L is HOMO-LUMO gap energy band. All units are in eV.

Table 3.
Selected anthocyanidins' energy results using DFT (B3LYP/6311+g(d,p)) in gas phase and with different solvents.

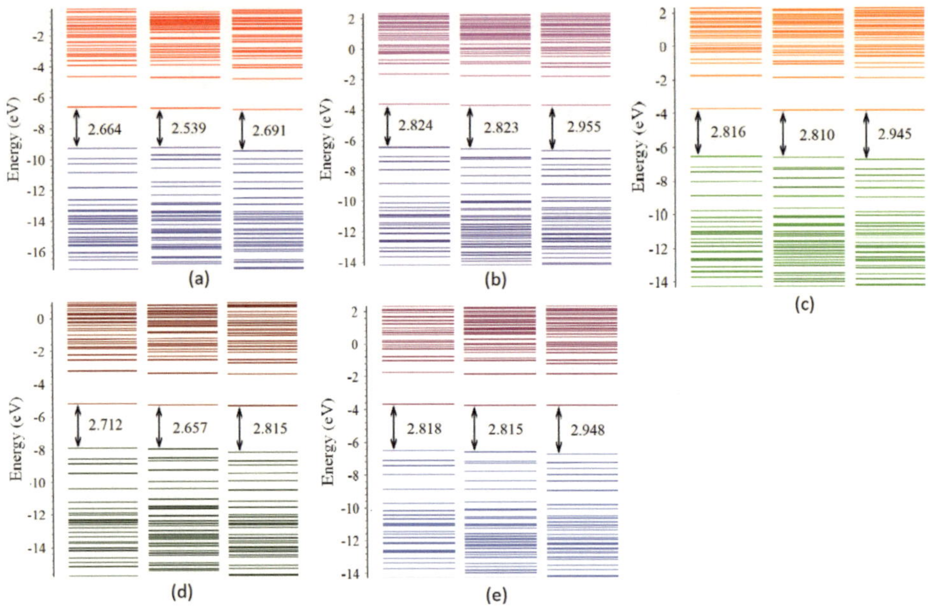

Figure 3.
Molecular orbitals for selected anthocyanidins cyanidin, malvidin, and peonidin corresponding to (a) gas phase, (b) water, (c) ethanol, (d) n-hexane, and (e) methanol. H-L gap energy units are shown in eV.

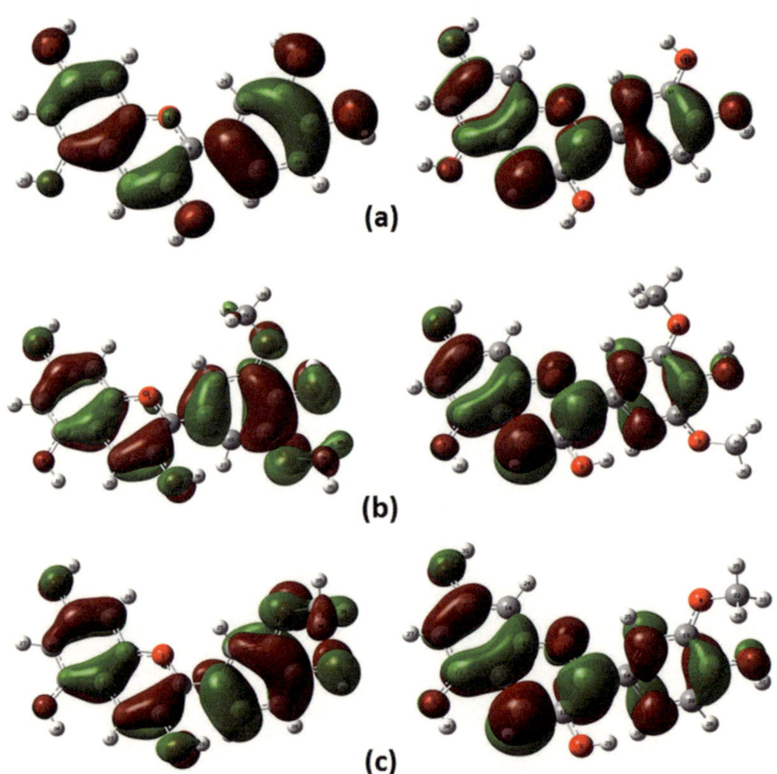

Figure 4.
HOMO and LUMO molecular orbital charge distributions using B3LYP/6-311 + g(d,p), corresponding to: (a) cyanidin, (b) malvidin, and (c) peonidin.

When solvents either water, ethanol, n-hexane, or methanol are added, H-L values shift slightly for all selected pigments. Malvidin in its gas phase has a lower value for gap energy, and with addition of solvents, H-L increases in all cases but malvidin with n-hexane is the narrower. Solvent addition has a more noticeable effect for water solvent if compared with ethanol and methanol.

Overall, H-L values are similar in magnitude for all selected pigments when using either solvent water, ethanol, n-hexane, or methanol. The H-L shift in all cases is <10% if compared with their H-L values for its respective gas phase. Malvidin and peonidin presented the bigger shift with 11 and 10%, respectively, with the exception of malvidin using n-hexane which had a shift of 5%. Among selected solvents, water caused the bigger H-L shift and n-hexane caused the smaller shift. Energy gap of anthocyanidins has few variations with ~0.3 eV as the mean difference between variants. Overall, planarity and the relative angle among rings have small contribution to gap energy results and predominates their family common features to determine the H-L parameter.

Intramolecular reorganization energies were calculated to find the required energy for the molecule to go from neutral to ionized state (as cation if charge is lost and anion if charge is accepted). Also, these calculations help understand the inverse process when the ionized molecule becomes neutral and these two different processes relate to the charge transfer process.

Values as low as possible are desirable for reorganization energies so the available energy is used in the charge transfer process instead of using the energy in reorganization processes in such a way that λ should be as low as possible in order to avoid wasting solar energy instead of taking advantage of sunlight during the energy transferring process. Overall, solvent addition helps the pigment decrease λ, and display similar values for water, ethanol, and methanol. Solvent n-hexane also helps decrease λ values but with less impact than the other solvents. From selected anthocyanidins, cyanidin presented lower electron reorganization energy (λ_e) using solvent water but ethanol and methanol had similar values.

For the hole reorganization energy (λ_h), again cyanidin had the lower values but now with methanol solvent followed by water and methanol with near values but not as close as for λ_h. Hole extraction potential (HEP) and the electron extraction potential (EEP) were calculated, and the results overall present the higher values for gas phase and the value for each case is decreased when any of the selected solvents are added.

When n-hexane solvent decreases around 8 eV and with water, ethanol, and methanol as solvents HEP is around 6 eV. For EEP, a similar effect occurs but the values decrease in less than around 1 eV when water, ethanol, and methanol solvents are used and around 0.5 eV when n-hexane is used. The three variant anthocyanidins had similar values in gas phase with <0.1 eV of difference.

Reorganization energies show that malvidin is the best choice for sensitization applications. Electron energy λ_e indicates clearly that cyanidin with methanol is the best choice followed by water and ethanol.

For hole energy λ_h also cyanidin with the same solvents is the best choice; this behavior with λ values may be attributed to its molecular planarity. EEP and HEP are not as clear as λ; in the case of these two parameters, malvidin with solvent water is the best choice but only with slight differences for the same solvent in other molecules like cyanidin and peonidin.

4.3 Chemical properties calculated from DFT results

Conceptual DFT was used to calculate the chemical properties of these three selected anthocyanidin variants. Chemical property results are shown in **Table 4**.

Pigment	Solvent	IP	EA	χ	η	ω	S
$(C_{15}H_{11}O_6)^+$	Gas phase	10.642	5.154	7.898	2.744	11.439	0.364
	Water	6.322	3.802	5.062	1.26	10.165	0.793
	Ethanol	6.443	3.838	5.141	1.302	10.147	0.768
	n-Hexane	8.608	4.522	6.565	2.043	10.549	0.49
	Methanol	6.382	3.825	5.104	1.278	10.189	0.782
$(C_{15}H_{13}O_5)^+$	Gas phase	10.614	5.296	7.955	2.659	11.899	0.376
	Water	6.406	3.878	5.142	1.264	10.462	0.791
	Ethanol	6.528	3.921	5.224	1.304	10.469	0.767
	n-Hexane	8.647	4.633	6.64	2.007	10.983	0.498
	Methanol	6.486	3.906	5.196	1.29	10.466	0.775
$(C_{15}H_{13}O_6)^+$	Gas phase	10.869	5.34	8.105	2.765	11.879	0.362
	Water	6.545	3.881	5.213	1.332	10.199	0.751
	Ethanol	6.67	3.922	5.296	1.374	10.209	0.728
	n-Hexane	8.85	4.652	6.751	2.099	10.859	0.477
	Methanol	6.627	3.908	5.267	1.359	10.205	0.736

Values include ionization potential (IP), electron affinity (EA), electronegativity (χ), chemical hardness (η), electrophilicity index (ω), and chemical softness (S), all of them in eV.

Table 4.
Chemical property results for selected anthocyanidins.

Ionization potential (IP) is the needed energy to extract an electron from a neutral molecule in order to form a cation. This property is related with the stiffness of the electronic cloud. In regard to reactivity, the cloud is more reluctant to participate in electron transfer. Then, a lower ionization potential value is desirable so there is a higher molecular potential to serve as an electron donor. The molecule with the lower IP was malvidin in its gas phase but with solvent addition, IP decreased in all cases. Although water, ethanol, and methanol cause a similar effect in IP magnitude, it was water used as solvent in cyanidin, the variant with the lower IP value among all variants. IP in gas phase was around 11 eV for selected anthocyanidins and when water, ethanol, and methanol were used, IP decreased to values around 6 eV.

Solvent n-hexane also had a decreasing effect in IP values but the values were observed around 8 eV. Cyanidin using water and methanol presented lower IP values and other molecules like malvidin also presented their lower values with water and methanol.

Selected anthocyanidins in gas phase had EA values around 5 eV and with solvents water, ethanol, and methanol, values decreased to around 3 eV while n-hexane effect decreased the EA to around 4 eV. Regarding electronegativity (χ), it is calculated to estimate the capacity of molecules to attract electron pairs. The highest the χ value, the highest its suitability to act as a charge acceptor.

In general, selected anthocyanidins had χ values around 8 eV, and with solvents like water, ethanol, and methanol this value decreased to around 5 eV while n-hexane solvent effect was less with values around 6 eV.

Overall, the chemical properties estimated display some similarity among calculated values which may be attributed to molecular resemblance such as relative angle at ring B, and the differentiator relates to the small structural differences as well as their molecule constituents.

4.4 Excited states for absorption energy calculation using TDDFT

Excited states were calculated using the TDDFT scheme as implemented in Gaussian09 using the B3LYP/6311+g(d,p) theoretical method for selected anthocyanidins. B3LYP has been reported as an efficient hybrid functional that has been compared with several other functionals with good results [46–50, 52] to process different anthocyanins and anthocyanidins. For any DSSC to be effective, its absorption spectrum must match the solar irradiation spectrum. The absorption property of the dye determines its light harvesting capability and thus affects the performance of dye sensitizers in DSSCs [53–57].

Our calculations showed that there is a slight difference with experimental values due to solvent effects and variation contributed by measuring methodologies [52, 58–60]. Two main regions in the anthocyanidin UV-Vis spectra have been reported in the literature, the first located between 260 and 280 nm and the second is located at the visible region between 490 and 550 nm. A third peak appears at 310–360 nm [59]; our discussion will focus on the principal peak located in the visible region.

Molecule	Solvent	State	ΔE (eV)	λ (nm)	Transition	Contribution	f
$(C_{15}H_{11}O_6)^+$	Gas phase	1	2.546	487.1 (522*)	H → L	67%	0.507
	Water	1	2.524	491.2	H → L	68%	0.619
					H-1 → L	17%	
					H-2 → L	12%	
	Ethanol	1	2.528	490.4	H → L	68%	0.629
					H-1 → L	15%	
					H-2 → L	12%	
	n-Hexane	1	2.473	501.4	H → L	69%	0.686
	Methanol	1	2.524	491.3	H → L	68%	0.622
					H-1 → L	17%	
					H-2 → L	12%	
$(C_{15}H_{13}O_5)^+$	Gas phase	1	2.312	536.4 (542*)	H → L	60%	0.24
	Water	1	2.434	509.3	H-1 → L	30%	0.604
	Ethanol	1	2.481	499.8	H → L	68%	0.591
					H-2 → L	17%	
	n-Hexane	1	2.376	521.9	H → L	70%	0.627
	Methanol	1	2.431	510.1	H → L	61%	0.601
$(C_{15}H_{13}O_6)^+$	Gas phase	1	2.401	516.3 (532*)	H → L	67%	0.288
					H-1 → L	11%	
	Water	1	2.509	494.2	H → L	69%	0.53
	Ethanol	1	2.564	483.6	H → L	67%	0.515
	n-Hexane	1	2.465	503	H → L	69%	0.535
	Methanol	1	2.505	494.9	H → L	69%	0.527

Table 5.
Excited state absorption results for selected anthocyanidins using TD-DFT.

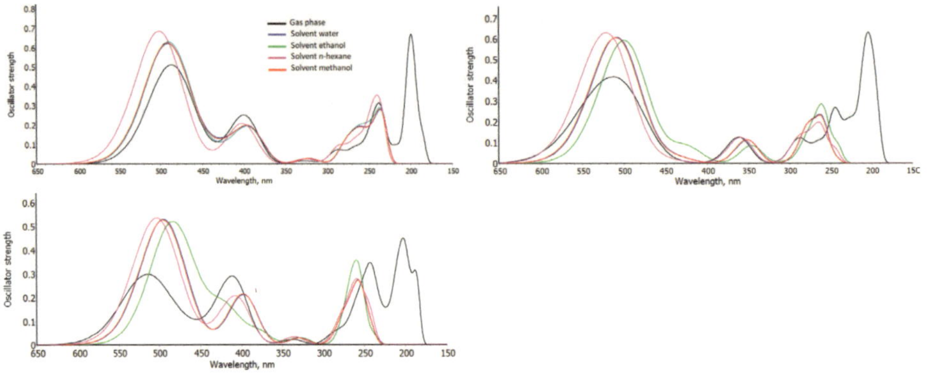

Figure 5.
Anthocyanidin excited state spectra from results using the TD-DFT scheme for gas phase and solvents water, ethane, n-hexane, and methane corresponding to: (a) cyanidin, (b) malvidin, and (c) peonidin.

In a general view of absorption results, selected anthocyanidins in gas phase had absorption wavelength between 479.1 and 536.4 nm so, all selected molecules work in the visible part of the electromagnetic spectrum. Cyanidin works in the blue region and displays lower values calculated for wavelength. Malvidin has higher values while cyanidin presents a similar value. These results suggest that there is an effect caused by the small relative angle at B ring considering that these molecules are the simplest regarding their constituents. Addition of solvent shifts the absorption spectrum by increasing its wavelength by <5 nm in the case of water, ethanol, and methanol. For n-hexane solvent, absorption spectrum shifts the wavelength by slightly more than 10 nm.

First excited state values using TDDFT to calculate absorption data are displayed in **Table 5** and absorption spectrum is shown in **Figure 5**. The visible and near-UV regions are the most important for photon-to-current conversion to obtain the microscopic information about the electronic transitions and their corresponding MO properties.

5. Conclusions

Ground state geometries were analyzed using a well-known theoretical methodology, and an analysis on their relative angles comparing dihedrals within individual rings provides insight into the different planarity characteristics between rings and establishes that functionalization with OCH_3 is an important feature for the structural and energy gap differences. Molecular orbitals are analyzed and compared with our results from prior research for TiO_2, the more widely used photocatalyst. These results mainly with MO analysis show that there is good compatibility between the semiconducting oxide and these pigments if they are to be used as dye sensitizers. Malvidin in its gas phase has a lower value for gap energy and with addition of solvents, gap energy increases in all cases but malvidin with n-hexane is the narrower. Conceptual DFT results show that cyanidin and malvidin may have good charge transfer. Furthermore, excited state data display the absorption capabilities of the selected pigments and confirm that cyanidin and malvidin, in that order, may be the best choices for dye sensitization applications.

Acknowledgements

This work was financed by CONACyT (Mexican Science and Technology National Council) through 2015 CONACyT SEP-CB (Basic Science-Public

Education Ministry) project fund 258553/CONACyT/CB-2015-01. MAFH thanks CONACYT for a postdoctoral scholarship (2014). Thanks to the Scientific Computational Laboratory at FCQ-UJED for computational resources. Thanks to Academic Group UJED-CA-129 for valuable discussions.

Conflict of interest

Authors state that this research was completed without any conflicts of interest related with funding to develop the present work.

Author details

Diana Barraza-Jiménez[1], Azael Martínez-De la Cruz[2], Leticia Saucedo-Mendiola[1], Sandra Iliana Torres-Herrera[3], Adolfo Padilla Mendiola[1], Elva Marcela Coria Quiñones[1,4], Raúl Armando Olvera Corral[1], María Estela Frías-Zepeda[1,5] and Manuel Alberto Flores-Hidalgo[1]*

1 Department of Chemical Sciences, Juarez University of Durango State, Durango, México

2 Graduate Studies Division, Faculty of Mechanical and Electrical Engineering, Autonomous University of Nuevo Leon, San Nicolás de los Garza, NL, Mexico

3 Faculty of Forestry Science, Juarez University of Durango State, Durango, México

4 TecNM/Durango Institute of Technology, Durango, Mexico

5 CIIDIR-IPN, Durango, México

*Address all correspondence to: maflores.hidalgo02@gmail.com

IntechOpen

© 2019 The Author(s). Licensee IntechOpen. This chapter is distributed under the terms of the Creative Commons Attribution License (http://creativecommons.org/licenses/by/3.0), which permits unrestricted use, distribution, and reproduction in any medium, provided the original work is properly cited.

References

[1] Mathew S, Yella A, Gao P, Humphry-Baker R, Curchod BFE, Ashari-Astani N, et al. Dye-sensitized solar cells with 13% efficiency achieved through the molecular engineering of porphyrin sensitizers. Nature Chemistry. 2014;**6**:242-247

[2] O'Regan B, Grätzel M. A low-cost, high-efficiency solar cell based on dye-sensitized colloidal TiO_2 films. Nature. 1991;**353**:737-740

[3] Grätzel M. Conversion of sunlight to electric power by nanocrystalline dye-sensitized solar cells. Journal of Photochemistry and Photobiology A. 2004;**164**:3-14

[4] Grätzel M. Dye-sensitized solar cells. Journal of Photochemistry and Photobiology C. 2003;**4**:145-153

[5] Komiya R et al. Improvement of the conversion efficiency of a monolithic type 8 dye-sensitized solar cell module. In: Technical Digest, 21st International Photovoltaic Science and Engineering Conference 2 C-5O-08. 2011

[6] Yella A et al. Porphyrin-sensitized solar cells with cobalt (II/III)-based redox electrolyte exceed 12 percent efficiency. Science. 2011;**334**:629-634

[7] Shrestha M et al. Dual functionality of BODIPY chromophore in porphyrin-sensitized nanocrystalline solar cells. Journal of Physical Chemistry C. 2012;**116**:10451-10460

[8] Nattestad A et al. Highly efficient photocathodes for dye-sensitized tandem solar cells. Nature Materials. 2010;**9**:31-35

[9] Yamaguchi T, Uchida Y, Agatsuma S, Arakawa H. Series-connected tandem dye-sensitized solar cell for improving efficiency to more than 10%. Solar Energy Materials & Solar Cells. 2009;**93**:733-736

[10] Murayama M, Mori T. Dye-sensitized solar cell using novel tandem cell structure. Journal of Physics D. 2007;**40**:1664-1668

[11] Kubo W, Sakamoto S, Kitamura T, Wada Y, Yanagida S. DSSC: Improvement of spectral response by tandem structure. Journal of Photochemistry and Photobiology A. 2004;**164**:33-39

[12] Ito S et al. Optimization of the dye-sensitized solar cell with anthocyanin as photosensitizer. Thin Solid Films. 2008;**516**:4613-4619

[13] Imahori H, Umeyama T, Ito S. Large π-aromatic molecules as potential sensitizers for highly efficient dye-sensitized solar cells. Accounts of Chemical Research. 2009;**42**:1809-1818

[14] Bessho T, Zakeeruddin SM, Yeh CY, Diau EWG, Grätzel M. Highly efficient mesoscopic dye-sensitized solar cells based on donor-acceptor-substituted porphyrins. Angewandte Chemie, International Edition. 2010;**49**:6646-6649

[15] Chien C-Y, Hsu B-D. Optimization of the dye-sensitized solar cell with anthocyanin as photosensitizer. Solar Energy. 2013;**98**:203-211

[16] Cherepy NJ, Smestad GP, Grätzel M, Zhang JZ. Ultrafast electron injection: Implications for a photoelectrochemical cell utilizing an anthocyanin dye-sensitized TiO_2 nanocrystalline electrode. The Journal of Physical Chemistry. B. 1997;**101**:9342-9351

[17] Aduloju KA, Shitta MB. Dye sensitized solar cell using natural dyes extracted from red leave onion.

International Journal of Physical Sciences. 2012;7:709-712

[18] Calogero G, Di Marco G. Red sicilian orange and purple eggplant fruits as natural sensitizers for dye-sensitized solar cells. Solar Energy Materials and Solar Cells. 2008;**92**:1341-1346

[19] Chang H, Lo YJ. Pomegranate leaves and mulberry fruit as natural sensitizers for dye-sensitized solar cells. Solar Energy. 2010;**84**:1833-1837

[20] Furukawa S, Iino H, Iwamoto T, Kukita K, Yamauchi S. Characteristics of dye sensitized solar cells using natural dye. Thin Solid Films. 2009;**518**:526-529

[21] Luo P, Niu H, Zheng G, Bai X, Zhang M, Wang W. From salmon pink to blue natural sensitizers for solar cells: *Canna indica* L., salvia splendens, cowberry and *Solanum nigrum* L. Spectrochimica Acta Part A: Molecular and Biomolecular Spectroscopy. 2009;**74**:936-942

[22] Calogero G, Yum JH, Sinopoli A, Di Marco G, Grätzel M, Nazeeruddin MK. Anthocyanins and betalains as light-harvesting pigments for dye-sensitized solar cells. Solar Energy. 2012;**86**:1563-1575

[23] Kumar K, Chowdhury A. Use of novel nanostructured photocatalysts for the environmental sustainability of wastewater treatments. In: Reference Module in Materials Science and Materials Engineering. Elsevier; 2018

[24] Grätzel M. Recent advances in sensitized mesoscopic solar cells. Accounts of Chemical Research. 2009;**42**(11):1788-1798

[25] Kamat PV, Das S, Thomas KG, George MV. Ultrafast photochemical events associated with the photosensitization properties of a squaraine dye. Chemical Physics Letters. 1991;**178**(1):75-79

[26] Kamat PV. Photoelectrochemistry in semiconductor particulate systems. 14. Picosecond charge-transfer events in the photosensitization of colloidal TiO_2. Langmuir. 1990;**6**(2):512-513

[27] Holkar CR, Jadhav AJ, Pinjari DV, Mahamuni NM, Pandit AB. A critical review on textile wastewater treatments: Possible approaches. Journal of Environmental Management. 2016;**182**:351-366

[28] Ge X, Timrov I, Binnie S, Biancardi A, Calzolari A, Baroni S. Accurate and inexpensive prediction of the color optical properties of anthocyanins in solution. The Journal of Physical Chemistry. A. 2015;**119**:3816. DOI: 10.1021/acs.jpca.5b01272

[29] Chemical Entities of Biological Interest (ChEBI). Online Molecular Database [Internet]. 2018. Available from: https://www.ebi.ac.uk/chebi/ [Accessed: June 20, 2018]

[30] Woodford JN. A DFT investigation of anthocyanidins. Chemical Physics Letters. 2005;**410**:182

[31] Buseta PB, Colleter JC, Gadret M. Structure du chlorure d'apigéninidine monohydrate. Acta Crystallographica. Section B. 1974;**30**:1448

[32] Ueno K, Saito N. Cyanidin bromide monohydrate (3,5,7,3′,4′-pentahydroxyflavylium bromide monohydrate). Acta Crystallographica. Section B. 1977;**33**:114

[33] Meyer M. Ab initio study of flavonoids. International Journal of Quantum Chemistry. 2000;**76**:724

[34] Skrede G, Wrolstad RE, Durst RW. Changes in anthocyanins and polyphenolics during juice processing of highbush blueberries (*Vaccinium*

corymbosum L.). Journal of Food Science. 2000;**65**:357-364

[35] Veberic R, Jakopic J, Stampar F, Schmitzer V. European elderberry (*Sambucus nigra* L.) rich in sugars, organic acids, anthocyanins and selected polyphenols. Food Chemistry. 2009;**114**:511-515

[36] Pedreschi R, Cisneros-Zevallos L. Phenolic profiles of andean purple corn (*Zea mays* L.). Food Chemistry. 2007;**100**:956-963

[37] Dai Y, Rozema E, Verpoorte R, Choi YH. Application of natural deep eutectic solvents to the extraction of anthocyanins from *Catharanthus roseus* with high extractability and stability replacing conventional organic solvents. Journal of Chromatography. A. 2016;**1434**:50-56

[38] Myjavcova R, Marhol P, Kren V, Simanek V, Ulrichova J, Palikova I, et al. Analysis of anthocyanin pigments in *Lonicera* (Caerulea) extracts using chromatographic fractionation followed by microcolumn liquid chromatography-mass spectrometry. Journal of Chromatography. A. 2010;**1217**:7932-7941

[39] Maran JP, Priya B, Manikandan S. Modeling and optimization of supercritical fluid extraction of anthocyanin and phenolic compounds from *Syzygium cumini* fruit pulp. Journal of Food Science and Technology. 2014;**51**:1938-1946

[40] Feuereisen MM, Gamero Barraza M, Zimmermann BF, Schieber A, Schulze-Kaysers N. Pressurized liquid extraction of anthocyanins and biflavonoids from *Schinus terebinthifolius* raddi: A multivariate optimization. Food Chemistry. 2017;**214**:564-571

[41] Pap N, Beszédes S, Pongrácz E, Myllykoski L, Gábor M, Gyimes E, et al. Microwave-assisted extraction of anthocyanins from black currant marc. Food and Bioprocess Technology. 2012;**6**:2666-2674

[42] Celli GB, Ghanem A, Brooks MS. Optimization of ultrasound-assisted extraction of anthocyanins from haskap berries (*Lonicera caerulea* L.) using response surface methodology. Ultrasonics Sonochemistry. 2015;**27**:449-455

[43] Choi YH, Verpoorte R. Green solvents for the extraction of bioactive compounds from natural products using ionic liquids and deep eutectic solvents. Current Opinion in Food Science. 2019;**26**:87-93

[44] Becke AD. Density-functional thermochemistry. III. The role of exact exchange. The Journal of Chemical Physics. 1993;**98**:5648

[45] Frisch MJ et al. Gaussian 09, Revision D.01. Wallingford, CT: Gaussian, Inc; 2013

[46] Terranova U, Bowler DR. Δ self-consistent field method for natural anthocyanidin dyes. Journal of Chemical Theory and Computation. 2013;**9**:3181

[47] Fan W, Deng W. Incorporation of thiadiazole derivatives as π-spacer to construct efficient metal-free organic dye sensitizers for dye-sensitized solar cells: A theoretical study. Communications in Computational Chemistry. 2013;**1**:152

[48] Armas R, Miguel M, Ovideo J, Sanz JF. Coumarin derivatives for dye sensitized solar cells: A TD-DFT study. Physical Chemistry Chemical Physics. 2012;**14**:225

[49] Lopez JB, Gonzalez JC, Holguin NF, Sanchez JA, Mitnik DG. Density functional theory (DFT) study of triphenylamine-based dyes for their use as sensitizers in molecular photovoltaics.

International Journal of Molecular Sciences. 2012;**13**:4418

[50] Xu J, Zhang H, Liang G, Wang L, Xu W, Ciu W, et al. DFT studies on the electronic structures of indoline dyes for dye-sensitized solar cells. Journal of the Serbian Chemical Society. 2010;**75**:259

[51] Leonid S. Chemissian. V4. **43**:2005-2016

[52] Sanchez-Bojorge NA, Rodriguez-Valdez LM, Glossman-Mitnik MD, Flores-Holguin N. Theoretical calculation of the maximum absorption wavelength for cyanidin molecules with several methodologies. Computational and Theoretical Chemistry. 2015;**1067**:129

[53] Liu Z. Theoretical studies of natural pigments relevant to dye-sensitized solar cells. Journal of Molecular Structure: THEOCHEM. 2008;**862**:44

[54] Sanchez-de-Armas R, San-Miguel MA, Oviedo J, Sanz J. Direct vs. indirect mechanisms for electron injection in DSSC: Catechol and alizarin. Computational & Theoretical Chemistry. 2011;**975**:99

[55] Mitnik DG. Computational molecular characterization of coumarin-102. Journal of Molecular Structure: THEOCHEM. 2009;**911**:105

[56] Mohammadi N, Wang F. First-principles study of Carbz-PAHTDDT dye sensitizer and two carbz-derived dyes for dye sensitized solar cells. Journal of Molecular Modeling. 2014;**20**:2177

[57] Megala M, Rajkumar BJM. Theoretical study of anthoxanthin dyes for dye sensitized solar cells (DSSCs). Journal of Computational Electronics. 2016;**15**:557

[58] Harborne JB. Spectral methods of characterizing anthocyanins. The Biochemical Journal. 1958;**70**:22

[59] Brouillard R, Harborne JB, editors. The Flavonoids: Advances in Research Since 1980. London: Chapman and Hall Ltd; 1988. p. 525

[60] Harborne JB. Comparative Biochemistry of Flavonoids. London, New York: Academic Press; 1967

Chapter 11

Solvent Effects in Supramolecular Systems

Raffaello Papadakis and Ioanna Deligkiozi

Abstract

Today it is well-established that solvents demonstrate an important role in chemistry. Solvents are able to affect the reactivity, as well as the electronic, optical, and generally physicochemical properties of compounds in solution. Taking this into account, in this chapter we analyze the importance of solvent polarity in phenomena closely related to supramolecular systems as well as the aptitude of various supramolecules to interact with solvent molecules and thus to give rise to chromic effects such as solvatochromism. Main focus is placed on mechanically interlocked molecules, e.g., rotaxanes, catenanes, etc., exhibiting solvent-controlled shuttling movements, switching, and/or solvatochromism. The effect of solvents in various supramolecular architectures is a further focus of this chapter.

Keywords: solvents, medium responsiveness, supramolecular recognition, solvatochromism, supramolecules, mechanically interlocked molecules, shuttle movements, supramolecular architectures

1. Introduction

Fundamental nature's operations are dominated and regulated by noncovalent interactions. Solvation demonstrates a key role in all these processes as it drastically influences the energetics of host-guest (Ho-G) interactions as well as the supramolecular recognition phenomena. In many occasions solvents can influence and modulate the supramolecular structure of complex systems through various possible interactions with solutes.

In recent years a rapidly increasing interest and development in the field of supramolecular engineering have been observed. Specifically within the scope of modern materials and chemical science, the conducted research is continuously growing [1]. Fundamentally, the target is to create molecular systems by design, intending to regulate the interactions of complex building blocks in the solid and liquid state and to obtain desirable complex systems exhibiting multifunctional properties. Since the recent awarding of the Nobel Prize of chemistry to Jean-Pierre Sauvage, Fraser Stoddart, and Ben Feringa, this area has gained plenty of scientific attention [2]. The microenvironment-dependent complexation of supramolecular complexes provides a conventional tool for creating a vast variety of mechanically interlocked molecules, supramolecular architectures, and molecular machines. Rotaxanes and catenanes, which are at the heart of the development of "molecular machine" chemistry [3], are of principal importance. Under this framework, molecular machines that have been studied until today include molecular motors [4], shuttles [5], muscles [6], pumps [7], elevators [8], etc. Supramolecular chemistry showed an impulsive interest in

IntechOpen

molecular-engineered compounds, whereby complexes are formed from small molecular building blocks held together by reversible intermolecular noncovalent interactions such as van der Waals interactions, hydrogen bonding, electrostatic, π-π stacking, and hydrophobic interactions. Their design, control, and function compose new relevant interdisciplinary key enabling areas (KEA) of science and technology. In all these solvents and solvation are demonstrating a fundamentally important role. Ordinarily, solvents are categorized into two main categories cited as polar and nonpolar, whereby their efficacy is often characterized by their dielectric constants. Solvents with a dielectric constant of less than 15 are usually regarded to be nonpolar. Nonpolar solvents contain bonds between atoms with similar electronegativities, such as carbon and hydrogen. Polar solvents have large dipole moments and they comprise bonds between atoms with very different electronegativities, such as oxygen and hydrogen. The aforementioned solvents are additionally divided into polar aprotic and polar protic. The solvation efficacy in a predefined medium pays a key role in thermodynamics and kinetics especially in supramolecular host-guest interactions. This is profoundly correlated to changes in solubility, stability constant, reactivity, redox potential, and some spectral parameters. Host-guest association behaviors can essentially be controlled only by applying diverse solvent system, thus altering by demand their solvation properties. Hence, the solvation environment plays a dynamic role for supramolecular solutes, being able thus to affect the thermodynamics of complex systems. In general, the interactions involved in supramolecular systems are quite weaker than covalent bonds, and thus they can be highly controlled and reversible. Acknowledging the importance of the above, this chapter discusses the impact of solvents in various types of supramolecular systems.

2. Supramolecular recognition

2.1 General aspects

One of the key roles of solvents which are dominant in supramolecular chemistry is their role in supramolecular recognition. This is essential for systems consisting of a host (Ho) and a guest (G) (**Figure 1**). In solution solvent molecules can interact with Ho and G molecules through various types of noncovalent weak interactions, and this process readily affects the mutual interactions between the Ho and G counterparts. Consequently, the thermodynamics of their binding can be significantly altered simply by changing solvents. The most dominant interactions in Ho-G-S system (where S is a solvent) can be electrostatic (i.e., ion-ion, ion-dipole, dipole-dipole, or dipole-induced dipole interactions), H-bonding, van der Waals, or π-π interactions. Inevitably, the type of developed interactions is influenced by the physicochemical properties of the Ho, G, and S molecules [9].

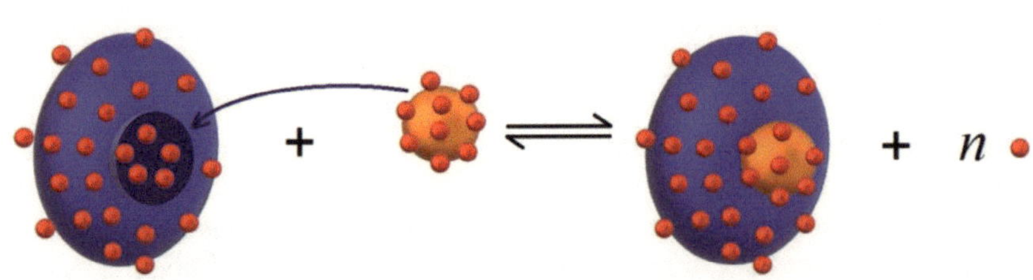

Figure 1.
Illustration of the equilibrium of the binding between a solvated Ho (blue) and a solvated G (orange) molecule involving the release of solvent molecules (red).

From a thermodynamic perspective, the simple equilibrium of **Figure 1** is described by the equilibrium (association) constant which is connected to the thermodynamic activities of the Ho, G, and (Ho-G) species, i.e., $K_a = a_{[Ho\text{-}G]}/a_{[Ho]}.\ a_{[G]}$. The association constant is in turn related to the changes in standard Gibbs free energy (ΔG^o), enthalpy (ΔH^o), and entropy (ΔS^o) upon binding of G to Ho, i.e., $RT \ln K_a = \Delta G^o = \Delta H^o - T\Delta S^o$, where R and T correspond to the ideal gas constant and temperature, respectively. Solvents are capable of influencing the above-described equilibrium by affecting the terms ΔH^o and ΔS^o [9, 10]. When it comes to Supramolecular Complexation (SC), a general classification of solvents is based on their aptitude to undertake to self-organization. Self-organized (structured) solvents are in general relatively polar solvents, e.g., water, alcohols, amides, etc., whereas the nonstructured are generally less polar, e.g., hydrocarbons, haloalkanes, etc. One might assume that the importance of nonstructured solvents would be minor in supramolecular complexation due to the weaker interaction with the Ho and G molecules; the situation however is much different. Of course highly structured/polar solvents like water exhibit major effects on the complexation of Ho and G. Surprisingly though, dramatic differences in the stabilization of a (Ho-G) complex can be observed when shifting from a haloalkane like chloroform to an aromatic solvent like benzene.

The tremendous impact of solvent polarity on the supramolecular assembling is easily manifested through the following example by Nishimura and coworkers [11]. In their work by employing dynamic covalent chemistry, they managed to develop a complementary capsule-guest supramolecular system (**Figure 2**) behaving very differently in two deuterated solvents of interest: $CDCl_3$ and C_6D_6.

Specifically, the thermodynamics of the supramolecular recognition were different in these two solvents with K_a values differing by three orders of magnitude ($K_a(C_6D_6)/K_a(CDCl_3) = 1150$). It was also found that the supramolecular recognition effect was in both solvent cases enthalpy driven. Yet, ΔH(kcal/mol) was determined to be −18.6 in deuterated benzene and −2.7 in deuterated chloroform which corresponds to a significant thermodynamic solvent effect. Interestingly, Kang and Rebek some years earlier designed and synthesized a dimeric supramolecular guest corresponding to various carboxylic acids such as 1-adamantanecarboxylic acid [12]. Working in the same two deuterated solvents, they discovered a reversed solvent effect (compared to that of Nishimura and coworkers). The supramolecular recognition effect in that case was found to be entropically favored with however two orders of magnitude difference in K_a: ($K_a(CDCl_3)/K_a(C_6D_6) = 243$). Through these stimulating examples, it is easily made understood that solvents have a drastic effect on supramolecular binding/recognition effects [9].

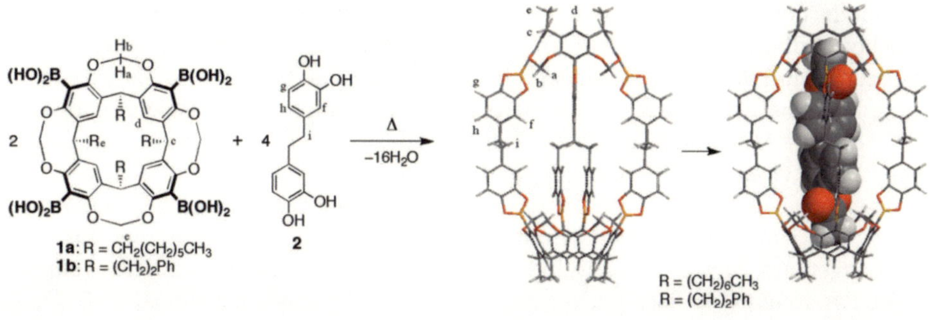

Figure 2.
Supramolecular assembly exploiting dynamic covalent chemistry. Reprinted with permission from Nishimura and coworkers [11].

Large solvent effects are also encountered in supramolecular complexation (SC) involving ionic and neutral, e.g., hydrophobic, entities. These effects are largely dependent of the nature of the target guest molecule for a given host molecule. Noteworthy, ionic and neutral SC often exhibits opposite solvent polarity dependencies. Two characteristic such examples are the SC of aromatic hydrophobic molecules by a cyclophanes and that of potassium ions by the crown ether 18-crown-6. For instance, Smithrud and Diederich observed five orders of magnitude higher association constant in water compared to the solvent carbon disulfide for a cyclophane/pyrene SC system [13]. The hydrophobic 3D cyclophane developed by Smithrud and Diederich involved a large cavity accessible to solvent molecules, and the huge K_a determined in water was attributed to the solvophobic effect. In simple words pyrene prefers to be encapsulated in the hydrophobic cavity of the cyclophane instead of interacting with water. The effect becomes less and less important as one moves from water to apolar solvents [13]. The opposite effect is observed for the SC of potassium ions by ether 18-crown-6 [14]. In that case the association constant becomes larger in solvents of lower polarity, e.g., $\log K_a$ (H_2O) = 2.0, whereas $\log K_a$ (acetone) = 6.0 [14]. Interestingly, the $\log K_a$ exhibits a linear dependence to the surface tension of the medium as well as to other parameters/properties of solvents [15, 16].

The above-described examples are fundamental for the development of complex supramolecular systems with possibilities of external control and the design of molecular machines. Focus of the next section is the effect of solvents on some characteristic molecular machines and switches.

3. Nano-mechanical motions affected by solvents

3.1 Rotaxanes and catenanes

Rotaxanes, pseudorotaxanes, and catenanes are prominent members of the supramolecular family of compounds. They may be composed of both organic and inorganic (macro)molecules mechanically linked together. One of the latter parts may exhibit the ability to move in relation to the rest of the other parts. Due to this effect, many of these systems have been proposed for molecular machinery applications. In the case of rotaxanes bulky substitutes, the so-called stoppers are integrated in these systems preserving the stability of the supramolecular assembly. Pseudorotaxanes consist of the same structural units as rotaxanes, but do not include the aforementioned stoppering bulky substituents. In contrast, catenanes consist of two or more macrocyclic molecules tied together, forming chain-like supramolecular assemblies (**Figure 3**). The stability in all these systems is achieved through various types of weak interactions such as van der Waals forces, hydrophobic effects, hydrogen bonds, donor acceptor interactions, etc. Today a large number of scientific works have been published following the pioneering synthesis of the first rotaxane by Wasserman in 1960 [17]. Additional scientific support has been provided by a stream of publications by pioneering researchers such as Luttringhaus [18], Wasserman [17], Harrison [19], and Schill [20]. All of them dealt with the creation of functional molecular devices of high complexity and specialization.

Today multiple supramolecular structures have been created through the inclusion of a variety of linear axle-like molecules in the cavities of macromolecules such as crown ethers, cyclophanes, cyclodextrins (CD), etc. In this way numerous supramolecular systems exhibiting diverse one-, two-, and three-dimensional (1D, 2D, and 3D) architectures have been reported. The methodologies leading to catenanes and rotaxanes after assembling and pseudorotaxane formation are illustrated in the schematic representation of **Figure 3**.

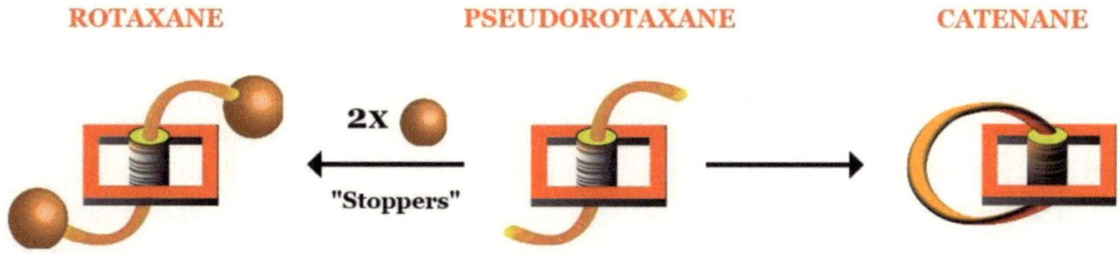

Figure 3.
From pseudorotaxanes to rotaxanes and catenanes.

3.2 Molecular shuttles

Interlocking a part of a linear molecule of a rotaxane into the cavity of a macrocycle molecule is associated with a series of complex interaction phenomena. An important function that many of the aforementioned systems can undergo is that of molecular shuttling. This often happens when a macrocycle trapped onto a linear component (axle) is capable of moving reversibly between two or more Regions on the axle (often called stations), in response to external stimuli (e.g., electrochemical stimuli, irradiation, heating/cooling, and/or solvent polarity changes) [21].

As already mentioned the main forces that hold these supramolecular structures together are relatively weak, and therefore the systems can undergo the described shuttling movement under mild external changes in a fully controlled manner [21]. Across all the potential driving forces, all the kind of energy inputs, and all aforementioned parameters, it is noteworthy that a simple change in solvent polarity can be harvested in order to induce a controllable molecular machine function (**Figure 4**). In this section the stimulating role of solvents on the function of interlocked systems is reviewed.

One of the key/pioneering contributions in the field of solvent effects on the (multi)functional behavior of rotaxanes has been made by Leigh and coworkers. By applying chemistry similar to that occurring in natural systems and specifically

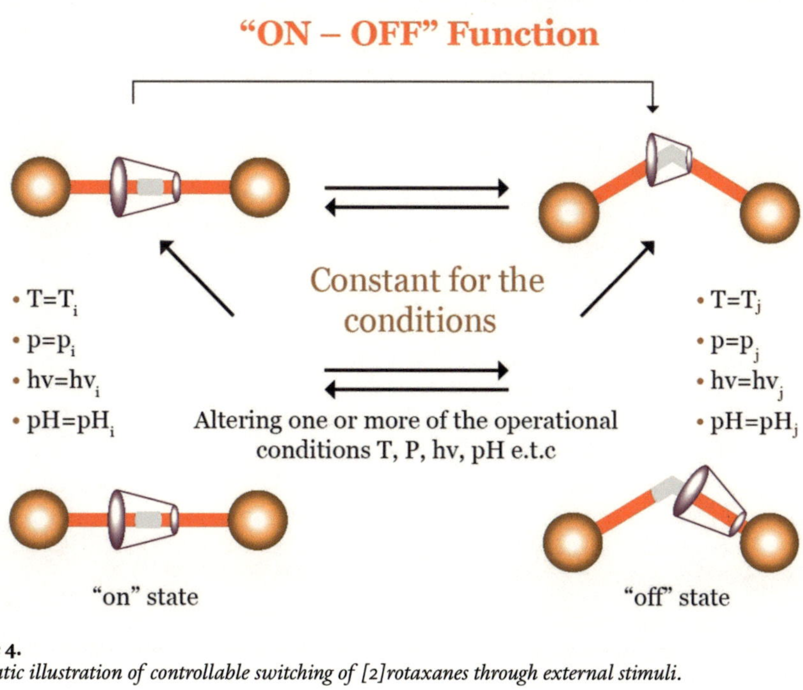

Figure 4.
Schematic illustration of controllable switching of [2]rotaxanes through external stimuli.

in peptides, they managed to synthesize the rotaxanes of **Figure 5A** [22]. In both cases of rotaxanes of **Figure 5A**, the linear component consists of a glycylglycine chain and two diphenylmethane end groups (stoppers). The stabilization of these [2]rotaxanes is achieved through the development of hydrogens bonds between amide hydrogen of the macrocycle molecule with the carbonyl groups of the linear compound and vice versa. The resulted bonds are very stable when the rotaxanes are dissolved in non polar solvents such as $CHCl_3$. However, when they are dissolved in polar solvents such as DMSO which can specifically interact with parts of these molecules, these bonds become unstable, and this results to a different molecular configuration for each of the two [2]rotaxanes. This solvent-driven feature is essential for triggering the switching ability of this supramolecular complex, thus functioning as a molecular machine, and has been a stimulating example for a number of later scientific works.

In 2003 Da Ros et al. published a [2]rotaxane which performs a solvent-induced shuttling movement as shown in **Figure 5B** [23]. This [2]rotaxane consists of fullerene C60 group behaving as both a stoppering unit and a photoactive group. The amphiphilic nature of the rotaxane thread was used to shuttle the macrocycle from close to the fullerene spheroid (in nonpolar solvents) to far away (in polar solvents). The rotaxane is based on hydrogen bond-directed assembly of a benzylic amide macrocycle around a dipeptide thread, solvent-switchable molecular shuttles in a similar fashion to the work by Leigh et al. [22]. In nonpolar solvents, e.g., CH_2Cl_2 or $CHCl_3$, the macrocycle forms hydrogen bonds with the peptide residue. In polar aprotic solvents such as DMSO, the hydrogen bonding between the macrocycle >NH group and the peptide carbonyl group is disrupted by the competing solvent interactions, and thus the macrocycle selectively stops over the alkyl chain [23].

In 2005 Gschwind and coworkers published a series of [2]rotaxanes, containing a phenol-involving linear part, amide-involving macrocycles, and triphenylmethane-stoppering units [24]. The dumbbell molecule 1 of **Figure 6** offers three diamide stations to the macrocyclic molecule in the protonated form of the [2] rotaxane. It was found that electrostatic interactions can modulate exceptionally well the speed of the mechanical motion between a fast- and a slow-motion state as a response to a reversible external solvent-provided stimulus. The electrostatic interactions in these rotaxanes are controllably regulated through solvent effects

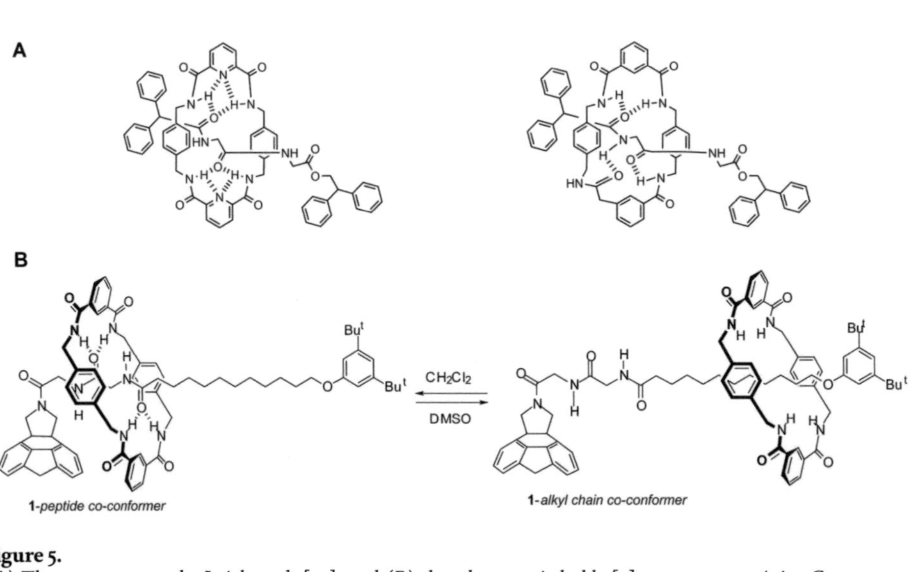

Figure 5.
(A) The two rotaxanes by Leigh et al. [22] and (B) the solvent-switchable [2]rotaxane containing C_{60} stoppering unit by Da Ros et al. Reprinted with permission from Da Ros et al. [23].

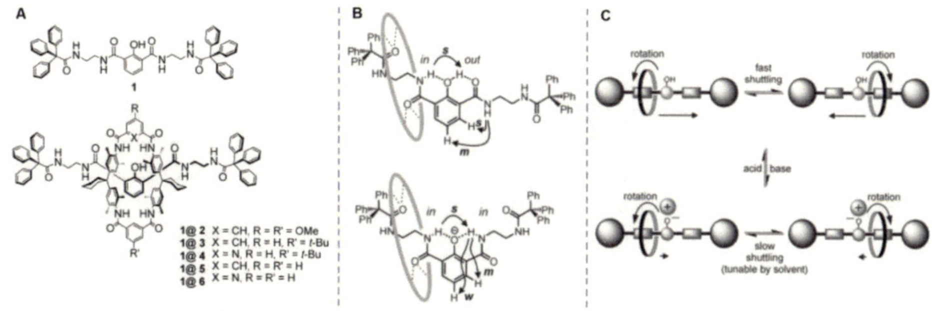

Figure 6.
(A) Various [2] rotaxanes by Gschwind and coworkers [24]. (B) Interactions in a [2]rotaxane. (C) Cartoon representation illustrating the dynamic processes in the acid-/base-regulated switching of [2]rotaxanes depicted in (A). Reprinted with permission from Gschwind and coworkers [24].

induced by altering the proportion of polar solvent in a binary solvent mixture. For example, when different amounts of DMSO are added to dichloromethane, solvent-driven shuttling modifications occur (**Figure 6C**). It was further found that the molecular wheel shuttling in deprotonated rotaxanes is hindered by the counter-cation held through electrostatic forces close to the anion at the axle-center region. Thus, the shuttling speed can easily be regulated by addition of acids and bases enabling a fast- and a slow-motion mode parallel to the on-off switching function.

Cai and coworkers have a long-standing interest in the effects of solvents in the shuttling movements in mechanically interlocked compounds [25–27]. In 2012 they reported a [2]rotaxane molecular shuttle controlled by solvent. The rotaxane involved α-cyclodextrin (α-CD), dodecamethylene, and bipyridinium moieties as shown in **Figure** 7 [26]. Cai et al. discovered that the molecular shuttling in this [2]rotaxane can be driven by both solvent and temperature changes. They indeed demonstrated the shuttling process of α-CD along the linear thread in solvents of different polarities such as DMSO and H_2O. The energy barrier in water was shown to be 4.0 kcal/mol higher than in DMSO. Water interacts favorably with the bipyridinium moieties, however, negligibly with the alkyl chain, and this yields to a higher free energy barrier in the case of water.

4. Solvatochromic supramolecular systems

4.1 Generalities

Solvatochromism is a well-studied phenomenon occurring in many diverse systems. It is described as the change in color (*χρώμα*, Greek word for color) induced by solvents. In a broader context, the term solvatochromism covers changes in the electronic (UV-Vis), FTIR, Raman, or EPR spectra induced by solvents [28]. A

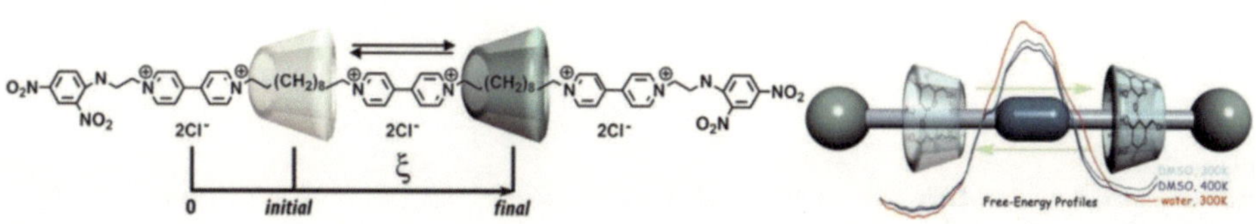

Figure 7.
Solvent-induced shuttle movement in a [2]rotaxane. Reprinted with permission from Cai and coworkers [26].

vast number of solvatochromic compounds have been reported to date exhibiting large structural diversity [29]. The most frequently studied class of solvatochromic dyes involves dyes bearing a **D-π-A** structure where **A** is an electron-withdrawing moiety, **D** is an electron-donating group, and **π** is a conjugated system (often aromatic), separating **A** and **D**. **D-π-A** dyes have recently received much attention as they can be used in hi-tech applications including materials with nonlinear optical (NLO) properties [30, 31], chromotropic sensors and molecular switches [32, 33]. They serve also in many cases as multifunctional building blocks for supramolecular architectures, e.g., in rotaxanes [34]. Some examples of such dyes are depicted in **Figure 8**. The common characteristic of the compounds **I–IV** is that their **D** part (an iodine anion in **I**, a phenolate in **II**, a carbanion in **III**, and an iron(II) cation in **IV**) is capable of transferring an electron pair (in **II** and **III**) or a single electron (in **I** and **IV**) to the electron-deficient positively charged pyridinium ring. The π-system through which the charge transfer occurs is either the aromatic backbone of pyridine itself (the cases of dyes **I**, **III**, and **IV**) or another π-system in conjugation with the pyridine ring (the case of dye **II**). This charge transfer (CT) is induced by light. The required energy of light for the CT transition depends strongly on solvent polarity [28]. Noteworthy, solvent polarity can affect CT energy in various ways. When the increase of medium polarity leads to a drop of the CT energy of a dye, the corresponding effect is called positive solvatochromism. In those cases bathochromic shifts in the electronic spectra of the compound are induced by an increase in solvent polarity (**Figure 9A**). When the opposite effect is observed, the observed phenomenon is called negative solvatochromism (**Figure 9B**). The main focus of this section is the solvatochromism in supramolecular systems.

4.2 Supramolecules involving solvatochromic entities

When conducting a deep literature search, it is easily made obvious that there are not many examples of solvent-switchable supramolecular structures such as rotaxanes and catenanes exhibiting also solvatochromism. As mentioned above, solvatochromic supramolecular assemblies exhibit a strong change in position and sometimes the intensity of their absorption spectra, which is achieved by changing

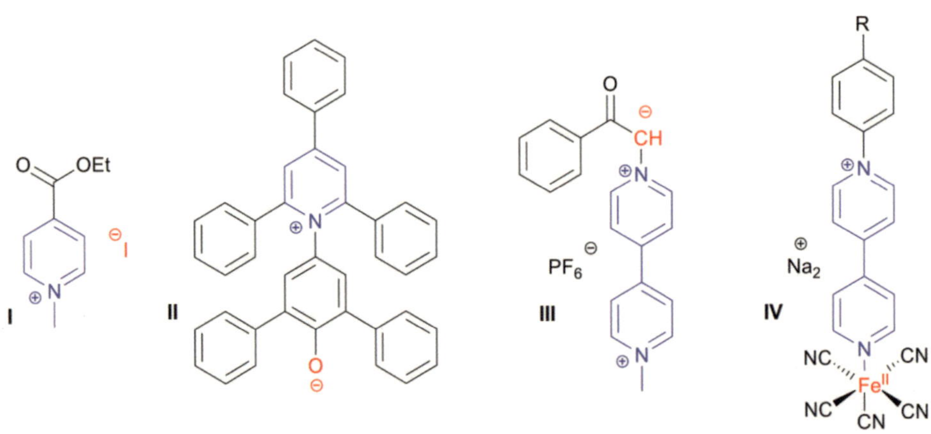

Figure 8.
*Various pyridinium-based solvatochromic **D-π-A** dyes. **I**, Kosower salt [28]; **II**, Reichardt's betaine [28]; **III**, betaine of N-aryl and N'-phenacetyl-4,4'-bipyridines [35, 36]; and **IV**, pentacyanoferrate(II) complexes with N-aryl4,4'-bipyridines (monoquats) ligands (R = OMe, Me, H, Cl, Br, CN) [37, 38].*

Figure 9.
Plots of CT energy as a function of solvent polarity corresponding to (A) positive and (B) negative solvatochromism (G corresponds to ground and E to excited state).

the polarity of the solvent (solvatochromic effect). This phenomenon is pronounced in electron acceptor-donor systems.

An interesting example combining solvent-controlled shuttle movement in a rotaxane (see previous section) and solvatochromic behavior was reported by Günbaş et al. in 2011 (**Figure 10**) [39]. The solvatochromic behavior of their [2]rotaxane and its dumbbell-like precursor molecule was investigated in a variety of solvents of different polarities. It was observed that both compounds exhibited solvatochromic shifts in their absorption spectra when increasing the polarity of the solvent. Spectroscopic data showed a wavelength shift of 575 nm in toluene to 621 nm in DMSO for the molecule which corresponds to a positive solvatochromic shift. The observed values were attributed to the pyrrolidine group. For the [2]rotaxane, however, the solvatochromic changes were smaller. The absorbance was shifted from 608 to 621 nm when the solvent was changed from the nonpolar toluene to the highly polar DMSO. In general, both in the case of the [2]rotaxane and the dumbbell molecule, solvatochromic shifts were observed, indicating that polar solvent interacts stronger with the molecules and also stabilizes the excited state. Of course, this effect is stronger for the dumbbell molecule than for [2]rotaxane, an effect which could be attributed to interactions developed among the chromophore linear molecule and the macrocycle molecule.

In 2007, Toma and his scientific team published a paper in which they disclosed the solvatochromic properties of a [2]rotaxane, which involved a β-cyclodextrin

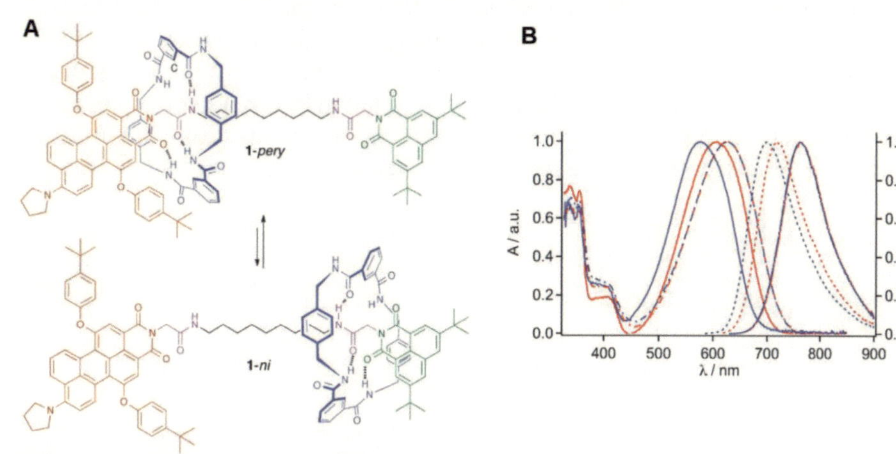

Figure 10.
(A) Shuttle movement in a [2]rotaxane and (B) solvatochromic shifts observed. Reprinted with permission from Günbaş et al. [39].

(β-CD) macrocycle [40]. The linear molecule of their rotaxane consisted of trans-1,4-di-[(4-pyridyl) ethylene] benzene (**Figure 11**) and trans ferrocyanide(II) anions ligated by the pyridyl groups of the linear molecule which act as stoppering groups. The UV-Vis absorption spectral analysis of $\{[Fe^{II}(CN)]_2(BPEB)\}$ (dumbbell) and the corresponding β-CD-involving rotaxane indicated that the dumbbell molecule exhibited two absorption bands, one around 352 nm and the other at 454 nm.

During the addition of β-CD to the linear molecule solution, a wavelength shift was observed denoting the formation of the [2] rotaxane, which was attributable to the metal-to-ligand charge transfer (MLCT). The formation of [2]rotaxane resulted in a decrease in the energy of MLCT, i.e., a bathochromic shift from 454 to 479 nm. Commonly, the reaction of the iron complex (II) with N-heterocyclic substituents results in deep chromatic shifts of MLCT when the final products are dissolved in less polar solvents. In these systems the hydrophobic forces increase the solvatochromic effect. Thus, for the [2]rotaxane, the low-wavelength shift values are attributed to the inclusion of the β-CD cavity and to the selective solubilization of the rotaxane. This behavior could be related to the stabilization of the energy levels of the complex between β-CD and the ligand leading to a decrease in the energy of MLCT [41].

Various other similar examples of solvation effects on rotaxane have been reported in the literature; one of these examples is the [2]rotaxane reported by Baer and Macartney in 2000 [41]. Cyclodextrin (CD) inclusion complexes with linear guest parts have been studied extensively using a variety of spectroscopic techniques. There have also been several reports of cyclodextrin (α- or β-CD)-based rotaxanes, polyrotaxanes, and catenanes using linear parts (L) containing biphenyl, stilbene, and azobenzene dyes. These [2]rotaxanes can be formed rapidly by the addition of $[Fe^{II}(CN)_5]^{3-}$ stoppering units, and a cyclodextrin macrocyclic unit can be threaded by the linear Skeleton. It has been proven that such [2]rotaxanes exhibit intense metal-to-ligand charge transfer (MLCT) transition bands in the visible spectrum. This transition as mentioned above is prone to changes in energy and intensity induced by solvents (solvatochromism). Different ligands acting as the axial parts for series of rotaxanes have been exploited so far such as *trans*-1,2-bis(4-pyridyl) ethylene ligand (BPE, **Figure 12**) which can nicely provide a conjugated bridging of inner sphere and intervalence electron transfer between transition metal centers [42]. The 4,4′-azopyridine ligand (AZP, **Figure 12**) has been employed as a bridging ligand as well mainly for ruthenium amines and porphyrins [43, 44]. The APA and PCA, i.e., 4-acetylpyridine azine and 4-pyridinecarboxaldehyde azine, respectively, are interesting examples of azines which have also been reported as building blocks of [2],rotaxanes of this class [45] (structures depicted in **Figure 12** correspond to some prominent examples of such ligands/linear skeletons). Spectroscopic studies conducted in order to validate the maximum wavelength difference for CD-free

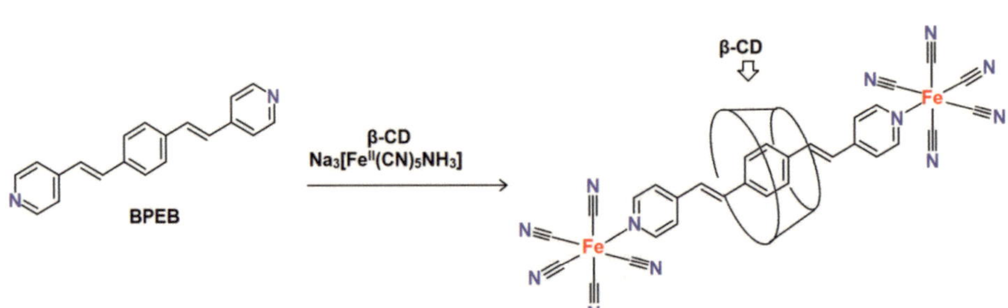

Figure 11.
Pioneering solvatochromic [2]rotaxane synthesis by Toma and coworkers [40].

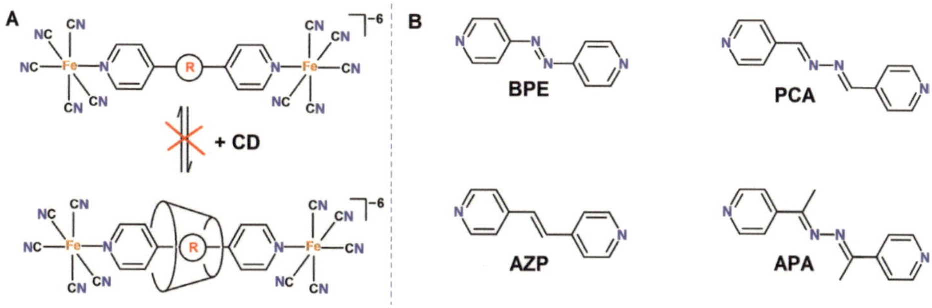

Figure 12.
Pyridine-involving ligands BPE, PCA, AZP, and APA (left) and the corresponding dumbbells and cyclodextrin-involving [2]rotaxanes (right).

dumbbells and CD-involving rotaxanes resulted in the following **λmax(nm) {Fe(CN)$_5$L$_{3-}$/ (Fe(CN)$_5${L,CD}$_{3-}$)}**: (*i*) BPE, 460/496 (α-CD) and 478 (β-CD); (*ii*) AZP, 596/698 (α-CD) and 644 (β-CD); (*iii*) PCA 508/536 (α-CD) and 518 (β-CD); and (*iv*) APA 448/456 (α-CD) and 458 (β-CD). It is evident in all cases that the spectra are affected by the presence of α- or β-CD, resulting in bathochromic shifts in the MLCT band in the visible spectrum. This effect constitutes an important response (of high sensitivity) to the polarity of their environment. This effect is much connected to solvatochromism. Indeed the aptitude of both non-rotaxanated dumbbells and the corresponding rotaxanes to yield solvatochromic shifts is very large [41].

In 2012 Deligkiozi et al. reported the synthesis of a [2]rotaxane consisting of a fully conjugated arylazo-based linear part entrapped in α-cyclodextrin and stoppered by bulky dinitrophenyl end groups [46]. Recording the UV-Vis spectra of both compounds, a broadband in the region 300–400 nm was observed, which was attributed to the π-π* transition of the group (**—**N = N**—**). Comparing the spectra of the [2]rotaxane and the dumbbell precursor, a bathochromic shift was observed. Specifically, the maximum wavelength of the dumbbell precursor was positioned at 337 nm, whereas that of the [2]rotaxane was centered at 351 nm. This shift is attributed to the interaction of the α-CD cavity and the (**—**N = N**—**) group of the dumbbell-like compound, which causes a decrease in the energy difference between the ground and excited states of the azo-compound leading to bathochromism. Both compounds were found to undergo *E-Z* reversible isomerizations, and in the case of the [2]rotaxane, light-induced shuttle movement was reported [46]. Interestingly the same complexes were found to be photoconductive in the solid state, and this was accredited to the extended π-conjugation in these molecules [47, 48]. These findings enforced the same group to develop a system involving the same π-conjugated backbone, however involving strong electron-donating groups (pentacyanoferrate(II) stoppering groups). This led to the formation of strong **D-π-A** systems (**Figure 13**) [49]. More recently, Papadakis et al. exploited the solvatochromism of two CD-containing [2]rotaxanes and their CD-free linear precursor in order to investigate preferential solvation (PS) effects in water/ethylene glycol (EG) mixtures [50]. Pentacyanoferrate(II) groups which served as strong electron donors facilitated charge transfer to the viologen electron-deficient parts. It was proven that the pentacyanoferrate(II) units were able to trigger an intense solvatochromic behavior in such systems in neat solvents, solvent mixtures, and other types of aqueous media. In order to study the solvatochromic behaviors, aqueous/ethylene glycol (EG) mixtures were used as the media, and alteration of the polarity was achieved through changing solvent/cosolvent mole ratio. The medium-responsive behavior of these compounds (**Figure 13**) is mainly pronounced in very polar media such as water, aqueous EG mixtures, and neat EG (see **Figure 13**).

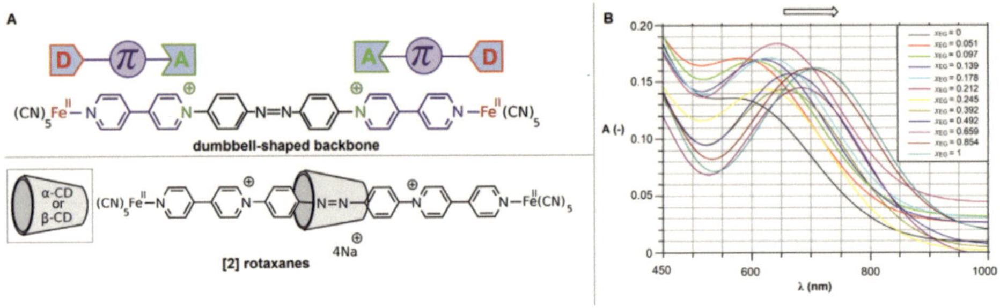

Figure 13.
A) The dumbbell-shaped push-pull backbone developed by Deligkiozi et al. (top) and the corresponding cyclodextrin involving [2]rotaxanes (bottom). B) The bathochromic shifts of the MLCT band of the [2] rotaxane with α-cyclodextrin depicted in A observed in aqueous ethylene glycol mixtures. Reprinted with permission from: [34] and [50].

In some cases rotaxanes and catenanes can also exhibit solvent-dependent emission of light. A representative example is that by Baggerman et al. involving [2]rotaxanes and [3]rotaxanes bearing a tetraphenoxy perylene diimide core [51]. In their work Baggerman et al. observed the influence of hydrogen bonding developed between the amide and the wheel macrocycle of these rotaxanes on the optical behavior of the chromophore (perylene). Specifically, they showed that both absorption and fluorescence spectra are bathochromically shifted upon rotaxanation. All systems including the wheel-free axle (WFA) exhibited fluorosolvatochromism with red shifts of up to 47 nm (WFA case) and a reduced fluorosolvatochromism when going to the [2] rotaxanes and [3] rotaxanes [51]. On the other hand, Boer et al. very recently exploited the solvent-dependent excimer and exciplex emissive behavior of naphthalene diimide metallomacrocycles and catenanes and thus managed to perform a solution speciation of the metallosupramolecular complexes and their solvent-dependent nature [52].

4.3 Supramolecular solvatochromism

Solvatochromism derived by an interaction of solvent molecules with a supramolecular system is characterized by various authors as "supramolecular solvatochromism." In many cases the systems involve transition metals coordinated to ligands forming supramolecular architectures in which solvent molecules can be trapped or simply interact with parts of the system giving rise to different responses, e.g., in their electronic spectra.

A characteristic early such example is that reported by Lee and Kimizuka in 2002 [53]. In their studies they developed a lipid-packaged 1D supramolecular complex bearing platinum. The anionic lipids acted as counter anions of the positively charged Pt complex. The electronic spectra of the as described supramolecular complex were found to be readily influenced by solvent polarity and that the packaging of the complex is vital for the overall medium-responsive properties. Some years later, Kuroiwa et al. reported on the supramolecular solvatochromism of lipid-packaged, mixed-valence linear platinum complexes (**Figure 14**) which were investigated in dispersions employing the solvents $CHCl_3$, chlorocyclohexane, and methylcyclohexane [54]. It was found that solid samples were all indigo-colored but displayed supramolecular thermochromism, attributed to heat-induced dissociation and concomitant recovery of coordination chains. The reassembled supramolecular complexes exhibited color changes depending on the solvent employed, and the CT energy measured was found to decrease as the polarity of the organic medium

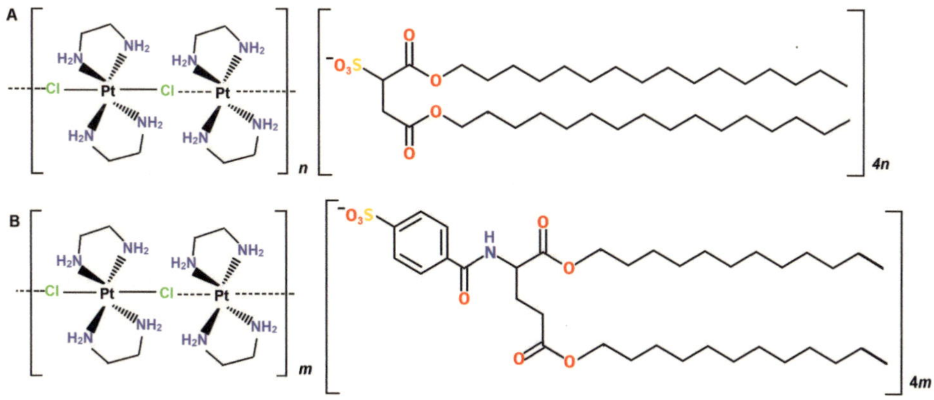

Figure 14.
Structure of the two Pt(IV) complexes of Kuroiwa et al. [54].

increases following the sequence: methylcyclohexane, benzene, chlorocyclohexane, $CHCl_3$, and 1,2-dichloroethane.

Such approaches gain more and more the attention of materials scientist as it could be envisioned that through mixing optical sensors and solvatochromic species or even aggregachromic compounds, the effects of various stimuli on the rheology of viscoelastic gels (VEGs) could be facilitated and this is considered as an important step before the development of new products based on VEGs becomes reality [55]. More recently, Nikolayenko et al. reported the supramolecular solvatochromic behavior of a dinuclear copper(II)-involving metallacycle [56]. The authors exploited the intense solvatochromic behavior of the Cu^{II} complex and the capacity of the macrocycle to trap small solvent molecules like tetrahydrofuran, diethyl ether, and pentane at temperatures well above their boiling points. The latter effect is attributed to the suitable guest shape and size which drastically limit lattice diffusion. Solvent exchange was found to induce intense color changes (**Figure 15B**) and sizable shifts in the visible region of the diffuse reflectance spectra (**Figure 15B**). The high intensity of the supramolecular solvatochromic effect is furthermore excellently illustrated through microphotographs of the variable colors of crystals.

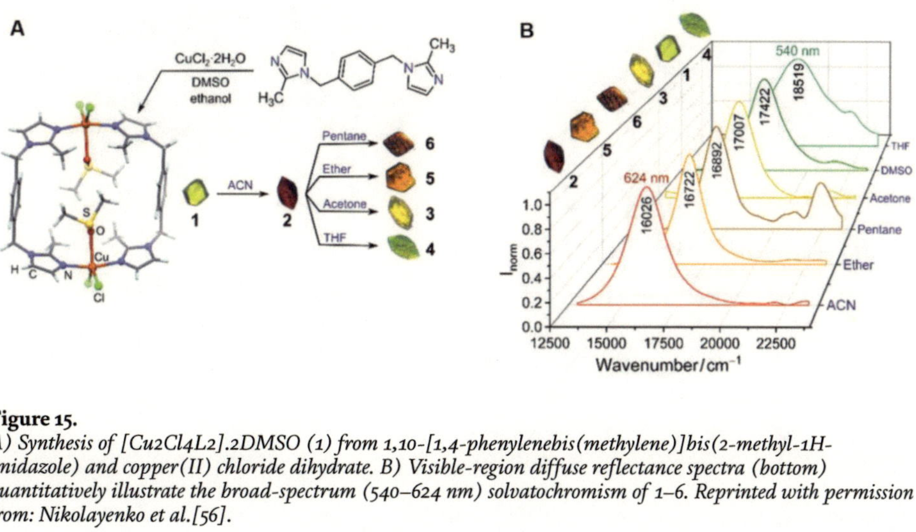

Figure 15.
A) Synthesis of [Cu2Cl4L2].2DMSO (1) from 1,10-[1,4-phenylenebis(methylene)]bis(2-methyl-1H-imidazole) and copper(II) chloride dihydrate. B) Visible-region diffuse reflectance spectra (bottom) quantitatively illustrate the broad-spectrum (540–624 nm) solvatochromism of 1–6. Reprinted with permission from: Nikolayenko et al.[56].

5. Solvent effects on supramolecular architecture

In the previous sections, the role of solvents in supramolecular functions and processes has been discussed. As described, in those cases solvents are capable of influencing the thermodynamics of supramolecular binding processes as well as the energetics of supramolecular systems so as to induce solvatochromism or some nano-mechanical functions like molecular shuttle movements in rotaxanes and catenanes or other conformational changes.

In this section focus is placed on the effect of solvents in supramolecular architectures. Their regulating role in metallosupramolecular solids has been thoroughly investigated in recent years [57] as well as their aptitude to affect crystal growth and assembly and dynamic transformations. Many solvents especially those bearing N, O, or S atoms exhibit aptitude to coordinate (i.e., to specifically interact) with metals in various coordination complexes, and thus they are capable of forming new complexes. In these complexes solvent molecules can demonstrate a stabilizing role as building blocks and variation of the solvent can lead to alternative molecular architectures. A very nice example illustrating this ability of solvents is that of the assembly of the ligand 1,4-benzene dicarboxylic acid (*bda*) with Mg^{II} in the solvents dimethyl acetamide (DMA), EtOH, and dimethylformamide (DMF). In the case of DMA, polymeric 2D layers (**Figure 16a**) of the type $[Mg_3(bda)_3(DMA)_4]_n$ are obtained. The situation is very different when EtOH or DMF are used as solvents.

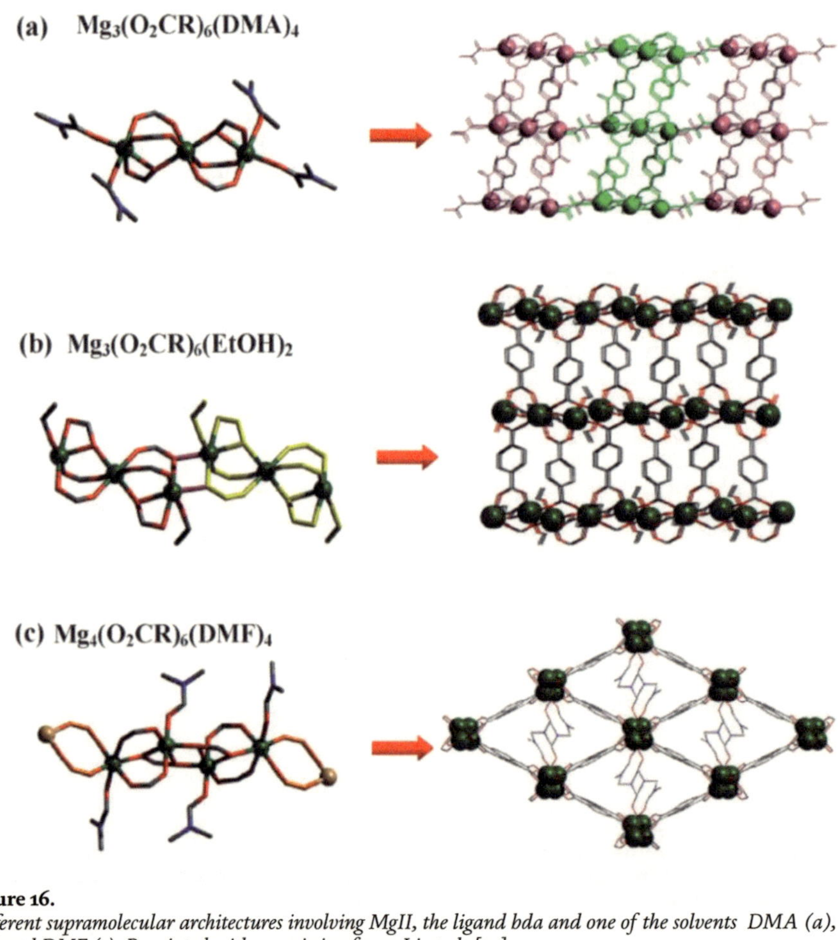

Figure 16.
Different supramolecular architectures involving MgII, the ligand bda and one of the solvents DMA (a), EtOH (b), and DMF (c). Reprinted with permission from: Li et al. [57].

In the latter two cases, 3D frameworks are obtained instead of the following: $[Mg_3(bda)_3(EtOH)_2]_m$ and $[Mg(bda)(DMF)]_k$, respectively (**Figure 16b,c**). Note that n, m, and k correspond to the number of repeated units in each case [58–60].

Another important feature of the solvents which readily affects the supramolecular architecture of coordination complexes is the steric effect they might introduce. Noro et al. reported the steric effect of different solvents with sp^1-, sp^2-, or sp^3-hybridized coordinating atoms on the assembly of $Cu^{II}(PF_6)_2$ and the ligand 1,2-bis(4-pyridyl)ethane (*bpe*) [61]. As depicted in **Figure 17A**, solvents with sp^1- or sp^2-hybridized coordinating atoms such as MeCN (sp^1 N) and DMF or acetone (sp^2 O carbonyl atom) are able to directly ligate Cu^{II}. The coordination of the solvents in these two cases happens axially with a Jahn-Teller distortion. On the other hand, alcohols, THF, and dioxane (which all have sp^3 O atoms) preferably form H-bonds with the H-atoms of coordinated water molecules already coordinated at the axial positions. This stimulating divisibility is achieved through the variation of the steric effect introduced by different solvents.

Synergistic solvent effects can also drastically influence the structure of supramolecular coordination polymers illustrated in 3D pillar-layered coordination polymers prepared by Wang et al. [62]. These complexes comprise Co^{II} metal centers and *meso-α,β*-bi(4-pyridyl)glygol (*bpg*) and azide (N_3^-) ligands. Their general formula is {[Co(bpg)(azide)2]S*x*}*n* (S represents a solvent and *x* the corresponding stoichiometric number), and they are prepared through a reaction of Co^{II}, azide, and *bpg* in the following solvent mixtures: MeOH–H_2O, DMSO–H_2O, and DMF–H_2O. The use of different solvent systems was found to drastically influence the 2D [Co(azide)2]*m* layers involved in the 3D networks. In the case of the protic mixture MeOH-water, a square net was observed without incorporation of solvent. On the other hand from DMSO–H_2O and DMF–H_2O, the aforementioned 2D layers obtained were of honeycomb and *kagomé* geometry, respectively, and solvent molecules were incorporated in the 3D polymers (DMSO and DMF, respectively) (see **Figure 17B**) [62].

Moreover, the protic or aprotic nature of the solvent can have a significant impact on the crystal structures of coordination compounds. A characteristic example is that of $Mn(OAc)_2$ and its complexation with the ligand 3-(2-pyridyl)-5-(4-pyridyl)-1,2,4-triazole (*ppt*). The use of binary solvent mixtures involving a protic and an aprotic solvent (viz., DMF-EtOH and toluene-MeOH) or neat MeCN led to three different 2D systems according to Lin et al. [63]. What is noteworthy is that in the case of MeCN (aprotic) as a solvent, the obtained supramolecular system did not incorporate any solvent molecule: $[Mn(ppt)_2]_n$. Strikingly, in the case of the aforementioned binary solvent mixtures, two supramolecular 2D polymers involving the corresponding

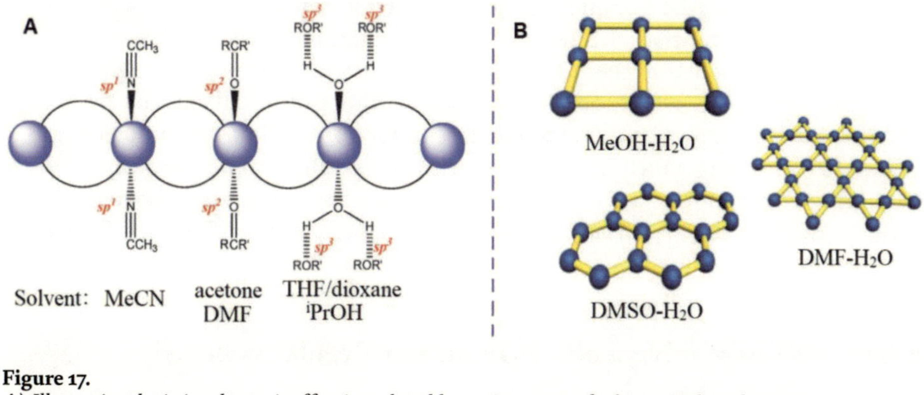

Figure 17.
A) Illustration depicting the steric effect introduced by various types of solvents in the polymeric supramolecular architecture involving CuII and bpe. B) Three different 2D polymeric [Co(azide)2]m layers involved in the complexes obtained by Wang et al. [62] Reprinted with permission from: Li et al. [57].

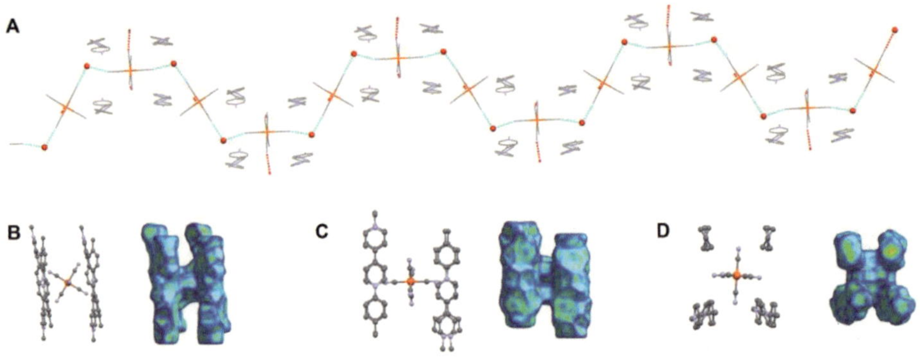

Figure 18.
(A) The zigzag $(^{\cdots}HOH^{\cdots}HCF^{\cdots})_n$ *polymers involved in a viologen/HCF CTC reported by Papadakis et al. (hydrogen atoms have been omitted). A cluster consisting of four viologen molecules around a HCF anion and the corresponding Hirshfeld surface representations shown along the b-(B), a-(C) and c-(D) axis. Reprinted with permission from: Papadakis et al. [65].*

solvents were obtained: $[Mn(ppt)_2(DMF)_{1/2}(H_2O)_3]_m$ and $[Mn(ppt)_2(toluene)_{1/2}(MeOH)_{3/2}]_k$, respectively. This interesting phenomenon was attributed to the fact that when only an aprotic solvent is utilized, the neighboring 2D polymer grid-sheets in $[Mn(ppt)_2]_n$ are stacked in a staggered mode, and this leads to a very compact 3D structure leaving out the solvent molecules. This effect is avoided when a protic solvent is employed as cosolvent [63].

In all above examples, the coordination of solvent molecules to the metal centers was found to affect the structure of the supramolecular coordination systems. Secondary interactions, which were already mentioned, mainly account to the H-bonding of a coordinating solvents and non-coordinating cosolvent. There are several cases however where secondary interactions alone can lead to stabilized 3D supramolecular coordination structures. The supramolecular charge transfer complexes (CTCs) of viologens with various electron donors have been reported long ago [64]. Such CTCs involving $[Fe^{II}(CN)_6]^{4-}$ (HCF) as a strong electron donor have been given some attention; however, their supramolecular structure has been scarcely investigated so far. An example pertaining to this category of supramolecular CTCs was recently reported by Papadakis et al. [65]. As depicted in **Figure 18**, the nonsymmetric dicationic viologen molecules tend to aggregate around the anionic HCF donor. However, the stability of a crystalline structure of such a CTC is achieved only if water is introduced in the reaction mixture. Water is found to readily form H-bonds with CN groups of HCF, and this results in the formation of a zigzag 2D polymer of the type: $(^{\cdots}HOH^{\cdots}HCF^{\cdots})_n$. The 3D supramolecular structure comprises the described 2D polymers and cationic channels of viologens perpendicular to these 2D polymers (**Figure 18**). Attempts to remove (by drying) or replace water (employing even another protic solvent) in these structures failed. Apparently the stabilizing interaction in these supramolecular systems is H-bonding which is stronger when H_2O is utilized. The importance of H_2O and the formation of a CTC in a similar fashion have been also reported earlier by Abouelwafa et al. [66]. In the aforementioned example, a symmetric viologen was utilized instead [66].

6. Applications of solvent effects in supramolecular systems

Except of publications that are indicative of scientific activity for furthering the knowledge base, patents are indicative of technology development for commercial

Keyword	Simple	Total
Supramolecular complexes	4682	16,727
Solvatochromism	296	806
Rotaxane/catenane	165	587
Solvatochromic Rotaxanes/catenanes	4	7

Table 1.
Summary of the patent search results.

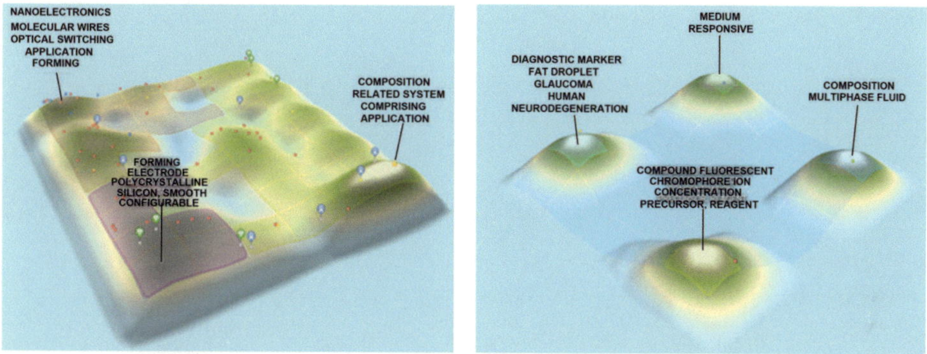

Figure 19.
Patent search results: rotaxane/catenane (left) and solvatochromic rotaxane/catenane (right).

or market potential. On this basis a patent search has been conducted so as to map the transition between science, technology, and market. For this purpose, we used the same parameters for our patent data extraction through Patsnap, a patent search engine and analytic portal. Using this patent search engine, an extended patent search in available patent libraries such as the European Patent Office, USPTO, and FPO was conducted. From this analysis it is evident that the sector supramolecular chemistry has started to grow significantly in recent years. The patent search followed a four-level approach starting by available patents using the following keywords: supramolecular complexes, solvatochromism, rotaxane/catenane, and solvatochromic rotaxane/catenane (**Table 1**). This refinement drives us to the specific market and indicates us the freedom to operate since by this procedure we conclude on the presence of very few patents that are mainly filled in the United States and China. The corresponding landscape is shown in **Figure 19**.

7. Conclusion

In this chapter, we reviewed some important examples of how solvents can influence supramolecular processes and structures. The impact of solvents on supramolecular binding is the starting point through which it is made clear that altering solvent polarity can drastically influence supramolecular binding processes. Solvents can also affect the relative supramolecular conformations of complex systems allowing the development of medium responsive molecular machines and switches. Moreover, solvents can give rise to supramolecular solvatochromic phenomena leading to optical changes and this observation has significantly assisted in the development of solvent (environment) sensing supramolecular systems. Last but not least, various metal ion-involving supramolecular architectures can be drastically influenced by specific solvents or solvent/cosolvent mixtures. This key effect enables the design of new supramolecular architectures encompassing modulation possibilities based on either the coordination of solvents to metal centers or the secondary interactions between them. In any case, the role of solvents in supramolecular chemistry is obviously enormous. Thus, it is highly important that solvent effects are taken seriously into account when designing new supramolecular systems.

Author details

Raffaello Papadakis[1,2]* and Ioanna Deligkiozi[1]

1 Laboratory of Organic Chemistry, School of Chemical Engineering, National Technical University of Athens (NTUA), Athens, Greece

2 Department of Chemistry—Ångström, Uppsala University, Uppsala, Sweden

*Address all correspondence to: rafpapadakis@gmail.com

IntechOpen

© 2019 The Author(s). Licensee IntechOpen. This chapter is distributed under the terms of the Creative Commons Attribution License (http://creativecommons.org/licenses/by/3.0), which permits unrestricted use, distribution, and reproduction in any medium, provided the original work is properly cited. (cc) BY

References

[1] Baroncini M, Casimiro L, de Vet C, Groppi J, Silvi S, Credi A. Making and operating molecular machines: A multidisciplinary challenge. ChemistryOpen. 2018;7:169-179. DOI: 10.1002/open.201700181

[2] Sauvage JP. From chemical topology to molecular machines (Nobel Lecture). Angewandte Chemie International Edition. 2017;**56**:11080-11093. DOI: 10.1002/anie.201702992

[3] Pezzato C, Cheng C, Stoddart JF, Astumian RD. Mastering the non-equilibrium assembly and operation of molecular machines. Chemical Society Reviews. 2017;**46**:5491-5507. DOI: 10.1039/C7CS00068E

[4] Van Leeuwen T, Lubbe AS, Štacko P, Wezenberg SJ, Feringa BL. Dynamic control of function by light-driven molecular motors. Nature Reviews Chemistry. 2017;**1**:0096. DOI: 10.1038/s41570-017-0096

[5] Silvi S, Venturi M, Credi A. Artificial molecular shuttles: From concepts to devices. Journal of Materials Chemistry. 2009;**19**:2279-2294. DOI: 10.1039/B818609J

[6] Jiménez MC, Dietrich-Buchecker C, Sauvage J-P. Towards synthetic molecular muscles: Contraction and stretching of a linear rotaxane dimer. Angewandte Chemie International Edition. 2000;**39**:3284-3287. DOI: 10.1002/1521-3773(20000915)39:18<3284::AID-ANIE3284>3.0.CO;2-7

[7] Bruns C, Stoddart JF. The Nature of the Mechanical Bond: From Molecules to Machines. Hoboken: Wiley; 2016. ISBN: 9781119044123. DOI: 10.1002/9781119044123

[8] Badjić JD, Balzani V, Credi A, Silvi S, Stoddart JF. A molecular elevator. Science. 2004;**303**:1845-1849. DOI: 10.1126/science.1094791

[9] Rekharsky M, Inoue Y. Solvation effects in supramolecular recognition. In: Supramolecular Chemistry: From Molecules to Nanomaterials. Hoboken, NJ: Wiley; 2012. DOI: 10.1002/9780470661345.smc009

[10] Kanagaraj K, Alagesan M, Inoue Y, Yang C. Solvation Effects in Supramolecular Chemistry. Part of Comprehensive Supramolecular Chemistry II Vol 1. Amsterdam: Elsevier; 2017. DOI: 10.1016/B978-0-12-409547-2.12481-3

[11] Nishimura N, Yoza K, Kobayashi K. Guest-encapsulation properties of a self-assembled capsule by dynamic boronic ester bonds. Journal of the American Chemical Society. 2009;**132**:777-790. DOI: 10.1021/ja9084918

[12] Kang J, Rebek J. Entropically driven binding in a self-assembling molecular capsule. Nature. 1996;**382**:239-241. DOI: 10.1038/382239a0

[13] Smithrud DB, Diederich F. Strength of molecular complexation of apolar solutes in water and in organic solvents is predictable by linear free energy relationships: a general model for solvation effects on apolar binding. Journal of the American Chemical Society. 1990;**112**:339-343. DOI: 10.1021/ja00157a052

[14] Izatt RM, Pawlak K, Bradshaw JS, Bruening RL. Thermodynamic and kinetic data for macrocycle interaction with cations, anions, and neutral molecules. Chemical Reviews. 1995;**95**:2529-2586. DOI: 10.1021/cr00039a010

[15] Marcus Y. Effect of ions on the structure of water: Structure making and breaking. Chemical Reviews.

2009;**109**:1346-1370. DOI: 10.1021/cr8003828

[16] Marcus Y. Ion Solvation. Chichester: John Wiley & Sons, Ltd; 1985

[17] Wasserman E. The preparation of interlocking rings: A catenane. Journal of the American Chemical Society. 1960;**82**:4433-4434. DOI: 10.1021/ja01501a082

[18] Luttringhaus A, Cramer F, Prinzbach H, Henglein F, Dithiolen MCVL. Versuche zur Darstellung sich umfassender Ringe mit Hilfe von Einschlußverbindungen. Liebigs Annalen der Chemie. 1958;**613**:185-198. DOI: 10.1002/jlac.19586130120

[19] Harrison IT, Harrison S. Synthesis of a stable complex of a macrocycle and a threaded chain. Journal of the American Chemical Society. 1967;**89**:5723-5724. DOI: 10.1021/ja00998a052

[20] Schill G. Catenanes, Rotaxanes, and Knots. Academic Press: NewYork; 1971

[21] Feringa BL, Browne WR. Molecular Switches. Weinheim: Wiley VCH; 2011. ISBN: 978-3-527-63441-5

[22] Leigh DA, Murphy A, Smart JP, Slawin AMZ. Glycylglycine rotaxanes—The hydrogen bond directed assembly of synthetic peptide rotaxanes. Angewandte Chemie International Edition. 1997;**36**:728-732. DOI: 10.1002/anie.199707281

[23] Da Ros T, Guldi DM, Morales AF, Leigh DA, Prato M, Turco R. Hydrogen bond-assembled fullerene molecular shuttle. Organic Letters. 2003;**5**: 689-691. DOI: 10.1021/ol0274110

[24] Ghosh P, Federwisch G, Kogej M, Schalley CA, Haase D, Saak W, et al. Controlling the rate of shuttling motions in [2]rotaxanes by electrostatic interactions: a cation as solvent-tunable brake. Organic & Biomolecular Chemistry. 2005;**3**:2691-2700. DOI: 10.1039/B506756A

[25] Du S, Fu H, Shao X, Chipot C, Cai W. Water-controlled switching in rotaxanes. Journal of Physical Chemistry C. 2018;**122**:9229-9234. DOI: 10.1021/acs.jpcc.8b01993

[26] Liu Y, Chipot C, Shao X, Cai W. How does the solvent modulate shuttling in a pillararene/imidazolium [2]rotaxane? Insights from free energy calculations. Journal of Physical Chemistry C. 2016;**120**:6287-6293. DOI: 10.1021/acs.jpcc.6b00852

[27] Liu P, Chipot C, Shao X, Cai W. Solvent-controlled shuttling in a molecular switch. Journal of Physical Chemistry C. 2012;**116**:4471-4476. DOI: 10.1021/jp2114169

[28] Reichardt C. Solvatochromic dyes as solvent polarity indicators. Chemical Reviews. 1994;**94**:2319-2358. DOI: 10.1021/cr00032a005

[29] Reichardt C, Welton T. Solvents and Solvent Effects in Organic Chemistry. Weinheim: Wiley-VCH; 2011

[30] Coe BJ, Harries JL, et al. Pentacyanoiron(II) as an electron donor group for nonlinear optics: Medium-responsive properties and comparisons with related pentaammineruthenium(II) complexes. Journal of the American Chemical Society. 2006;**128**:12192-12204. DOI: 10.1021/ja063449m

[31] Bureš F, Pytela O, Diederich F. Solvent effects on electronic absorption spectra of donor-substituted 11,11,12,12-tetracyano-9, 10-anthraquinodimethanes (TCAQs). Journal of Physical Organic Chemistry. 2009;**22**:155-162. DOI: 10.1002/poc.1443

[32] Pinheiro C, Lima JC, Parola AJ. Using hydrogen bonding-specific interactions to detect water in aprotic

solvents at concentrations below 50 ppm. Sensors and Actuators B. 2006;**114**:978-983. DOI: 10.1016/j.snb.2005.08.013

[33] Yoon G. Dielectric properties of glucose in bulk aqueous solutions: Influence of electrode polarization and modeling. Biosensors and Bioelectronics. 2011;**26**:2347-2353. DOI: 10.1016/j.bios.2010.10.009

[34] Deligkiozi I, Voyiatzis E, Tsolomitis A, Papadakis R. Synthesis and characterization of new azobenzene-containing bis pentacyanoferrate(II) stoppered push–pull [2]rotaxanes, with α- and β-cyclodextrin. Towards highly medium responsive dyes. Dyes and Pigments. 2015;**113**:709-722. DOI: 10.1016/j.dyepig.2014.10.005

[35] Papadakis R, Deligkiozi I, Tsolomitis A. Spectroscopic investigation of the solvatochromic behavior of a new synthesized non symmetric viologen dye: Study of the solvent-solute interactions. Analytical and Bioanalytical Chemistry. 2010;**397**:2253-2259. DOI: 10.1007/s00216-010-3792-7

[36] Papadakis R, Deligkiozi I, Tsolomitis A. Synthesis and characterization of a group of new medium responsive non-symmetric viologens. Chromotropism and structural effects. Dyes and Pigments. 2012;**95**:478-484. DOI: 10.1016/j.dyepig.2012.06.013

[37] Papadakis R, Tsolomitis A. Study of the correlations of the MLCT Vis absorption maxima of 4-pentacyanoferrate-40′-arylsubstituted bispyridinium complexes with the Hammett substituent parameters and the solvent polarity parameters ETN and AN. Journal of Physical Organic Chemistry. 2009;**22**:515-521. DOI: 10.1002/poc.1514

[38] Papadakis R. Preferential solvation of a highly medium responsive pentacyanoferrate(II) complex in binary solvent mixtures: Understanding the role of dielectric enrichment and the specificity of solute–solvent interactions. Journal of Physical Chemistry B. 2016;**120**:9422-9433. DOI: 10.1021/acs.jpcb.6b05868

[39] Günbaş DD, Zalewskia L, Brouwer AM. Solvatochromic rotaxane molecular shuttles. Chemical Communications. 2011;**47**:4977-4979. DOI: 10.1039/c0cc05755j

[40] Toma SH, Toma HE. Self assembled rotaxane and pseudo-rotaxanes based on β-cyclodextrin inclusion compounds with trans-1,4-Bis[(4-pyridyl)ethenyl] benzene-pentacyanoferrate(II) Complexes. Journal of the Brazilian Chemical Society. 2007;**18**:279-283. DOI: 10.1590/S0103-50532007000200006

[41] Baer AJ, Macartney DH. α- and β-cyclodextrin rotaxanes of μ-Bis(4-pyridyl)bis[pentacyanoferrate(II)] Complexes. Inorganic Chemistry. 2000;**39**:1410-1417. DOI: 10.1021/ic990502h

[42] Creutz C. Mixed valence complexes of d^5-d^6metal centers. Progress in Inorganic Chemistry. 1983;**30**:1. DOI: 10.1002/9780470166314.ch1

[43] Launay JP, Marvaud V. Control of intramolecular electron transfer by protonation: Dimers and polymers containing ruthenium II/III and 44′ azopyridine. AIP Conference Proceedings. 1992;**262**:118. DOI: 10.1063/1.42663

[44] Launay JP, Touriel-Pagis M, Lipskier JF, Marvaud V, Joachim C. Control of intramolecular electron transfer by a chemical reaction. The 4,4′-azopyridine/1,2-bis(4-pyridyl) hydrazine system. Inorganic Chemistry. 1991;**30**:1033-1038. DOI: 10.1021/ic00005a029

[45] Kesslen EC, Euler WB. Synthesis and characterization of pyridine

end-capped oligoazines. Tetrahedron Letters. 1995;**36**:4725-4728. DOI: 10.1016/0040-4039(95)00888-J

[46] Deligkiozi I, Papadakis R, Tsolomitis A. Synthesis, characterisation and photoswitchability of a new [2] rotaxane of α-cyclodextrin with a diazobenzene containing π-conjugated molecular dumbbell. Supramolecular Chemistry. 2012;**24**:333-343. DOI: 10.1080/10610278.2012.660529

[47] Deligkiozi I, Papadakis R, Tsolomitis A. Photoconductive properties of a π-conjugated α-cyclodextrin containing [2]rotaxane and its corresponding molecular dumbbell. Physical Chemistry Chemical Physics. 2013;**15**:3497-3503. DOI: 10.1039/C3CP43794A

[48] Papadakis R, Deligkiozi I, Li H. Photoconductive interlocked molecules and macromolecules, IntechOpen: Rijeka, In book: Photodetectors. 2018. DOI: 10.5772/intechopen.79798

[49] Deligkiozi I, Voyiatzis E, Tsolomitis A, Papadakis R. Synthesis and characterization of new azobenzene-containing bis pentacyanoferrate(II) stoppered push-pull [2]rotaxanes, with alpha- and beta-cyclodextrin. Towards highly medium responsive dyes. Dyes and Pigments. 2015;**113**:709-722. DOI: 10.1016/j.dyepig.2014.10.005

[50] Papadakis R, Deligkiozi I, Nowak KE. Study of the preferential solvation effects in binary solvent mixtures with the use of intensely solvatochromic azobenzene involving [2]rotaxane solutes. Journal of Molecular Liquids. 2019;**274**:715-723. DOI: 10.1016/j.molliq.2018.10.164

[51] Baggerman J, Jagesar DC, Vallée RAL, Hofkens J, De Schryver FC, Schelhase F, et al. Fluorescent perylene diimide rotaxanes: Spectroscopic signatures of wheel–chromophore interactions. Chemistry A European Journal. 2007;**22**:1291-1299. DOI: 10.1002/chem.200601014

[52] Boer SA, Cox RP, Beards MJ, Wang H, Donald WA, Bell TDM, et al. Elucidation of naphthalene diimide metallomacrocycles and catenanes by solvent dependent excimer and exciplex emission. Chemical Communications. 2019;**55**:663-666. DOI: 10.1039/C8CC09191A

[53] Lee CS, Kimizuka N. Solvatochromic nanowires self-assembled from cationic, chloro-bridged linear platinum complexes and anionic amphiphiles. Chemistry Letters. 2002;**31**:1252-1253. DOI: 10.1246/cl.2002.1252

[54] Kuroiwa K, Oda N, Kimizuka N. Supramolecular solvatochromism. Effect of solvents on the self-assembly and charge transfer absorption characteristics of lipid-packaged, linear mixed-valence platinum complexes. Science and Technology of Advanced Materials. 2006;**7**:629-634. DOI: 10.1016/j.stam.2006.09.011

[55] Kawai T, Hashizume M. Stimuli-Responsive Interfaces. Chapter 5. Singapore: Springer Verlag; 2017

[56] Nikolayenko VI, Wyk LM, Munro OQ, Barbour LJ. Supramolecular solvatochromism: Mechanistic insight from crystallography, spectroscopy and theory. Chemical Communications. 2018;**54**:6975-6978. DOI: 10.1039/c8cc02197j

[57] Li CP, Du M. Role of solvents in coordination supramolecular systems. Chemical Communications. 2011;**47**:5958-5972. DOI: 10.1039/c1cc10935a

[58] Rood JA, Noll BC, Henderson KW. Cubic networks and 36 tilings assembled from isostructural trimeric magnesium aryldicarboxylates. Main Group Chemistry. 2006;**5**:21-30. DOI: 10.1080/10241220600815718

[59] Davies RP, Less RJ, Lickiss PD, White AJP. Framework materials

assembled from magnesium carboxylate building units. Dalton Transactions. 2007:2528-2535. DOI: 10.1039/B705028C

[60] Williams CA, Blake AJ, Wilson C, Hubberstey P, Schröder M. Novel metal–organic frameworks derived from group II metal cations and aryldicarboxylate anionic ligands. Crystal Growth and Design. 2008;**8**: 911-922. DOI: 10.1021/cg700731d

[61] Noro S, Horike S, Tanaka D, Kitagawa S, Akutagawa T, Nakamura T. Flexible and shape-selective guest binding at CuII axial sites in 1-dimensional Cu^{II}–1,2-Bis(4-pyridyl) ethane coordination polymers. Inorganic Chemistry. 2006;**45**: 9290-9300. DOI: 10.1021/ic0609249

[62] Wang XY, Wang L, Wang ZM, Gao S. Solvent-tuned azido-bridged Co^{2+} layers: Square, honeycomb, and kagomé. Journal of the American Chemical Society. 2006;**128**:674-675. DOI: 10.1021/ja055725n

[63] Lin JB, Zhang JP, Zhang WX, Xue W, Xue DX, Chen XM. Porous manganese(II) 3-(2-pyridyl)-5-(4-pyridyl)-1,2,4-triazolate frameworks: Rational self-assembly, supramolecular isomerism, solid-state transformation, and sorption properties. Inorganic Chemistry. 2009;**48**:6652-6660. DOI: 10.1021/ic900621c

[64] Nakahara A, Wang JH. Charge-transfer complexes of methylviologen. Journal of Physical Chemistry. 1963;**67**:496-498. DOI: 10.1021/j100796a503

[65] Papadakis R, Deligkiozi I, Giorgi M, Faure B, Tsolomitis A. Supramolecular complexes involving nonsymmetric viologen cations and hexacyanoferrate(II) anions. A spectroscopic, crystallographic and computational study. RSC Advances. 2016;**6**:575-585. DOI: 10.1039/c5ra16732a

[66] Abouelwafa AS, Mereacre V, Balaban TS, Anson CE, Powell AK. Photo- and thermally-enhanced charge separation in supramolecular viologen–hexacyanoferrate complexes. CrystEngComm. 2010;**12**:94-99. DOI: 10.1039/B915642A

Chapter 12

Solvent Effect on a Model of S_NAr Reaction in Conventional and Non-Conventional Solvents

Paola R. Campodónico

Abstract

In this chapter some theoretical and experimental reports in order to elucidate solvent effects (preferential solvation and iso-solvation effects, respectively) over nucleophilic aromatic substitution reactions as reaction model were examined. Solvent effects phenomena are introduced to predict their mechanism highlighting the hydrogen bond role mainly in ionic liquids, a new generation of solvents that can be designed in order to improve the reactivities of the reacting pair and intermediate species through of the potential energy surface (PES). Then, the preferential solvent effect may be defined as the difference between local and bulk compositions of the solute with respect to the various components of the solvent; usually mixtures of solvents and iso-solvation effect indicate the composition of a mixture in which the probe under consideration is solvated by approximately an equal number of cosolvent molecules in the solvent mixture.

Keywords: solvent effects, preferential solvation, S_NAr reaction, ionic liquids, iso-solvation effect

1. Introduction

Studies suggest that nucleophilic aromatic substitution (S_NAr) reactions are significantly affected by the reaction media, because it involves the stabilization of species associated to the potential energy surface (PES) determining selectivity, reaction rates, and mechanisms [1–3]. In this chapter an integrative analysis based on experimental and theoretical results as an input to perform a deeper analysis based on preferential solvation and iso-solvation effects, respectively, is described. This chapter is organized by summarizing the main achievements on solvent effects based on S_NAr reactions considering that these reactions have widely been studied in water, conventional organic solvents (COS), and more recently in ionic liquids (IL) and mixtures of them [1–11]. Note that the most discussed articles are based on kinetic responses in order to evaluate the solvent effect over this reaction which has been used as model.

2. Nucleophilic aromatic substitution reactions

Nucleophilic substitution is an addition-elimination (A_ND_N) process that depending on the nature of the substrate, the attacking nucleophile, and the solvent

IntechOpen

effect may lead to a nucleophilic substitution (NS) product or a S_NAr product or both [12–16]. A S_NAr reaction occurs in activated aromatic compounds bearing good leaving groups (LG). In general, it is widely accepted that the mechanism of the S_NAr reactions involves the formation of a σ-complex (also called Meisenheimer complex (MC)) that occurs after the nucleophilic attack step at the ipso atom of aromatic moiety. Next, the departure of LG with re-aromatization of the aromatic ring closes the set of steps to give the desired product. Commonly, the LG departure step is faster than the nucleophilic attack; therefore, the addition of the nucleophile to the ring moiety appears as the rate-limiting step in these processes [13, 17–24]. In the last time, a concerted reaction mechanism could be prevailing [25–27]. A lot of work has been carried out to clarify whether the concerted mechanism is an exception or the dominant pathway in these processes [28, 29].

2.1 Hydrogen Bond in S_NAr Reactions

Bernasconi et al. [23, 30] postulated the existence of an intramolecular hydrogen bond between a hydrogen atom of the nucleophilic center in the nucleophile and the orto-NO_2 group of the substrate in order to explain the reactivity trends in orto-halonitrobenzenes to respect para-halonitrobenzenes toward amines. Ormazábal-Toledo et al. [31] carried out computational studies about the role of HB effects along the intrinsic reaction coordinate profile, demonstrating that it promotes the activation of both the substrate and nucleophile, respectively. Note that the analysis was performed in transition state (TS) structures, because the reactant states hide most of the information about specific interactions that characterize the S_NAr reactions. Recently, Gallardo-Fuentes and Calfumán et al., respectively, showed that the HB not only determines the reactivities, but also it could be involved in concerted routes in S_NAr reactions [1, 26, 32].

3. Room temperature ionic liquids

Ionic liquids or room temperature ionic liquids (RTILs) are defined as molten salts (composed entirely of cations and anions) that melt below 100°C [33] with remarkable physicochemical properties: non-flammable, non-corrosive, nonvolatile, and bulk physical constant, which can be tuned by the combination of different cations and anions [34–38]. RTILs are composed by bulky organic cations usually imidazolium or pyridinium derivatives substituted with alkyl chains and an inorganic or organic anion (usually a halide, tetrafluoroborate, hexafluorophosphate, and others). The high combinatorial flexibility has converted these materials into "designer solvents" or "task-specific" solvents [33, 35, 38] whose properties can be specified to suit the requirements of a particular reaction [2, 4, 12, 39]. For these reasons, RTILs have gained importance in the solvent effects field being recognized as very promising reaction media with green features.

3.1 S_NAr reactions in ionic liquids

A series of reaction have been studied in RTILs and mixtures of them with water or COS. The criteria to select the RTILs were based on the following: (i) the solubility of substrates and nucleophiles; (ii) to have a reasonable number of anions and cations to assess anion and cation effects; and (iii) to ensure that these RTILs do not interfere with the reaction [12]. Solvent effects in RTILs are a complex problem, because the solute-solvent interactions will be masked by the leading solvent-solvent interactions that are coulombic in nature. Some strategies to study solvent effects in RTILs consider a reasonable large number of these and to evaluate their performance by fixing the anion and varying the cation and vice versa. For instance, a complete study based on the reaction of DNBSCl with piperidine was performed in 17 RTILs considering water as a solvent reference. This study identified three groups of RTILs showing 1-ethyl-3-methylimidazolium dicyanamide (EMIMDCN) as the best solvent considering all the studied RTILs and 21 COS [12, 40]. EMIMDCN shows a catalytic behavior attributed to its high polarizability given by the dicyanamide anion (DCN^-), which presents a highly rich π electron density, and its size could be in relationship with steric hindrance effects. Note that ethylammonium nitrate (EAN) was the only protic RTIL that decreased the reaction time in comparison to water and EMIMDCN, respectively. Then, a comparative study of the reaction of DNBSCl and propylamine in COS and RTILs, respectively, was performed in order to analyze the nature of the nucleophile. Piperidine showed to be more active than propylamine in polar solvents with the ability to donate and accept HB, while the reaction of propylamine was favored only with solvent that can accept HB, being the best COS solvent: N,N-dimethylformamide (DMF). Note that, in all the studied RTILs for propylamine, the reactivities of the reactions were lower than piperidine. This response was attributed to the capacity of the RTILs to donate/accept HB in agreement with the COS behavior [12]. On the other hand, propylammonium nitrate (PAN) was able to emulate the HB behavior of water toward the reaction between 4-chloroquinazoline and aniline [39]. PAN could be donating an HB by the ammonium moiety of PAN toward the substrate emulating an electrophilic solvation suggested in aqueous media improving its reactivity toward the nucleophile. These results are in agreement with the report of Welton et al. [2] based on the task-specific design of RTILs in order to optimize those properties that enhanced the reaction reactivities. Harper et al. reported the main role of the RTIL structure on the reaction rates of S_NAr reactions [41, 42].

3.2 Binary mixtures based on ionic liquids

The use of RTILs or ionic liquid binary mixtures could give variations in the structure of the ionic lattice of neat ILs after mixing [43–45]. This fact may have significant repercussions on the nature and strength of the interactions that contribute mainly to coulomb interactions that determine the 3D structure of ILs [46, 47]. Studies of binary mixtures with common anions, for instance, the same cation but different anions, have shown how the presence of random co-networks or block co-networks depends on the size of the anions [4, 47, 48]. Seddon suggests the use of IL mixtures to expand the range of room temperatures in ILs [49]. Initially, the hygroscopic nature of the ILs was a problem; however the high capacity of the ILs to solubilize water opens a wide spectrum of reaction media, mainly based on the role of the hydrogen bond (HB) and electrostatic interactions between molecules in the mixture. Reports have shown that the addition of COS to ILs may affect significantly the density, viscosity, and conductivity with respect to pure ILs. For instance, the direct relationships between the viscosity of the IL/COS mixtures with the solvent dielectric constant (ε) of the COS pure [50, 51]. It may be attributed to the difference in the ion-dipole interactions between the ions and solvents. The addition of water to ILs may change the molecular structure of pure ILs probably due to HB between the water molecules and the anions of the ILs [52, 53]. Sanchez et al. studied solvent mixtures between 1-butyl-3-methyl imidazolium tetrafluoroborate (BMIMBF4)/water at different molar fractions, observing on the studied range of compositions, a border line located close to $\chi = 0.2$. Before this value indicates that the added RTIL promotes the reactivity of the substrate by preferential solvation. After this value, the rate coefficients remain approximately constant. At low concentrations the water begins to break down the 3D structure of the ILs, which then goes on to form ionic clusters as the concentration of water increases until eventually ion pairs form, which are the dominant species in the aqueous solution [10, 46, 48].

4. Interactions and solvent effects

It is well known that the chemical reactivity is determined by the ability of the solvent to interact with solute, intermediates, and transition state (TS) structures along the reaction pathway [1–3]. The main difference between COS and RTILs are the electrostatic solvent-solvent interactions between cation-anion and cation-cation interactions [33]. These interactions in the COS are moderate dipole-dipole interactions; in the RTILs they become the leading term (ion-ion interactions) that are expected to outweigh the target solute-solvent interactions. Solute-solvent interactions contain the relevant information about catalysis, stabilizing/destabilizing effects affecting the electrophile/nucleophile pair (solute), TS structures, and the intermediate in a polar process [1]. Solvent effects can be split into two types: non-specific interactions and specific interactions, including all the possible interactions that can occur between solvent and the electrophile/nucleophile pair [5, 52].

4.1 Preferential solvation

Solvent effects can be split into two types: non-specific and specific interactions, including all the possible interactions that can occur between the solvent and solute [5, 39, 54]. Then, preferential solvent may be defined as the difference between local and bulk composition of the solute with respect to the various components of the solvent, usually mixtures of solvents [5, 55–57]. The "bulk of the solvent" is treated as the external shell, and it can be described using the classic theories of Kirkwood-Onsager, models of solvation based on reaction field theory or molecular dynamic [55, 58, 59]. Then, in a binary mixture of protic solvents, the "preferential solvation" may be cast into the form of specific solute-solvent interactions described as local solvation, which may be defined as a "first solvation shell." The local solvation may be classified as electrophilic or nucleophilic, respectively [17, 60–63]. Electrophilicity and nucleophilicity concepts are related to electron-deficient (electrophile) and electron-rich (nucleophile) species [39, 64, 65]. These concepts are based on the valence electron theory of Lewis [66] and the general acid–base theory of Brønsted and Lowry [67, 68] and introduced by Ingold in 1934. Then, for a mixture of polar solvents, the "electrophilic solvation" represents the specific interaction through a HB with the hydrogen atom of the solvent, whereas "nucleophilic solvation" describes a specific interaction through a HB between an acidic hydrogen atom of the solute and the heteroatom of the solvent [5, 60–63]. Mancini et al. have reported preferential solvation of 1-halo-2,4-dinitrobenzenes with amines in mixtures of dichloromethane with polar protic/polar aprotic solvents [7, 44, 45, 69, 70]. Ormazabal-Toledo et al. [5] reported an integrated experimental and theoretical study of 2,4,6-trinitrophenyl ether with a series of secondary alicyclic (SA) amines in ethanol/water mixtures at different compositions. In it only piperidine was sensitive to preferential solvation at high proportion of water. Piperidine increases its rate coefficient values suggesting a stabilization of the MC by HB displayed by the presence of more water molecules in the first shell at these proportions of water in the studied mixtures. This result shows that the environment of the MC changes for different solvent compositions. Then, for the remaining amines the environment showed to be similar being it attributed to polar nature of the substituent at position 4, suggesting that their kinetic responses are independent of the bulk properties of the reaction media. On the other hand, Alarcón-Espósito et al. [1] studied the reaction between 1-chloro and 1-fluoro-2,4-dinitrobenzenes, respectively (ClDNB and FDNB, respectively) with morpholine (a SA amine) in acetonitrile (MeCN)/water mixtures at different compositions. Experimental results were complemented with a theoretical analysis in order to study the bulk and specific interactions of solute-solvent in mixtures of a COS (MeCN) and water. Note that both solvents display significant HB abilities. Then, in 90% vol. MeCN substrates both displayed the maximum value of the rate coefficient constants. On the other hand, the exploration of the PES using the "super-molecule method" revealed that the solvation of the TS structure associated to the rate determining step (RDS) of the reaction mechanism expressed in the mode water/MeCN outweighs over MeCN/water, suggesting a preferential solvation in favor of the aqueous phase. The super-molecule model introduces solvent molecules explicitly around the solute. This model provides a detailed synopsis of the field of solvation sufficient to describe the interactions between the solute and solvent [71].

4.2 Iso-solvation

The concept of iso-solvation has been introduced to indicate the composition of a mixture in which the probe under consideration is solvated by an approximately an equal number of cosolvent molecules in the solvent mixture [48]. This effect has been extensively observed in COS mixtures [72–74]. Alarcón-Espósito et al. [48] studied the reaction between ClDNB with morpholine in a series of mixtures of ILs involving imidazolium cations. Iso-solvation effects were observed in the following mixtures: 1-ethyl-3-methyl imidazolium thiocyanate/1-ethyl-3-methyl imidazolium dicyanamide (EMIMSCN/EMIMDCN), 1-butyl-3-methyl imidazolium dicyanamide/1-butyl-3-methyl imidazolium tetrafluoroborate (BMIMDCN/$BMIMBF_4$), $BMIMBF_4$/1-butyl-3-methyl imidazolium hexafluorophosphate ($BMIMPF_6$), and $BMIMPF_6$/1-butyl-3-methyl imidazolium tris(pentafluoroethyl) trifluorophosphate (BMIMFAP), respectively. Iso-solvation regimes correspond to a solvent composition regime where the solute is being solvated by approximately the same number of different solvent molecules in the mixture. These results showed that for significant changes in composition, the rate coefficients remain approximately constant. On the other hand, for the solvent mixture $BMIMBF_4$/$BMIMPF_6$ at 0.9 molar fraction of $BMIMBF_4$, a slightly better kinetic response is observed than the pure $BMIMBF_4$ and $BMIMPF_6$. Another interesting result was observed in the mixture of EMIMSCN/EMIMDCN; an increasing proportion of EMIMSCN with respect to EMINDCN results in a decrease of the rate coefficient within the range 0.1–0.75 in molar fraction of EMIMSCN. This result could be expressed as a competition between the anions toward the reaction center driven by the basicity of the reaction media.

4.3 Polarity and solvent effects

Experimentally, the most common way to measure the polarity of a solvent is through its (bulk) dielectric constant (ε). The concept of polarity has been defined as the sum of all possible intermolecular interactions between the solvent and the solute, including specific interactions, for instance, HB effects, dipole-dipole, dipole-induced dipole, electron pair acceptor-electron pair donor, and acid-base interactions [1, 33]. Gazitúa et al. [12, 40] studied the solvation patterns of 21 COS and water over the reaction between 2,4-dinitrophenylsulfonyl chloride (DNBSCl) with SA amines in order to determine the solvent polarity effect on the reaction mechanism. Note that solvent polarity became relevant only in the reactions that proceeded by the non-catalyzed route. On this way, water and tetrahydrofuran

(THF) have a key role due to its ambiphilic character as an HB donor/acceptor that promotes a nucleophilic activation at the nitrogen center of the piperidine (nucleophile).

5. Solvent models

5.1 Kamlet-Taft model

Solvent effect studies have been focused mainly on the polarity of the reaction medium as a determinant of chemical reactivity properties. Experimentally, the most common way to measure the polarity of a solvent is through the ε. However, the measure requires that the reaction medium will be non-conductive, which does not happen in the RTLIs. The concept of polarity has been defined as the sum of all possible intermolecular interactions between the solvent and a solute, excluding those interactions that lead toward chemical reactions of the solute and including Coulombic interactions, HB interactions, dipole-dipole, dipole-induced dipole, electron pair acceptor-electron pair donor, and acid-base interactions [1, 33]. There are many empirical solvent polarity scales [75–83] that attempt to give quantitative estimates of solvent polarity, some of those were applied to RTLIs [84]. However, the high number of interactions in non-conventional reaction media cannot be incorporated in a measurement or polarity scale. The most used solvent polarity scale is the one developed by Kamlet and Taft [78] based on solvatochromism properties that show specific and non-specific interactions. Solvatochromism is solvent dependence of the electronic spectrum of chromophore. Intensity, position, and shape of absorption bands of dissolved chromophores are influenced by the change in solvents or mixture of solvents, according to their electronic and molecular structure, due to the different stabilization of their electronic ground and excited states. Therefore, any solvent-dependent property (SDP) in solution is normally expressed as a linear solvation free energy relationship (LSER) as follows:

$$SDP = SDP_0 + a\,\alpha_s + b\,\beta_s + s\,\pi_s^* \qquad (1)$$

where *SDP* corresponds to any kinetic property, namely, selectivity or reaction rate coefficients, which is modeled as a linear combination of two H-bond terms, one for hydrogen bond donor ($a\,\alpha_s$) and hydrogen-bond acceptor ($b\,\beta_s$) and a dipolarity/polarizability term ($s\,\pi_s^*$), with SDP_0 a constant describing intrinsic properties of the solute [84]. In this approach, empirical solvatochromic parameters are introduced to describe specific HB interaction, ion-dipole, dipole-dipole, dipole-induced dipole, solvophobic, dispersion London, and possible π-π and p-π stacking effects. The reason is that while empirical solvatochromic parameters in COS work reasonably well, for RTILs they consistently fail. The main reason seems to be the transferability of the response of a particular probe chromophore from some known SOC to RTILs. This transferability would warrant that the polarities of the RTILs and the SOC are the same and that the appropriate value of the parameter can then safely be assigned to the RTIL. The second implicit assumption is that the effect of transferring from a SOC to an RTIL is the same for all probes. They conclude that it is important to consider the nature of the chromophore as well as the solvent when establishing reliable solvent polarity parameters, mainly when this chromophore is transferred from a neutral molecular solvent to an RTIL. The main message, however, is that it cannot be a priori established if one solvent polarity scale with respect to another one is right or wrong: “each scale will turn useful in a given set of circumstances and in other ones they will not” [1, 12].

5.2 Theoretical models of solvation

In pure conventional solvents, the determination of properties and type of interactions has been reasonably achieved with the use of non-specific solute-solvent interactions, based on continuum dielectric models [85, 86]. In RTILs, the results are both scarce and not yet systematized [87, 88]. The super-molecule model provides a detailed synopsis of the field of solvation [71].

5.3 Gutmann's donor and acceptor numbers

Donor (DN) and acceptor (AN) numbers proposed by Gutmann [89–91] are used to describe acid-base solvent properties in RTILs based on a reformulation by Schmeisser et al. [92, 93]. On the original definition of Gutmann, DN and AN are a quantitative measure of Lewis basicity and acidity of a solvent, generally a nonaqueous media [4, 91]. These numbers can be measured by calorimetrical technique and by using the chemical shift in ^{31}P NMR spectra [92, 94]. In COS these parameters are used in order to describe the ability of solvents to donate or accept electron pairs or at least electron density to the substrate. Then, DN represents a measure for the donor properties of a solvents, and AN is a measure of the electrophilic properties of a solvent. DN parameter in RTILs shows a strong dependence on the anionic component of the RTIL; however, AN is dependent on both anionic and cationic moieties of the RTIL [92].

Alarcón-Espósito et al. [4] studied three families of RTILs, based on 1-ethyl-3-methyl imidazolium ($EMIM^+$), 1-butyl-3-methyl imidazolium ($BMIM^+$), and 1-butyl-1-methyl-pyrrolidinium ($BMPyr^+$) cations, respectively, with a wide series of anions in order to evaluate both models of solvent effects toward the reaction between 1-chloro-2,4-dinitrobenzene with morpholine by kinetic experiments. The first approximation of solvent effects was attributed to an "anion effect." This effect appears to be related to the anion size, polarizability, and its HB ability toward the substrate. The comparison between rate constants and Kamlet-Taft solvatochromic model systematically failed. However, the anion effect was confirmed by performing a comparison of the rate constants and DN emphasizing the main role of the charge transfer from the anion to the substrate.

6. Conclusions

In this chapter some theoretical and experimental reports in order to elucidate solvent effects over nucleophilic aromatic substitution reactions were examined. Solvent effects are introduced over mechanistic behaviors highlighting the HB role mainly in RTILs, a new generation of solvents that can be designed in order to improve the reactivities of the reacting pair and intermediate species through the PES. For instance, (i) solvent polarity could be modulating the reaction pathways differently; (ii) the ability of the solvent to establish HB could drive the reaction mechanism opening the possibility of preferential solvation; (iii) in mixtures of solvents and depending on the constituents of them could be affecting the reaction rate by solvent structural organization, viscosity, and HB interactions; and (iv) in ionic liquids the solvent effect could be attributed to an anion effect being it related to the size and HB abilities of the anions.

Acknowledgements

This work was supported by Fondecyt Grant 1150759 (PC), Proyecto de Mejoramiento Institucional postdoctoral fellowship PMI-UDD from Instituto de Ciencias e Innovación en Medicina (ICIM-UDD). Postdoctoral fellowships by Fondecyt and project ICM-MINECOM, RC-130006-CILIS, granted by Fondo de Innovación para la Competitividad del Ministerio de Economia, Fomento y Turismo, all of them from Chile.

Author details

Paola R. Campodónico
Facultad de Medicina, Centro de Química Médica, Clínica Alemana Universidad del Desarrollo, Santiago, Chile

*Address all correspondence to: pcampodonico@udd.cl

IntechOpen

© 2019 The Author(s). Licensee IntechOpen. This chapter is distributed under the terms of the Creative Commons Attribution License (http://creativecommons.org/licenses/by/3.0), which permits unrestricted use, distribution, and reproduction in any medium, provided the original work is properly cited.

References

[1] Alarcón-Espósito J, Tapia RA, Contreras R, Campodónico PR. Changes in the SNAr reaction mechanism brought about by preferential solvation [Internet]. RSC Advances. 2015;**5**:99322-99328. DOI: 10.1039/c5ra20779g

[2] Newington I, Perez-Arlandis JM, Welton T. Ionic li-quids as designer solvents for nucleophilic aromatic substitutions. Organic Letters. 2007;**9**:5247-5250

[3] D'Anna F, Marullo S, Noto R. Aryl azides formation under mild conditions: A kinetic study in some ionic liquid solutions. The Journal of Organic Chemistry. 2010;**75**:767-771

[4] Alarcón-Espósito J, Contreras R, Tapia RA, Campodónico PR. Gutmann's donor numbers correctly assess the effect of the solvent on the kinetics of SN Ar reactions in ionic liquids. Chemistry. 2016;**22**:13347-13351

[5] Ormazabal-Toledo R, Santos JG, Ríos P, Castro EA, Campodónico PR, Contreras R. Hydrogen bond contribution to preferential solvation in SNAr reactions. The Journal of Physical Chemistry. B. 2013;**117**:5908-5915

[6] Um I-H, Min S-W, Dust JM. Choice of solvent (MeCN vs H2O) decides rate-limiting step in SNAr aminolysis of 1-fluoro-2,4-dinitrobenzene with secondary amines: Importance of Brønsted-type analysis in acetonitrile [Internet]. The Journal of Organic Chemistry. 2007;**72**:8797-8803. DOI: 10.1021/jo701549h

[7] Nudelman NS, Mancini PME, Martinez RD, Vottero LR. Solvents effects on aromatic nucleophilic substitutions. Part 5. Kinetics of the reactions of 1-fluoro-2,4-dinitrobenzene with piperidine in aprotic solvents [Internet]. Journal of the Chemical Society, Perkin Transactions. 1987;**2**:951. DOI: 10.1039/p29870000951

[8] Swager TM, Wang P. A negotiation between different nucleophiles in SNAr reactions. Synfacts. 2017;**13**:0148-0148

[9] Park S, Lee S. Effects of ion and protic solvent on nucleophilic aromatic substitution (SNAr) reactions. Bulletin of the Korean Chemical Society. 2010;**31**:2571-2573

[10] Marullo S, D'Anna F, Campodonico PR, Noto R. Ionic liquid binary mixtures: How different factors contribute to determine their effect on the reactivity. RSC Advances. 2016;**6**:90165-90171

[11] D'Anna F, Frenna V, Noto R, Pace V, Spinelli D. Study of aromatic nucleophilic substitution with amines on nitrothiophenes in room-temperature ionic liquids: Are the different effects on the behavior of para-like and ortho-like isomers on going from conventional solvents to room-temperature ionic liquids related to solvation effects? [Internet]. The Journal of Organic Chemistry. 2006;**71**:5144-5150. DOI: 10.1021/jo060435q

[12] Gazitúa M, Tapia RA, Contreras R, Campodónico PR. Mechanistic pathways of aromatic nucleophilic substitution in conventional solvents and ionic liquids. New Journal of Chemistry. 2014;**38**:2611-2618

[13] Bunnett JF, Zahler RE. Aromatic nucleophilic substitution reactions [Internet]. Chemical Reviews. 1951;**49**:273-412. DOI: 10.1021/cr60153a002

[14] Choi JH, Lee BC, Lee HW, Lee I. Competitive reaction pathways in the nucleophilic substitution reactions of aryl benzenesulfonates

with benzylamines in acetonitrile. The Journal of Organic Chemistry. 2002;**67**:1277-1281

[15] Um I-H, Hong J-Y, Kim J-J, Chae O-M, Bae S-K. Regioselectivity and the nature of the reaction mechanism in nucleophilic substitution reactions of 2, 4-dinitrophenyl X-substituted benzenesulfonates with primary amines. The Journal of Organic Chemistry. 2003;**68**:5180-5185

[16] Um I-H, Chun S-M, Chae O-M, Fujio M, Tsuno Y. Effect of amine nature on reaction rate and mechanism in nucleophilic substitution reactions of 2, 4-dinitrophenyl X-substituted benzenesulfonates with alicyclic secondary amines. The Journal of Organic Chemistry. 2004;**69**:3166-3172

[17] Mortier J. Arene Chemistry: Reaction Mechanisms and Methods for Aromatic Compounds. Chapter 7. New Jersey, USA: John Wiley & Sons; 2015. p. 175-193

[18] Buncel E, Norris AR, Russell KE. The interaction of aromatic nitro-compounds with bases. Quarterly Reviews, Chemical Society. 1968;**22**:123-146

[19] Bunnett JF. Some novel concepts in aromatic reactivity. Tetrahedron. 1993;**49**:4477-4484

[20] Miller J. 1918. Aromatic Nucleophilic Substitution. 1968. Available from: http://agris.fao.org/agris-search/search.do?recordID=US201300456768

[21] Bernasconi CF. Kinetic behavior of short-lived anionic .sigma. complexes. Accounts of Chemical Research. 1978;**11**:147-152

[22] Bunnett JF, Creary X. Nucleophilic replacement of two halogens in dihalobenzenes without the intermediacy of monosubstitution products. The Journal of Organic Chemistry. 1974;**39**:3611-3612

[23] Bernasconi CF, De Rossi RH. Influence of the o-nitro group on base catalysis in nucleophilic aromatic substitution. Reactions in benzene solution [internet]. The Journal of Organic Chemistry. 1976;**41**:44-49. DOI: 10.1021/jo00863a010

[24] Crampton MR, Emokpae TA, Howard JAK, Isanbor C, Mondal R. Leaving group effects on the mechanism of aromatic nucleophilic substitution (SNAr) reactions of some phenyl 2,4,6-trinitrophenyl ethers with aniline in acetonitrile [Internet]. Journal of Physical Organic Chemistry. 2004:65-70. DOI: 10.1002/poc.690

[25] Terrier F. Modern Nucleophilic Aromatic Substitution [Internet]. Chapter 1. Weinheim, Germany: Wiley-VCH; 2013:76. DOI: 10.1002/9783527656141

[26] Calfumán K, Gallardo-Fuentes S, Contreras R, Tapia RA, Campodónico PR. Mechanism for the SNAr reaction of atrazine with endogenous thiols: Experimental and theoretical study [Internet]. New Journal of Chemistry. 2017;**41**:12671-12677. DOI: 10.1039/c7nj02708g

[27] Kwan EE, Zeng Y, Besser HA, Jacobsen EN. Concerted nucleophilic aromatic substitutions. Nature Chemistry. 2018;**10**:917-923

[28] Neumann CN, Hooker JM, Ritter T. Concerted nucleophilic aromatic substitution with (19)F(-) and (18)F(-). Nature. 2016;**534**:369-373

[29] Neumann CN, Ritter T. Facile C–F bond formation through a concerted nucleophilic aromatic substitution mediated by the phenofluor reagent. Accounts of Chemical Research. 2017;**50**:2822-2833

[30] Bunnett JF, Morath RJ. The rates of condensation of piperidine with 1-chloro-2,4-dinitrobenzene in various solvents [Internet]. Journal of the American Chemical Society. 1955;**5**:5165-5165. DOI: 10.1021/ja01624a063

[31] Ormazábal-Toledo R, Contreras R, Tapia RA, Campodónico PR. Specific nucleophile-electrophile interactions in nucleophilic aromatic substitutions. Organic & Biomolecular Chemistry. 2013;**11**:2302-2309

[32] Gallardo-Fuentes S, Tapia RA, Contreras R, Campodónico PR. Site activation effects promoted by intramolecular hydrogen bond interactions in SNAr reactions. RSC Advances. 2014;**4**:30638-30643

[33] Hallett JP, Welton T. Room-temperature ionic liquids: Solvents for synthesis and catalysis. 2. Chemical Reviews. 2011;**111**:3508-3576

[34] Welton T. Room-temperature ionic liquids. Solvents for synthesis and catalysis. Chemical Reviews. 1999;**99**:2071-2084

[35] Earle MJ, Seddon KR. Ionic liquids. Green solvents for the future. Journal of Macromolecular Science Part A Pure and Applied Chemistry. 2000;**72**:1391-1398. Available from: https://www.degruyter.com/view/j/pac.2000.72.issue-7/pac200072071391/pac200072071391.xml

[36] Sheldon RA. Green solvents for sustainable organic synthesis: State of the art. Green Chemistry. 2005;**7**:267-278

[37] Rogers RD, Seddon KR. Chemistry. Ionic liquids—Solvents of the future? Science. 2003;**302**:792-793

[38] Seddon KR. Ionic liquids for clean technology. Journal of Chemical Technology & Biotechnology: International Research in Process, Environmental and Clean Technology. 1997;**68**:351-356

[39] Sánchez B, Calderón C, Tapia RA, Contreras R, Campodónico PR. Activation of electrophile/nucleophile pair by a nucleophilic and electrophilic solvation in a SNAr reaction [Internet]. Frontiers in Chemistry. 2018;**6**:509-517. DOI: 10.3389/fchem.2018.00509

[40] Gazitúa M, Tapia RA, Contreras R, Campodónico PR. Effect of the nature of the nucleophile and solvent on an SNAr reaction [Internet]. New Journal of Chemistry. 2018;**42**:260-264. DOI: 10.1039/c7nj03212a

[41] Tanner EEL, Hawker RR, Yau HM, Croft AK, Harper JB. Probing the importance of ionic liquid structure: A general ionic liquid effect on an S(N) Ar process. Organic & Biomolecular Chemistry. 2013;**11**:7516-7521

[42] Tanner EEL, Hawker RR, Yau HM, Croft AK, Harper JB. Probing the importance of ionic liquid structure: A general ionic liquid effect on an SN Ar process. Organic & Biomolecular Chemistry. 2013;**11**:7516-7521

[43] Dawber JG, Ward J, Williams RA. A study in preferential solvation using a solvatochromic pyridinium betaine and its relationship with reaction rates in mixed solvents. Journal of the Chemical Society, Faraday Transactions 1: Physical Chemistry in Condensed Phases. 1988;**84**:713-727

[44] Mancini PME, Terenzani A, Adam C, Vottero LR. Solvent effects on aromatic nucleophilic substitution reactions. Part 9. Special kinetic synergistic behavior in binary solvent mixtures [Internet]. Journal of Physical Organic Chemistry. 1999;**12**:430-440. DOI: 10.1002/(SICI)1099-1395(199906)12:6<430::AID-POC142>3.0.CO;2-W

[45] Mancini PM, Terenzani A, Adam C, Pérez A d C, Vottero LR.

Characterization of solvent mixtures: Preferential solvation of chemical probes in binary solvent mixtures of polar hydrogen-bond acceptor solvents with polychlorinated co-solvents [Internet]. Journal of Physical Organic Chemistry. 1999;**12**:713-724. DOI: 10.1002/(SICI)1099-1395(199909)12:9<713::AID-POC182>3.0.CO;2-0

[46] D'Anna F, Marullo S, Vitale P, Noto R. Binary mixtures of ionic liquids: A joint approach to investigate their properties and catalytic ability. ChemPhysChem. 2012;**13**:1877-1884

[47] Xiao D, Rajian JR, Hines LG, Li S, Bartsch RA, Quitevis EL. Nanostructural organization and anion effects in the optical Kerr effect spectra of binary ionic liquid mixtures [Internet]. The Journal of Physical Chemistry B. 2008;**112**:13316-13325. DOI: 10.1021/jp804417t

[48] Alarcón-Espósito J, Contreras R, Campodónico PR. Iso-solvation effects in mixtures of ionic liquids on the kinetics of a model SNAr reaction [Internet]. New Journal of Chemistry. 2017;**41**:13435-13441. DOI: 10.1039/c7nj03246c

[49] Holbrey JD, Seddon KR. Ionic liquids. Clean Products and Processes. 1999;**1**:223-236

[50] Mele A, Tran CD, De Paoli Lacerda SH. The structure of a room-temperature ionic liquid with and without trace amounts of water: The role of C■H···O and C■H···F interactions in 1-n-butyl-3-methylimidazolium tetrafluoroborate [Internet]. Angewandte Chemie International Edition. 2003;**115**:4364-4366. DOI: 10.1002/anie.200351783

[51] Saha S, Hamaguchi H-O. Effect of water on the molecular structure and arrangement of nitrile-functionalized ionic liquids. The Journal of Physical Chemistry. B. 2006;**110**:2777-2781

[52] Sánchez B, Calderón C, Garrido C, Contreras R, Campodónico PR. Solvent effect on a model SNAr reaction in ionic liquid/water mixtures at different compositions [Internet]. New Journal of Chemistry. 2018;**42**:9645-9650. DOI: 10.1039/c7nj04820c

[53] Crowhurst L, Mawdsley PR, Perez-Arlandis JM, Salter PA, Welton T. Solvent–solute interactions in ionic liquids [Internet]. Physical Chemistry Chemical Physics. 2003;**5**:2790-2794. DOI: 10.1039/b303095d

[54] Chiappe C, Pomelli CS, Rajamani S. Influence of structural variations in cationic and anionic moieties on the polarity of ionic liquids. The Journal of Physical Chemistry. B. 2011;**115**:9653-9661

[55] Ben-Naim A. Preferential solvation in two-and in three-component systems. Journal of Macromolecular Science, Part A Pure and Applied Chemistry. 1990;**62**:25-34. Available from: https://www.degruyter.com/view/j/pac.1990.62.issue-1/pac199062010025/pac199062010025.xml

[56] Covington AK, Newman KE. Approaches to the problems of solvation in pure solvents and preferential solvation in mixed solvents. Journal of Macromolecular Science, Part A Pure and Applied Chemistry. 1979;**51**:2041-2058

[57] Langford CH, Tong JPK. Preferential solvation and the role of solvent in kinetics. Examples from ligand substitution reactions. Accounts of Chemical Research. 1977;**10**:258-264

[58] Sengwa RJ, Khatri V, Sankhla S. Dielectric behaviour and hydrogen bond molecular interaction study of formamide-dipolar solvents binary mixtures. Journal of Molecular Liquids. 2009;**144**:89-96

[59] Kirkwood JG. Theory of solutions of molecules containing widely separated

charges with special application to zwitterions. The Journal of Chemical Physics. 1934;**2**:351-361

[60] Olah GA, Klumpp DA. Superelectrophilic solvation. Accounts of Chemical Research. 2004;**37**:211-220

[61] Bentley TW, William Bentley T, Llewellyn G, Ryu ZH. Solvolytic reactions in fluorinated alcohols. Role of nucleophilic and other solvation effects [Internet]. The Journal of Organic Chemistry. 1998;**63**:4654-4659. DOI: 10.1021/jo980109d

[62] Schadt FL, Bentley TW, Schleyer PR. The SN2-SN1 spectrum. 2. Quantitative treatments of nucleophilic solvent assistance. A scale of solvent nucleophilicities. Journal of the American Chemical Society. 1976;**98**:7667-7675

[63] Winstein S, Grunwald E, Jones HW. The correlation of solvolysis rates and the classification of solvolysis reactions into mechanistic categories. Journal of the American Chemical Society. 1951;**73**:2700-2707

[64] Ingold CK. The principles of aromatic substitution, from the standpoint of the electronic theory of valency. Recueil des Travaux Chimiques des Pays-Bas. 1929;**48**:797-812. Available from: https://onlinelibrary.wiley.com/doi/abs/10.1002/recl.19290480808

[65] Ingold CK. Principles of an electronic theory of organic reactions. Chemical Reviews. 1934;**15**:225-274

[66] Lewis GN. Valence and the Structure of Atoms and Molecules. T.M. Lowry, New York. USA: Chemical Catalog Company, Incorporated; 1923

[67] Brønsted JN. Some remarks on the concept of acids and bases. Recueil des Travaux Chimiques des Pays-Bas. 1923;**42**:718-728

[68] Lowry TM. The uniqueness of hydrogen. Journal of Chemical Technology and Biotechnology. 1923;**42**:43-47

[69] Mancini PME, Terenzani A, Adam C, Vottero LR. Solvent effects on aromatic nucleophilic substitution reactions. Part 7. Determination of the empirical polarity parameter ET (30) for dipolar hydrogen bond acceptor-co-solvent (chloroform or dichloromethane) mixtures. Kinetics of the reactions of halonitrobenzenes with aliphatic amines. Journal of Physical Organic Chemistry. 1997;**10**:849-860

[70] Martinez RD, Mancini PME, Vottero LR, Nudelman NS. Solvent effects on aromatic nucleophilic substitutions. Part 4. Kinetics of the reaction of 1-chloro-2,4-dinitrobenzene with piperidine in protic solvents. Journal of the Chemical Society, Perkin Transactions. 1986;**2**:1427-1431

[71] Beveridge DL, Schnuelle GW. Statistical thermodynamic consideration of solvent effects on conformational stability. Supermolecule-continuum model. The Journal of Physical Chemistry. 1974;**78**:2064-2069

[72] Baltzer L, Bergman N-Å, Drakenberg T, Raldow W, Nielsen PH. Solvation of the sodium ion in mixtures of methanol and dimethyl sulfoxide [Internet]. Acta Chemica Scandinavica. 1981;**35**:759-762. DOI: 10.3891/acta.chem.scand.35a-0759

[73] Frankel LS, Stengle TR, Langford CH. A study of preferential solvation utilizing nuclear magnetic resonance [Internet]. Chemical Communications (London). 1965;**17**:393. DOI: 10.1039/c19650000393

[74] Taha A, Ramadan AAT, El-Behairy MA, Ismail AI, Mahmoud MM. Preferential solvation

studies using the solvatochromic dicyanobis(1,10-phenanthroline) iron(II) complex [Internet]. New Journal of Chemistry. 2001;**25**:1306-1312. DOI: 10.1039/b104093f

[75] Buncel E, Rajagopal S. Solvatochromism and solvent polarity scales. Accounts of Chemical Research. 1990;**23**:226-231

[76] Kamlet MJ, Taft RW. The solvatochromic comparison method. I. The .beta.-scale of solvent hydrogen-bond acceptor (HBA) basicities. Journal of the American Chemical Society. 1976;**98**:377-383

[77] Kamlet MJ, Abboud JL, Taft RW. The solvatochromic comparison method. 6. The .pi.* scale of solvent polarities. Journal of the American Chemical Society. 1977;**99**:6027-6038

[78] Kamlet MJ, Abboud JLM, Abraham MH, Taft RW. Linear solvation energy relationships. 23. A comprehensive collection of the solvatochromic parameters, .pi.*, .alpha., and .beta., and some methods for simplifying the generalized solvatochromic equation. The Journal of Organic Chemistry. 1983;**48**:2877-2887

[79] Catalan J. Solvent effects based on pure solvent scales [Internet]. ChemInform. 2003;**34**:583-616. DOI: 10.1002/chin.200320297

[80] Cerón-Carrasco JP, Jacquemin D, Laurence C, Planchat A, Reichardt C, Sraïdi K. Solvent polarity scales: Determination of newET(30) values for 84 organic solvents [Internet]. Journal of Physical Organic Chemistry. 2014;**27**:512-518. DOI: 10.1002/poc.3293

[81] Barton AFM. Solvent scales [Internet]. In: CRC Handbook of Solubility Parameters and Other Cohesion Parameters. Chapter 8. 2017:178. DOI: 10.1201/9781315140575-8

[82] Reichardt C, Welton T. Solvents and Solvent Effects in Organic Chemistry. Weinheim, Germany: Wiley-VCH Verlag GmbH & Co. KGaA; 2011

[83] Reichardt C. Polarity of ionic liquids determined empirically by means of solvatochromic pyridinium N-phenolate betaine dyes [Internet]. Green Chemistry. 2005;**7**:339. DOI: 10.1039/b500106b

[84] Cerda-Monje A, Aizman A, Tapia RA, Chiappe C, Contreras R. Solvent effects in ionic liquids: Empirical linear energy-density relationships [Internet]. Physical Chemistry Chemical Physics. 2012;**14**:10041. DOI: 10.1039/c2cp40619e

[85] Rostov IV, Basilevsky MV, Newton MD. Advanced dielectric continuum models of solvation, their connection to microscopic solvent models, and application to electron transfer reactions [Internet]. Simulation and Theory of Electrostatic Interactions in Solution. AIP Conference Proceeding. 1999;**492**:331-351. DOI: 10.1063/1.1301535

[86] Contreras R, Aizman A. On the SCF theory of continuum solvent effects representation: Introduction of local dielectric effects [Internet]. International Journal of Quantum Chemistry. 1985;**27**:293-301. DOI: 10.1002/qua.560270307

[87] Kobrak MN, Li H. Electrostatic interactions in ionic liquids: The dangers of dipole and dielectric descriptions. Physical Chemistry Chemical Physics. 2010;**12**:1922-1932

[88] Göllei A. Dielectric characteristics of ionic liquids and usage in advanced energy storage cells [Internet]. Progress and Developments in Ionic Liquids. IntechOpen; 2017:451-473. Chapter 17. DOI: 10.5772/66948

[89] Baker SN, Baker GA, Bright FV. Temperature-dependent microscopic

solvent properties of "dry" and "wet" 1-butyl-3-methylimidazolium hexafluorophosphate: Correlation with ET(30) and Kamlet–Taft polarity scales [Internet]. Green Chemistry. 2002;**4**:165-169. DOI: 10.1039/b111285f

[90] Gutmann V, Wychera E. Coordination reactions in non aqueous solutions—The role of the donor strength [Internet]. Inorganic and Nuclear Chemistry Letters. 1966;**2**:257-260. DOI: 10.1016/0020-1650(66)80056-9

[91] Mayer U, Gutmann V, Gerger W. The acceptor number? A quantitative empirical parameter for the electrophilic properties of solvents [Internet]. Monatshefte für Chemie. 1975;**106**:1235-1257. DOI: 10.1007/bf00913599

[92] Schmeisser M, Illner P, Puchta R, Zahl A, van Eldik R. Gutmann donor and acceptor numbers for ionic liquids. Chemistry. 2012;**18**:10969-10982

[93] Schmeisser M, Illner P, Puchta R, Zahl A, van Eldik R. Cover picture: Gutmann donor and acceptor numbers for ionic liquids (Chem. Eur. J. 35/2012) [Internet]. Chemistry—A European Journal. 2012;**18**:10765-10765. DOI: 10.1002/chem.201290149

[94] Coleman S, Byrne R, Minkovska S, Diamond D. Investigating nanostructuring within imidazolium ionic liquids: A thermodynamic study using photochromic molecular probes. Journal of Physical Chemistry B. 2009;**113**:15589-15596